Chemistry – a practical approach

3

Chemistry – a practical approach

A. L. Barker

Head of Science
Reigate Grammar School

K. A. Knapp

Formerly of Reigate Grammar School

Macmillan Education
London and Basingstoke

First published 1978
Reprinted 1979, 1981, 1982, 1983, 1984

Published by
Macmillan Education Limited
Houndmills Basingstoke Hampshire RG21 2XS
and London
Associated companies in Delhi Dublin
Hong Kong Johannesburg Lagos Melbourne
New York Singapore and Tokyo

Printed in Hong Kong

British Library Cataloguing in Publication Data
Barker, A. L.
  Chemistry: a practical approach
  1. Chemistry
  I. Title II. Knapp, K.A.
  540   QD33

  ISBN 0-333-18222-7

Contents

Preface   ix

Acknowledgements   xi

A note to the pupil   xii

1   Some simple investigations – separations and purity   13
    Experiments: 1.1 The bunsen burner; 1.2 Separation of mixtures;
    1.3 Tests for purity; Questions

2   Classes of substance and types of change   29
    Experiments: 2.1 Elements, compounds and mixtures; 2.2 Evidence for
    chemical reactions—chemical and physical changes; Questions

3   The atmosphere and combustion   37
    Experiments: 3.1 Combustion; 3.2 The rusting of iron; 3.3 Respiration
    and photosynthesis; 3.4 The composition of the air; Questions

4   Water and hydrogen   48
    Experiments: 4.1 Water, the most common liquid in the world; 4.2 Tests
    for water; 4.3 The breaking and making of water; 4.4 Hydrogen;
    Questions

5   Acids, bases and salts   62
    Experiments: 5.1 An introduction to acids, alkalis and indicators;
    5.2 Acids; 5.3 Bases and alkalis; 5.4 Salts; 5.5 Some ionic considerations;
    5.6 Hydrolysis of salts; Questions

6   Atoms and molecules   86
    Experiments: 6.1 Evidence for the existence of atoms; 6.2 How big are
    atoms?; 6.3 The kinetic theory; 6.4 The chemical laws and Dalton's
    atomic theory; 6.5 Atomic structure; 6.6 Radioactivity; Questions

7   The periodic table   110
    Experiments: 7.1 Early attempts at classification of elements;
    7.2 Electronic arrangement and the periodic table; 7.3 Periodic variation
    in physical properties of the elements; 7.4 Trends in properties across the
    third period (sodium to argon); 7.5 Trends in properties going down
    groups; Questions

8   Bonding and structure   125
    Experiments: 8.1 The ionic (electrovalent) bond; 8.2 The covalent bond;
    8.3 Some differences between ionic and covalent compounds; 8.4 The
    shapes of covalent molecules; 8.5 Crystal structure; 8.6 Oxidation and
    reduction; Questions

9   Moles, formulae and equations   152
Experiments: 9.1 Masses of atoms and the mole; 9.2 Empirical formulae; 9.3 Masses of molecules, molecular formulae and percentage composition; 9.4 Chemical equations; 9.5 Ionic equations; 9.6 Calculating reacting masses from chemical equations; Questions

10   The molecular theory of gases   173
Experiments: 10.1 The gas laws; 10.2 Atomicity and molar volume of a gas; Questions

11   Electrochemistry   187
Experiments: 11.1 The electrochemical series; 11.2 Electrolysis; 11.3 Faraday's laws of electrolysis; 11.4 Obtaining electrical energy from chemical changes; Questions

12   Solvents and solubility   219
Experiments: 12.1 Solutions and suspensions; 12.2 Water of crystallisation; 12.3 Efflorescence, deliquescence and hygroscopic substances; 12.4 How solids dissolve in liquids; 12.5 The solubility of one liquid in another; 12.6 The solubility of gases in liquids; Questions

13   Rate of reaction   238
Experiments: 13.1 Measuring the rate of a reaction; 13.2 The effect of particle size on the rate of a reaction; 13.3 The effect of changes in concentration on the rate of a reaction; 13.4 The effect of change in temperature on the rate of a reaction; 13.5 The effect of a catalyst on the rate of a reaction; 13.6 The effect of light on the rate of a reaction; Questions

14   Reversible reactions   254
Experiments: 14.1 Reactions which can go both ways; 14.2 Dynamic equilibrium; 14.3 Varying the position of equilibrium – Le Chatelier's principle; 14.4 Reversible reactions in industry; 14.5 Further examples of reversible reactions; Questions

15   Energy changes in chemistry   270
Experiments: 15.1 Energy changes in chemistry; 15.2 Enthalpy of fusion and enthalpy of vaporisation; 15.3 Enthalpy of combustion; 15.4 Enthalpy of neutralisation; 15.5 Enthalpy of solution and enthalpy of hydration; 15.6 Enthalpy of precipitation; 15.7 Enthalpy of reaction and electrical energy; Questions

16   Carbon and silicon   290
Experiments: 16.1 Allotropes of carbon; 16.2 Reactions and uses of carbon; 16.3 Carbon hydrides; 16.4 Carbon dioxide; 16.5 Carbon monoxide; 16.6 Tetrachloromethane; 16.7 Silicon and the other group IV elements; Questions

17   Nitrogen and phosphorus   309
Experiments: 17.1 Nitrogen; 17.2 The oxides of nitrogen; 17.3 Ammonia; 17.4 Nitric acid; 17.5 Phosphorus; Questions

18 Oxygen and sulphur   333
    Experiments: 18.1 Oxygen and oxides; 18.2 Metal hydroxides;
    18.3 Hydrogen peroxide; 18.4 Sulphur; 18.5 Hydrogen sulphide; 18.6 Sul-
    phur dioxide; 18.7 Sulphur trioxide; 18.8 Sulphuric acid; Questions

19 The halogens   364
    Experiments: 19.1 Hydrogen chloride; 19.2 Chlorine; 19.3 A comparison
    of fluorine, bromine and iodine with chlorine; 19.4 A comparison of the
    hydrogen halides; Questions

20 The metals   382
    Experiments: 20.1 The reactivity series; 20.2 Potassium, sodium
    and lithium – the alkali metals; 20.3 Calcium and magnesium;
    20.4 Aluminium; 20.5 Zinc; 20.6 Iron; 20.7 Lead; 20.8 Copper;
    20.9 Differences between metals and non-metals; Questions

21 Organic chemistry   414
    Experiments: 21.1 Organic chemistry – an introduction; 21.2 The
    alkanes; 21.3 The alkenes; 21.4 The alkynes; 21.5 Ring compounds;
    21.6 Petroleum and coal; 21.7 Carbohydrates; 21.8 The alcohols;
    21.9 The carboxylic acids; 21.10 Esters and detergents; 21.11 Proteins;
    21.12 Man-made macromolecules; Questions

22 Chemical analysis   451
    22.1 Qualitative analysis. 22.2 Quantitative analysis; Questions

    Answers   460

    Appendix   463
    The periodic table; Table of relative atomic masses

    Index   469

Preface

This book has been written to satisfy the needs of modern chemistry courses leading to O level and similar examinations. Its aim is to present chemistry as an experimental science and to encourage an investigational approach to the subject.

In order to achieve these objectives a somewhat unusual layout has been adopted. Each chapter begins with an experimental section containing practical instructions followed by questions designed to emphasise the necessary observations and to provoke thought about their implications. Theoretical and factual sections, uncluttered by experimental details, appear later in the chapter. This separation of experiments and text results in a greater likelihood of pupils trying to deduce the answers to the questions posed rather than simply looking them up. However, the less confident ones have the security of knowing that most of the answers can be found further on in the chapter, even if their experiment is unsuccessful or their deductions are incorrect.

Although the experiments are referred to in the theory sections, the text has been written in such a way that it can be understood even if the experiments have not been performed. This is particularly useful where pupils have missed work through absence or if the teacher wishes to omit some of the experiments. The numbering system used throughout the book links clearly the experiments with the corresponding theory sections so that cross-referencing is easy. For example, section 13.5 covers catalysis, and the experiments concerned with this topic are numbered 13.5.a, 13.5.b, and so on. In addition, diagrams and tables are given the same numbers as the experiments or theory sections to which they refer, making the location of material a very simple task indeed. Each experiment is given a 'star rating', no stars indicating suitability for class, one star (★) for class or demonstration and two stars (★★) for demonstration only. Extra warnings are given for particularly dangerous experiments.

No attempt has been made to arrange the material in the form of a teaching scheme, although most of the work in the first few chapters is simpler than that which appears later on. All the material on a particular topic has been included in one chapter in order to simplify revision at the end of the course and to make reference to that topic easier. The individual sections are self-contained so that a wide variety of teaching schemes is possible. Since all the material is found in one volume no one teaching sequence is dictated. This format makes the book suitable for all stages of an O level course and allows flexibility to meet specific examinations and individual teachers' requirements.

The nomenclature used throughout is that recommended in *Chemical Nomenclature, Symbols and Terminology* published by the Association for Science Education (1972), and in the circular issued by the GCE examining boards, except in cases such as 'citric acid' where the systematic name would be meaningless at this level.

Our thanks are due to the examining boards listed on page xi for their permission to reproduce questions from recent O level papers, to the publishers for their many helpful suggestions and to our families for their patience and understanding during the past months. Finally, we freely admit that much of our work has been influenced by that of others. We are deeply indebted to the producers of the Nuffield Chemistry Scheme, contributors to the *School Science Review* and to the many chemistry teachers who, over the years, have handed on their ideas so that others might benefit from them.

September 1977.   A.L.B.
                                                       K.A.K.

Acknowledgements

The authors and publishers would like to acknowledge the following examination boards for permission to reprint questions from past O level (or in the case of the Scottish Board, O grade) examination papers in Chemistry. Individual questions are credited in the body of the text.

The Associated Examining Board for the General Certificate of Education (AEB)
University of Cambridge Local Examinations Syndicate (C)
University of London School Examinations Department (L)
Northern Ireland Schools Examinations Council (NI)
Oxford and Cambridge Schools Examination Board (O & C)
Oxford Delegacy of Local Examinations (O)
Scottish Certificate of Education Examination Board (SCE)
Southern Universities Joint Board (SUJB)
Welsh Joint Education Committee (W)

We are also grateful to the following photograph sources:

The Trustees of the British Museum (Natural History) p. 142; British Oxygen p. 408; Camera Press p. 299; J. Allan Cash p. 399; Design and Materials Limited p. 277; Distillers Company Limited p. 440; Filden & Mawson Architects p. 357; ICI Agricultural Division p. 325; Brian Kirkpatrick p. 231; The late Professor Dame Kathleen Lonsdale p. 145 (collection in care of Dr. H. J. Milledge); Erwin Müller p. 95; NASA p. 346; Northern Ireland Tourist Board p. 23; Rhodesian High Commission p. 54; Science Museum p. 101. Permission for the use of the cover transparency was kindly given by The Monsanto Company.

The publishers have made every effort to trace the copyright holders of all illustrations, but, where they have failed to do so they will be pleased to make the necessary arrangements at the first opportunity.

A note to the pupil

Chemistry is about *you* and everything around you. Chemical changes in your body keep you alive and chemical processes in industry help to provide you with food, clothing and shelter; in fact, a knowledge of chemistry is required at some stage in the production of just about everything with which you come into contact. Obviously then, chemistry is a very important subject, but what exactly *is* it? The purpose of this book is to help you to find out the answer to this question for yourselves.

If you look through the book you will see that each chapter is divided into two parts, the first consisting of a series of experiments for you to perform and the second giving you the necessary theory and background material on each topic. Your teacher will select the experiments which are relevant to your course and before beginning each one you should read carefully through the instructions and the questions which follow them, so that you know what to look for. When you have obtained your results you should try to provide answers to the questions *yourself*, and only consult the text as a last resort if you are unable to think of any answer at all. By acting in this way you will develop an understanding of the principles of chemistry and the subject will not become a mass of dull facts which have to be learned by heart.

Finally, before you embark upon your studies, a word of warning is necessary. There are many dangers in a chemistry laboratory; if you are to make your discoveries in safety you must obey the following simple rules.

1 Carry out the experiment instructions exactly and *never* mix chemicals or heat them up just to see what happens.

2 Treat all apparatus and chemicals with care and *on no account* put either in your mouth.

3 Always report accidents and spillage of chemicals to your teacher *immediately*.

4 When you leave the laboratory, always *wash your hands*.

1 Some simple investigations – separations and purity

The experiments in this chapter will introduce you to a number of techniques which you will use throughout your chemistry course. They are concerned with the separation of mixtures into their components and with simple ways of testing the purity of substances.

Before you begin, read the note on the page opposite.

Experiments

1.1.a To investigate the bunsen burner

The following experiments are designed to enable you to discover how a bunsen burner works and how to use it efficiently.

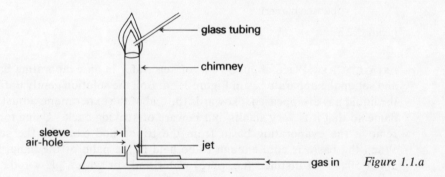

Figure 1.1.a

PROCEDURE (a) Close the air-hole at the bottom of the chimney by twisting the movable sleeve, turn the gas on and light the burner. Keep your head well away from the chimney as the flame may be bigger than you expect! Adjust the gas until the flame is about 10 cm high and then, using tongs, hold a piece of porcelain in it for about a minute. Remove the porcelain and examine it.
(b) Turn the gas full on, gradually open the air-hole, noting any changes in the flame, and repeat the experiment with the porcelain.
(c) Close the air-hole slightly and reduce the gas supply to give a 'softer' flame but make sure that a blue cone is still visible.
(d) With the flame as in (c) place a piece of nichrome wire across the top of the chimney and gradually raise it until it is out of the flame.
(e) With the flame as in (c) place one end of a 10 cm length of glass tubing in the inner cone of the flame. Hold the tube as shown in Figure 1.1.a and apply a lighted splint to the other end.

(f) Finally, with the air-hole fully open, observe the effect of *slowly* turning down the gas supply. Look for a flame in a different part of the bunsen and then turn off the gas completely.

Questions

1 What are the disadvantages of using the bunsen with the air-hole closed? Draw a diagram of the flame obtained.
2 Are there any disadvantages in using the bunsen with the air-hole wide open and the gas full on? Draw a diagram of the flame obtained.
3 Why is the bunsen burner usually used as in (c)? Under these circumstances, which is the hottest part of the flame and what does the inner cone contain? (Consult the results of (d) and (e).)
4 The effect you observed in (f) is known as 'striking back'. Can you explain what happened?

1.2.a To obtain salt from sea water

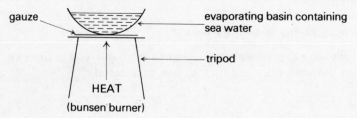

Figure 1.2.a

PROCEDURE   Place about 50 cm³ of sea water in an evaporating basin and set up the apparatus as in Figure 1.2.a. Boil the solution gently until all the liquid has disappeared. (Towards the end of the experiment, adjust the flame so that it is very small—but beware of striking back.) Using tongs, remove the evaporating basin from the tripod and examine the solid. When the basin is cool enough to be held in the palm of the hand, add about 50 cm³ of distilled (i.e. pure) water and stir with a glass rod.

Questions

1 Since substances cannot just disappear, where did the water go?
2 What do you think the solid residue in the basin was? Did it look pure?
3 What happened to the residue when the water was added?
4 What were the disadvantages, if any, of this method of separation?

1.2.b To find if temperature change affects dissolving

PROCEDURE   Add 10 g of copper(II) sulphate crystals to a Pyrex boiling tube containing 10 cm³ of distilled water and stir well with a glass rod. Boil the mixture gently for a few seconds. (When heating a tube containing liquid you should always shake it continuously, ensure that it is not pointing at anyone, and be ready to remove it from the flame if it boils too vigorously.) Cool the tube under the tap, note the results and repeat the experiment using 5 g of salt in place of the copper(II) sulphate.

Questions

1 Did any copper(II) sulphate dissolve at room temperature? How could you tell?
2 Did more copper(II) sulphate dissolve in hot water than in cold?
3 What happened on cooling the copper (II) sulphate solution?
4 Did temperature change affect the dissolving of salt very greatly?

1.2.c To prepare crystals of copper(II) sulphate from a solution of copper(II) sulphate in water

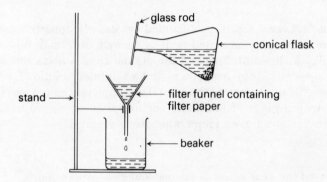

Figure 1.2.c

PROCEDURE   Place 50 cm³ of copper(II) sulphate solution (mass concentration 300 g dm⁻³*) in an evaporating basin and heat gently as shown in Figure 1.2.a. At frequent intervals, stir the solution with a glass rod, withdraw a drop of the solution on the end of the rod and place it on a clean watch glass. Look at this carefully to see if crystals are forming and continue the heating and testing with the glass rod until crystals are observed in a cooled drop. Using tongs, pour the solution into a 100 cm³ conical flask, stopper the flask and leave it to cool. When the mixture is completely cool, filter off the copper(II) sulphate crystals as shown in Figure 1.2.c. (Your teacher will show you how to fold the filter paper.) Wash the crystals with a little distilled water while they are still in the funnel and leave them to dry on a filter paper.

Questions

1 Why was the glass rod test carried out and why did crystals form in the drop on the watch glass as it cooled?
2 Would this method work with sea water? (Hint: check the results of Experiment 1.2.b.)
3 Why was the conical flask stoppered?
4 Why were the crystals washed with distilled water and why was only a little used?
5 Draw a diagram of one of the crystals.
6 What are the advantages and disadvantages, if any, of this experiment compared with Experiment 1.2.a?
7 Make a list of substances found at home which are crystalline in appearance.

*If you are not familiar with the minus sign in this notation, you may take it simply to mean 'per'. Thus g dm⁻³ means 'grams per cubic decimetre'.

1.2.d To prepare a large crystal of copper(II) sulphate

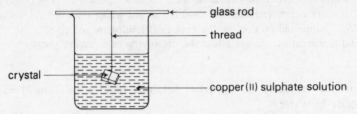

Figure 1.2.d

PROCEDURE Take a small, well-formed crystal of copper(II) sulphate from the previous experiment and tie it to a length of thread; this is best done by using a slip-knot. Suspend the crystal from a glass rod placed across the top of a 100 cm³ beaker and add a saturated solution (see page 225) of copper(II) sulphate until the crystal is completely immersed. Cover the beaker with a piece of thin cloth secured with an elastic band and leave the beaker at an even temperature for several days.

Questions

1 Is the crystal the same shape as the one you drew in Experiment 1.2.c?
2 Can you explain how the crystal grows?

1.2.e To obtain the solvent from a solution ★

So far, the experiments have been concerned with obtaining the solute from a solution but here distillation is used to recover the *solvent*.

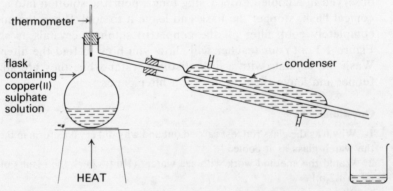

Figure 1.2.e

PROCEDURE Set up the apparatus as shown in Figure 1.2.e. Connect the water inlet to the top of the condenser, place the outlet in the sink and observe what happens when the water is turned on. Reverse the connections and decide which is the best arrangement for cooling the central tube. Heat the flask until only a small amount of liquid remains and note the temperature of the vapour when the mixture is boiling steadily. The *top* of the thermometer bulb should be level with the *bottom* of the side-arm so that the vapour passes over the bulb and an accurate reading of its temperature is obtained.

Questions

1 Which was the best water inlet and why? Make sure that your own diagram shows the water connections drawn correctly.
2 What was the temperature of the vapour? Do you think that the water you collected was pure? Give a reason for your answer.
3 Suggest some uses for distilled water.

1.2.f To obtain pure salt from rock salt

Rock salt is a mixture of two solids, sand and salt. Two techniques from earlier experiments, filtration and evaporation, are combined for this separation.

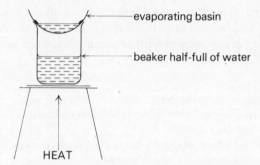

HEAT

Figure 1.2.f

PROCEDURE   Place about 1 cm depth of rock salt, previously ground to a coarse powder by means of a pestle and mortar, in a 100 cm³ beaker half-filled with distilled water. Warm (why?) and stir the mixture for about 2 minutes. Filter and collect the liquid in an evaporating basin; rinse the beaker with a little distilled water and pour the washings through the filter paper. When *all* the liquid has been collected, heat the evaporating basin until most of the water has been driven off, then transfer the evaporating basin to a steam bath (Figure 1.2.f) and evaporate to dryness.

Questions

1 Why did you rinse the beaker with a little water and pour the washings through the filter paper?
2 If two substances can be separated by this method, what can you say about their solubilities in water? Could you separate a mixture of sugar and salt in this way? Give a reason for your answer.
3 What are the advantages and disadvantages of this method of evaporation compared with that used in Experiment 1.2.a?
4 Rock salt could be mined in the same way as coal. Can you think of an easier way of getting the salt to the surface, involving no one going underground?

1.2.g To separate the dyes in ink

Inks usually consist of mixtures of coloured solids dissolved in a solvent.

These solids cannot be separated by filtration (why not?) and so the technique of paper chromatography is used.

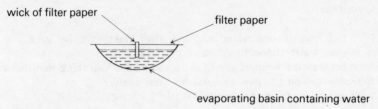

Figure 1.2.g

PROCEDURE Using a dropping pipette with a fine jet, place a small drop of black ink (e.g. Quink or Venus water colour felt tip ink) in the centre of a piece of filter paper and allow it to dry. Roll up a strip of filter paper about 3 cm wide to make a wick and push it through a hole in the centre of the ink spot. Set up the apparatus as in Figure 1.2.g, with the wick just dipping into the water and leave it until the water has reached the edge of the filter paper. Repeat the experiment with other substances of your own choice and keep the filter papers for reference.

Questions
1   Did the ink contain more than one dye?
2   Can you suggest an explanation for the separation which occurred as the water moved outwards across the filter paper?
3   Read the section on paper chromatography (page 24). Which dye was the least strongly adsorbed and which most strongly adsorbed?

1.2.h   To investigate the effect of heat on ammonium chloride
PROCEDURE Place a very small quantity of ammonium chloride in a test tube. Heat the *bottom* of the tube and observe closely until there is no further change.

Questions
1   Did the solid melt? Was any solid left in the bottom of the tube? Was any solid left *anywhere* in the tube?
2   The word used to describe the process which has taken place is 'sublimation'. In your own words, give a definition of sublimation.

1.2.i   To separate a mixture of salt and ammonium chloride

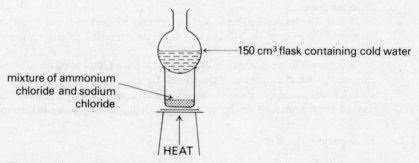

Figure 1.2.i

PROCEDURE Place a thin layer (2 mm) of the mixture in a 100 cm³ beaker and assemble the apparatus as shown in Figure 1.2.i. Heat the beaker gently until you think the separation is complete, remove the flask and examine it.

Questions

1   Why is filtration not used? Do you think that sublimation is widely used to separate mixtures? Give a reason for your answer. (Hint: can you think of any other solid which sublimes?)

2   Which of the solids, if any, do you think is (a) pure salt, (b) pure ammonium chloride, (c) a mixture of both?

1.2.j   To separate a mixture of cooking oil and water ★

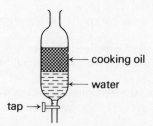

cooking oil

water

tap →

Figure 1.2.j

PROCEDURE Place the mixture in a separating funnel and wait until separation into two layers is complete as shown in Figure 1.2.j. Open the tap and allow the lower layer to run into a beaker. Close the tap when as much as possible of the lower layer, but none of the upper layer, is in the beaker and then tip the upper layer through the top of the funnel into another container.

Questions

1   Which of the two liquids has the greater density?

2   Can you think of another pair of liquids which could be separated in this way?

1.2.k   To separate a mixture of ethanol and water ★★

Many liquids mix with one another and cannot be separated by means of a separating funnel. For such mixtures fractional distillation is used.

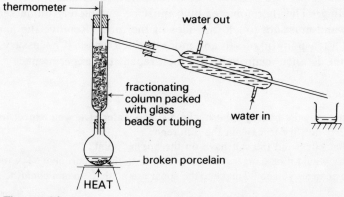

thermometer

water out

fractionating column packed with glass beads or tubing

water in

broken porcelain

HEAT

Figure 1.2.k

PROCEDURE Set up the apparatus as shown in Figure 1.2.k. One-third-fill the flask with a mixture of 1 part ethanol (i.e. the alcohol found in drinks) and 9 parts distilled water. Place a little of this mixture on a watch glass and try to ignite it by placing a burning taper close to the surface of the liquid. Heat the flask gently and observe what happens in the fractionating column. Watch the thermometer carefully and collect the distillate in five fractions: (a) up to 80 °C, (b) 80–85 °C, (c) 85–90 °C, (d) 90–95 °C, (e) over 95 °C. Try to ignite the various fractions and record your results.

Questions

1   Why should the thermometer bulb be kept in the position shown?
2   What is the purpose of the broken porcelain?
3   Which of the fractions were you able to ignite? Were the flames of the same size?
4   Do you think that the fractions which burned contained more or less ethanol than the starting material? Which fraction contained the most ethanol? (Ethanol boils at 78 °C and water at 100 °C.)
5   From your observations, try to explain how the fractionating column works.
6   Can you think of any substances used in everyday life which are obtained by fractional distillation?

1.3.a   To study the effect of a dissolved solid on the boiling point of a liquid

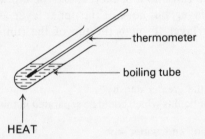

HEAT   Figure 1.3.a

PROCEDURE Half-fill a boiling tube with distilled water and clamp it in the position shown in Figure 1.3.a. Place a thermometer in the water, heat the tube gently and note the steady temperature obtained when the water is boiling. (The thermometer bulb must be completely immersed in the water and must not touch the sides of the tube.) Remove the flame, add about a teaspoonful of salt and more distilled water, if necessary, so that the tube is once again half-full, then repeat the experiment.

Questions

1   What was the boiling point of the distilled water? Did you expect to get this result? If not, try to explain the difference.
2   What effect did the salt have on the boiling point?
3   Why would it be dangerous to use ethanol in this experiment? Try to think of simple changes you could make to the apparatus so that ethanol *could* be used.

1.3.b To study the effect of a dissolved solid on the freezing point of a liquid

PROCEDURE Add some ice to distilled water in a 100 cm³ beaker and stir the mixture *carefully* with a thermometer. Record the steady temperature obtained (i.e. freezing point), and then add about a teaspoonful of salt, stir, and note what happens to the temperature.

Questions

1 What was the freezing point of the distilled water?
2 What effect did the salt have on the freezing point? Did you expect this result? If not, why did you think the result would be different?
3 Try to explain how the addition of salt causes ice on roads to melt, bearing in mind the results of this experiment.

1.3.c To study the effect of impurity on the melting point of a solid

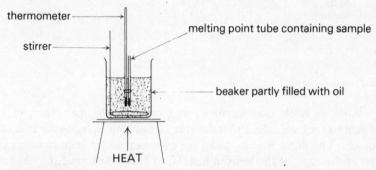

Figure 1.3.c

PROCEDURE Place a little powdered naphthalene, containing about 15% by mass of benzoic acid as impurity, on a watch glass. Seal the end of a melting point tube (a thin-walled capillary tube) in a bunsen flame, allow it to cool and press the open end of the tube into the mixture on the watch glass. Invert the tube and tap the closed end gently on the bench so that the solid which has entered goes to the bottom. Repeat, if necessary, until a depth of 3–4 mm is obtained. Using rubber bands, attach the tube to a thermometer (-5 to $105\,°C$) and set up the apparatus as shown in Figure 1.3.c. Heat the oil *gently*, stirring *continuously*, and note the temperature at which the solid melts. Observe the solid carefully for sharpness of melting point. Let the oil cool by $5\,°C$ (more, if the liquid in the tube does not solidify) and repeat the experiment with the same tube. Finally, repeat the whole procedure using a new tube containing pure naphthalene.

Questions

1 Did the impurity affect the melting point of naphthalene in the same way as salt affects the melting point of ice?
2 Which solid had the sharpest melting point, pure or impure naphthalene?

1.1 The bunsen burner

The bunsen burner (see Figure 1.1.a) was invented by Professor Robert Bunsen in Heidelberg in 1855 and is still used as a convenient source of heat in laboratories. It is usually used with the air-hole partly open. If the air-hole is closed, the gas burns with a yellow flame and a large amount of soot is produced. This is because insufficient air is supplied (see 21.2, page 428). When the air-hole is wide open, no soot is produced but the flame is fierce and noisy. This flame is difficult to see, and in direct sunlight it is completely invisible. Therefore, when the bunsen is on but not in use, the air-hole must always be closed.

When the air-hole is open, the flame is in three parts, as shown in Figure 1.1.

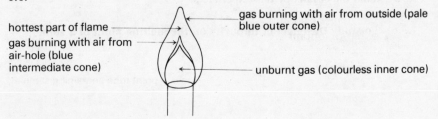

hottest part of flame

gas burning with air from air-hole (blue intermediate cone)

gas burning with air from outside (pale blue outer cone)

unburnt gas (colourless inner cone)

Figure 1.1 The bunsen flame

'Striking back' may occur if, with the air-hole open, the gas is turned down to such an extent that the rate of burning exceeds the rate of supply of gas. The flame moves down the chimney and is seen to come from the jet at the base of the bunsen instead of from the top of the chimney. The remedy is to turn off the gas and close the air-hole. (**Care**: the sleeve may well be hot!) The burner may then be relit and both gas and air adjusted to give the required flame, making sure that this time the air-hole is not opened too widely.

1.2 Separation of mixtures

Evaporation

When sea water is heated in an evaporating basin (Experiment 1.2.a), the level of the liquid goes down. This is because some of the water changes into vapour and passes into the atmosphere, the process being known as *evaporation*. Eventually a solid residue (salt) is left but when water is added to this, it dissolves again.

The salt is said to be *soluble* in water, whereas a solid which will not dissolve is *insoluble*. Sea water is a *solution* of salt (the *solute*) in water (the *solvent*). In fact, there are many other dissolved solids in sea water as well as the salt (see 4.1).

The effect of temperature change on dissolving

If copper(II) sulphate is added to cold water in a boiling tube (Experiment

1.2.b) a little of it dissolves, turning the liquid blue but the bulk of the solid remains undissolved at the bottom of the tube. The resulting solution is said to be *saturated* (12.1) since no more solute will dissolve under these conditions. However, if the temperature is raised, more copper(II) sulphate *does* dissolve, only to be 'thrown out' of solution again when the tube cools down. Most solids behave in this way, although for some, like salt, there is only a small change in solubility as the temperature is raised or lowered.

For most solids, then, *raising* the temperature *increases* the solubility in a given solvent and *lowering* the temperature *decreases* the amount of solute that will dissolve.

Crystallisation

The main disadvantage of evaporating all of the solvent from a solution in order to obtain the solute (Experiment 1.2.a) is that if there are any dissolved impurities present, they will be obtained with the solid at the end of the experiment. This problem is avoided if *crystallisation* is carried out.

The usual procedure is to evaporate the solution to crystallisation point, i.e. the point at which crystals of solute will form on cooling the solution to room temperature. Drops of solution are removed at intervals and allowed to cool; the formation of crystals in one of these drops indicates that the solution is at crystallisation point (see Experiment 1.2.c). The reason that crystals form on cooling is that the solute is less soluble in cold water than in hot so that as the temperature drops, the solution becomes saturated and then deposits crystals. Impurities are usually present to a small extent only; therefore, the solution should not become saturated with respect to them and they should remain dissolved even at

Crystals in nature – The Giant's Causeway, Northern Ireland

room temperature. Sometimes, however, crystallisation has to be repeated several times before purification is complete.

N.B. The flask in which the solution cools down must be stoppered to keep out dust and to prevent the solution from evaporating to dryness.

Crystals also form if a cold saturated solution is allowed to stand in the air for several days. Part of the solvent evaporates, causing some of the solute to come out of solution and be deposited as crystals. If this method is used, the crystals will form slowly and will generally be large and well-shaped.

If a single crystal of solute is placed in a saturated solution (Experiment 1.2.d) additional layers are deposited on the crystal and it is possible that after several days, only one large crystal will be observed in the solution. Dust must be excluded or the solute may start to crystallise around the dust particles and many small crystals will be obtained.

Distillation

If a liquid is required pure, then distillation is used (Experiment 1.2.e). For example, when a solution of copper(II) sulphate in water boils, the water changes to vapour, leaving the copper(II) sulphate behind. The vapour passes down a condenser where it is converted back to water and is collected as the *distillate* in a receiving flask. Water prepared in this way is called 'distilled water' and is colourless, odourless and tasteless. It has numerous uses, e.g. in making medicines, in steam irons and in car batteries.

Filtration

This method can be used to separate any substance which is soluble in a given solvent from one that is not. When water is added to rock salt (Experiment 1.2.f) the sandy impurities in the rock salt do not dissolve but the salt does. On filtering the mixture, the sand is left on the filter paper as the *residue* while the salt solution passes through and is collected in the evaporating basin as the *filtrate*. Salt is obtained from the filtrate by evaporating to dryness using a steam bath. This is slower than by direct heating with a bunsen but it usually gives better crystals and avoids loss by 'spitting'.

Paper chromatography

Ink usually consists of a mixture of coloured solids dissolved in a solvent. If a spot of ink is placed in the centre of a filter paper and treated as described in Experiment 1.2.g, then as the water passes over the ink and moves towards the edge of the filter paper, the various components in the ink separate into different bands of colour on the paper. The separation depends on differences in (a) the solubilities of the dyes in the water and (b) the tendencies of the dyes to stick to the surface of the paper (i.e. be *adsorbed*). Those dyes which have the greatest solubility in water and the least tendency to stick to the paper travel fastest and end up near to the edge of the paper. Those which are less soluble or stick to the paper better do not travel so far. Paper chromatography can easily be used to detect as little as one ten-millionth of a gram of a substance and has many applications in chemical analysis.

If several bands of colour are not obtained it may be that only one dye is present; alternatively, it can be that water is an unsuitable solvent for the material.

Sublimation

Sublimation is the direct conversion of a solid to a vapour on heating and of a vapour to a solid on cooling, without going through the liquid state.

If a mixture of ammonium chloride and salt is heated in a beaker the ammonium chloride sublimes and may be condensed on a cold surface (Experiment 1.2.i). The salt remains in the beaker but unless it is heated thoroughly it probably still contains some ammonium chloride which has not been driven off. Since only a few solids *sublime* this method of separation is very limited; other substances which can be purified in this way include iodine and anhydrous aluminium chloride.

Fractional distillation

If two liquids are *immiscible* (i.e. they separate into two distinct layers), then the simplest way to separate them is to use the separating funnel (Experiment 1.2.j). However, many liquids are *miscible* (i.e. they mix together completely and do not form layers); mixtures of liquids of this type may be separated by fractional distillation provided that their boiling points are different (Experiment 1.2.k).

When a mixture of two such liquids is heated to boiling in the apparatus shown in Figure 1.2.k the vapour evolved is richer than the liquid in the *more volatile* component (i.e. the one with the lower boiling point). This vapour condenses near the bottom of the fractionating column and the resulting liquid runs back towards the flask. More hot vapour passes from the flask to the column and thus the descending liquid is reheated and gives off vapour which is again richer in the more volatile component than it is itself. This process is repeated many times as the mixture passes up the column until eventually the vapour escaping to the condenser consists of the more volatile component only.

Fractionating columns are usually packed with glass beads or tubing to provide a large surface area on which condensation and evaporation may take place, thus improving the efficiency of the process. As the experiment proceeds the column gradually heats up from the bottom and vapour of the less volatile component climbs higher and higher until it, too, passes into the condenser. Careful observation of the thermometer enables fractions with different boiling point ranges to be collected.

You should now be able to understand why the first fraction collected in Experiment 1.2.k is almost pure ethanol and successive fractions contain progressively less of it, the last one being mainly water.

Important applications of fractional distillation include
  (a) the separation of liquid air into oxygen, nitrogen, etc. (18.1),
  (b) the separation of crude oil into petrol, paraffin, etc. (21.6) and
  (c) the manufacture of spirits (e.g. gin, whisky, brandy).

Summary

The chart shown in Figure 1.2 sums up what you have learned so far about the separation of solids and liquids from various mixtures.

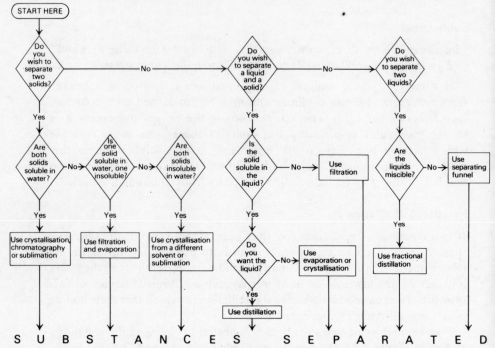

Figure 1.2 Flow diagram for the separation of a mixture

1.3   Tests for purity

A pure substance contains just one type of material. There are many cases where the use of pure substances is essential, one obvious example being in the preparation of medicines. Can you think of any others?

If the purity of a substance is to be tested, how can we do it? Most of the methods of chemical analysis used to detect the presence of impurities are too complicated for you to study at this stage, although chromatography is often used for this purpose. However, there *are* tests which you can carry out for yourself.

Pure substances usually have definite, sharp melting points and boiling points and the addition of only a trace of a contaminant will alter these values. Provided that the temperatures required are not too high, then, it is a simple matter to test for purity.

The effect of impurity on boiling point

The boiling point of pure water is normally stated to be 100°C (373 K). However, water only boils at this temperature at a pressure of 1 atmosphere (101325 Pa) and since the pressure of the air varies from this value, depending on the weather conditions, the boiling point of the water is rarely exactly 100 °C. In addition, the thermometer used may be slightly inaccurate. (To check this, try putting several thermometers in a beaker of water to see if they all give the same reading.)

Whatever value is obtained for the boiling point of pure water, the addition of an impurity,such as salt, causes it to *rise* (Experiment 1.3.a).

It would be dangerous to heat an inflammable liquid like ethanol in an open tube because of the fire hazard. You should, instead, use the apparatus in Figure 1.3. The anti-bumping granules ensure that the liquid boils smoothly; the vapour condenses in the long glass tube and then drips back into the boiling tube.

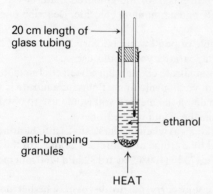

20 cm length of glass tubing

ethanol

anti-bumping granules

HEAT

Figure 1.3 Apparatus for finding the boiling point of an inflammable liquid

Measurement of boiling point often helps in the identification of a pure liquid. Suppose that you carried out chemical tests which led you to suspect that a liquid was ethanol; if you then found that it boiled at 78 °C, your suspicions would be confirmed.

Melting point

Pure water freezes at 0 °C but the addition of a trace of solute *lowers* this value. No doubt you know that salt is used to melt ice on roads in winter, and this may have led you to think that the salt raises the temperature in some way. However, when salt is added to ice and water the mixture becomes *colder* (Experiment 1.3.b). This may be explained by the following example.

Suppose that the temperature of the air were −5 °C. Any water on the roads would be frozen. However, if sufficient salt were added to lower the freezing point of the water to, say, −10 °C then the temperature of the air would be above the freezing point of the ice/salt mixture and the ice would melt. To do this it would have to absorb heat and thus the temperature of the mixture would fall.

Similarly, the presence of benzoic acid impurity *lowers* the melting point of naphthalene and makes it less sharp (Experiment 1.3.c).

Summary

The three experiments in this section illustrate the general rule that the presence of dissolved solid impurities raises the boiling point of a liquid and lowers the melting point of a solid, making it less sharp. (You may find it helpful to remember that when a substance becomes impure its melting point and boiling point move apart on the temperature scale.)

Questions on Chapter 1

1 Draw a diagram of the bunsen burner, labelling the various parts. Explain why the flame sometimes 'strikes back', and what you would do if this were to happen.

2 Name the method you would use to separate mixtures of the following substances: (a) sugar and powdered glass, (b) the coloured materials in a fruit juice, (c) petrol and water, (d) chalk and ammonium chloride. Describe *one* of the methods in detail.

3 How would you convert a sample of pond water into drinkable water? Include in your answer a diagram of the apparatus you would use.

4 You suspect that some potassium nitrate crystals have been contaminated with chalk dust. How would you confirm your suspicions? (Potassium nitrate is soluble in water, chalk is not.) Describe in detail the method you would use to obtain pure *crystals* of potassium nitrate from the contaminated ones.

5 Explain the following observations: (a) when black ink soaks into blotting paper the 'blot' is sometimes surrounded by a fringe of a different colour, (b) the boiling point of sea water is slightly above 373 K (100°C) at 101325 Pa (760 mm mercury) pressure.

6 Describe an experiment you could carry out to discover whether domestic paraffin is a pure liquid or a mixture of miscible liquids.

7 The supply of distilled water for your mother's steam iron has run out. She has been told that ice from the refrigerator may be melted and used in the iron but is not sure whether to use ice cubes or the 'frost' which collects on the ice box. Would either of these be of any use? Explain your answer.

8 You have found a plastic bottle of a colourless liquid in the laboratory. When you test it you find that it boils at 373 K (100°C). Why is it *not* safe to drink the liquid without further tests?

9 Describe how you would determine the melting point of benzoic acid, which is of the order of 393 K (120 °C). What would happen to the melting point if a little solid impurity were added?

10 (a) You are provided with a powdered mixture of calcium carbonate, sodium chloride and naphthalene (a solid hydrocarbon).

(i) Describe how, by the use of suitable *named* solvents, you could obtain pure samples of *each* of the three constituents of the mixture.

(ii) Outline a method by which you could determine the percentage by mass of naphthalene in the mixture.

(b) A given sample of a compound is known to have a melting point of between 60 °C and 80 °C but the exact melting point of the pure compound is unknown. Explain briefly how you could find out whether the sample is pure or impure.

(C)

2 Classes of substance and types of change

If you were asked to make a list of all the substances which exist you would run out of paper (or patience!) long before you finished the task. Thus, if you are going to study the properties of the materials which make up our world, it will simplify matters if you can find some means of grouping together those substances which have things in common. In this chapter you will learn how to place materials into three main classes and also how to divide the many changes which take place around us into just two types.

Experiments

2.1.a To investigate the effect of heat on copper(II) sulphate crystals

Experiment 1.2.c shows you how to purify copper(II) sulphate crystals. Now we will look at the *pure* solid and try to split it up into something *simpler*. The first problem is to find a way of telling whether a simpler substance has been formed. Perhaps measuring the mass before and after the experiment will give us a clue – would you expect the mass to increase or decrease if a simpler substance were formed?

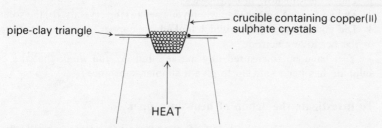

pipe-clay triangle

crucible containing copper(II) sulphate crystals

HEAT

Figure 2.1.a

PROCEDURE Find the mass of a crucible, half fill it with copper(II) sulphate crystals and find the mass again. Heat the crucible and contents on a pipe-clay triangle as shown in Figure 2:1.a for about 5 minutes, leave it until it is cool enough to be held in the palm of the hand (**care!**) and find the new mass. Empty out the residue, rinse the crucible with distilled water, dry it and once more find the mass.

Questions

1 Did the mass of the crucible or the copper(II) sulphate change?
2 Do you think that the copper(II) sulphate changed into a simpler or a more complicated substance? Give a reason for your answer.

2.1.b To investigate the effect of an electric current on copper(II) sulphate ★

The name copper(II) sulphate implies that the crystals consist of copper and something else. Did you obtain copper in the previous experiment? Perhaps the use of an electric current will help.

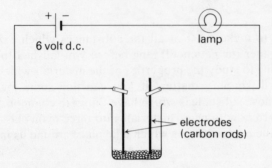

Figure 2.1.b

PROCEDURE Set up the circuit shown in Figure 2.1.b and, making sure the electrodes do not touch, dip them into about 1 cm depth of copper(II) sulphate crystals in a 100 cm³ beaker. Note whether a current flows (i.e. whether the lamp lights up), then remove the rods and examine them. Replace the crystals with about 50 cm³ of copper(II) sulphate solution and repeat the experiment but this time leave the rods in position for 1 minute. (**N.B.** On *no account* must mains electricity be used in experiments of this type.)

Questions

1 Did a current flow in both cases?
2 Did you notice any change when you used (a) the crystals, (b) the solution?
3 Did you obtain anything which looked like copper? If so, was it at the positive or the negative electrode?
4 You have not measured any masses, but do you *think* that the copper(II) sulphate has been split up to give a simpler substance?

2.1.c To investigate the action of heat on copper

PROCEDURE Place a piece of copper foil (a square of side 2 cm) on a watch glass and find the mass. Using tongs, hold the foil in the bunsen flame for about 2 minutes. Allow it to cool on the watch glass, note any change in appearance and find the mass again.

Questions

1 How did (a) the appearance, (b) the mass of the foil change?
2 Do you think that the copper changed into a simpler or a more complicated substance? Give a reason for your answer.
3 Find out the meaning of the word 'element'.

2.1.d A simple classification of elements

Elements form the first of the three main classes of substance mentioned at the beginning of the chapter. However, they themselves may be subdivided into two types – metals and non-metals.

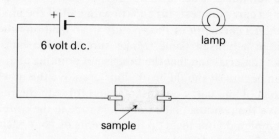

lamp

sample

Figure 2.1.d

PROCEDURE Copper is a typical metal; use a piece of copper foil (1 cm × 5 cm) for the following experiments. (a) Note its appearance, (b) try to bend it, (c) find if it conducts electricity by connecting it with crocodile clips in the circuit shown in Figure 2.1.d, (d) investigate its heat-conducting properties by holding one end in the bunsen flame and seeing if the other end becomes hot. Repeat the experiments with a lump of a typical non-metal, sulphur, but do not heat it (it will melt and burn). If you cannot bend it, try hitting it with a hammer. Check on the properties of other samples of metal foil, such as iron, aluminium, zinc and lead; test also some other non-metals, such as a carbon rod and a lump of silicon. Set out your results as in Table 2.1.d.

Name of substance	Metal or non-metal	Shiny or dull	Flexible or brittle	Good or bad conductor of electricity	Good or bad conductor of heat

Table 2.1.d Some typical properties of metals and non-metals

Questions

1 What are the general properties of (a) metals, (b) non-metals?
2 Do all elements fit exactly into the two classes? If not, give examples.
3 For each property in the table name a different metal which is used in everyday life because it has this property.
4 Can you think of any solid non-metallic elements used in everyday life?

2.1.e Some properties of aluminium and sulphur

PROCEDURE (a) Examine some flowers of sulphur (or very finely powdered roll sulphur) with a hand lens.
(b) Add a little of the powder to water in an evaporating basin and see if there is any change or smell produced.
(c) Heat a little of the sulphur *gently* in a test tube – stop heating if the sample melts.

(d) Repeat (a), (b) and (c) with some *fine* aluminium powder in place of the sulphur. (**Care:** aluminium powder is highly inflammable.)

(e) Mix some of the sulphur powder with about twice its own volume of aluminium powder and carry out tests (a) and (b) with half of the mixture.

(f) Place *not more than 1 cm depth* of the mixture in an ignition tube and, using tongs, heat it in a bunsen flame. Make sure that the bunsen is standing on fire-proof material and that the tube is not pointing at anyone. As soon as a reaction occurs (it should be obvious), remove the tube from the flame, allow it to cool and carry out tests (a) and (b) with the residue. (If you cannot remove the residue, place the whole tube in the water.) As soon as you have completed your tests tip the contents of your basin into a large beaker in the fume cupboard.

Questions

1   Was heat produced or absorbed on mixing the aluminium and sulphur, as far as you could tell?

2   Do you think more heat was produced than was absorbed on heating the mixture of aluminium and sulphur? Give a reason for your answer.

3   Could you see the separate elements in (a) the mixture, (b) the residue after heating?

4   Did the effect of water on (a) the mixture, (b) the residue differ from that on the separate elements?

5   Powdered roll sulphur is soluble in a liquid called methylbenzene (toluene) but aluminium is not. Bearing in mind that this liquid is highly inflammable, can you devise a method of separating the aluminium and sulphur mixture into its components? (Your teacher may demonstrate this experiment to you but you should *not* attempt it yourself.)

2.2.a   The effect of heat on various substances

Heat may bring about both physical and chemical changes. In this experiment you need to know the meaning of these terms and so you should read the relevant section of the text (2.2) before beginning.

PROCEDURE   Heat the following substances separately in ignition tubes: a small iodine crystal, 1 cm depth of sugar, zinc oxide, lead(II) nitrate, potassium manganate(VII) (potassium permanganate). Using tongs, heat 3-cm lengths of nichrome wire and magnesium ribbon. (Do not look directly at the flame when you heat the magnesium; use tinted glass if available.) Your teacher may demonstrate the effect of plunging a hot glass rod into a small pile of ammonium dichromate(VI) (ammonium dichromate) on an asbestos board. Look carefully for the formation of

Substance heated	New compound formed?	Difficult to reverse?	Heat produced?	Type of change
	√ or × or ?	√ or × or ?	√ or × or ?	chemical *or* physical

Table 2.2.a The effect of heat on various substances

what appears to be a new substance, production of heat and ease of reversibility. In each case, let the residue cool down before making your final observations. Record your results as in Table 2.2.a.

Questions

1 Did you find it easy to classify *all* of the changes in the experiment?
2 Can you think of another check or a measurement which you could make to help you come to your decision?
3 What sort of change, physical or chemical, takes place when an egg is boiled?
4 In all of the separation experiments in Chapter 1 the same type of change takes place. Which type is it, physical or chemical?

2.1 Elements, compounds and mixtures

Elements

Once we have a *pure* substance (i.e. one substance only), the methods used in Chapter 1 will not change its composition. However, if we convert it into a new substance having a smaller mass we will presumably have made it into something *simpler*, because some of the matter in it will have been removed. Of course, the easiest way of making the mass less would be to throw some of it away, but this would not be very helpful!

When solid copper(II) sulphate is heated (Experiment 2.1.a) its appearance changes and its mass decreases; it has been split into simpler materials, some of which have escaped into the air. Electric current will not pass through copper(II) sulphate crystals (Experiment 2.1.b) but it does decompose a solution of the crystals in water, giving copper at the negative electrode. If the mass of copper produced is compared with that of the copper(II) sulphate which has decomposed it is found that the mass of the copper is the smaller of the two. Therefore, copper must be simpler than copper(II) sulphate.

When copper is heated in air (Experiment 2.1.c), a black coating forms on it and its mass *increases*, i.e. it changes into something more complicated. Has the limit of simplicity been reached? One experiment is not enough to *prove* that this has happened but, in fact, whatever chemical processes are carried out, copper can never be made into anything simpler – it is an *element*.

> An **element** is a substance which cannot be split up into two or more simpler substances by chemical means, e.g. aluminium, sulphur, copper, oxygen.

There are over 100 elements grouped into two main classes, metals and non-metals, according to their properties. Throughout this book, differences between the classes are investigated and a final summary appears in Table 20.9. For the present, it is enough to know that metals are generally shiny, flexible and good conductors of heat and electricity, while non-metals are usually dull in appearance, brittle and bad conductors of heat and electricity. Not all elements fit the pattern exactly, e.g. carbon rods conduct electricity and yet carbon is a non-metal (Experiment 2.1.d).

Many of the non-metals are gases and very few are found free in everyday life.

Compounds and mixtures

A **compound** is a substance which consists of two or more elements chemically combined together, e.g. copper(II) sulphate crystals contain copper, sulphur, oxygen and hydrogen.

Compounds may be broken down by the action of heat or an electric current (Experiments 2.1.a and 2.1.b) but it is often difficult to separate all of the constituent elements. The composition of a given compound is found to be fixed, i.e. it always consists of the same elements combined in the same proportions by mass. In Experiment 2.1.e a compound (aluminium sulphide) is *synthesised* from a *mixture* of its elements (aluminium and sulphur).

Synthesis is the building up of a compound from simpler substances, often its elements.

A **mixture** consists of two or more elements or compounds which have not been chemically combined, e.g. water and ethanol, rock salt.

Heat is given out when the aluminium sulphide is formed and the compound reacts with water to give a gas which smells of bad eggs (hydrogen sulphide). The elements and mixture give no such reaction.

The main differences between compounds and mixtures are summarised in Table 2.1.

	Compound	Mixture
Composition	Fixed	Variable
Heat change	Heat usually produced or absorbed when compound made	Usually no heat change on making mixture
Appearance	Different from constituent elements	'Average' of constituents
Properties	Different from constituent elements	Similar to constituents
Separation	Cannot be separated into constituent elements by physical means (see 2.2)	Can be separated into constituents by physical means

Table 2.1 The properties of compounds and mixtures

The word 'CHAPS' should help you to remember the table.

2.2 Evidence for chemical reactions – chemical and physical changes

When a new substance is made we say that a **chemical change** has taken place or that a **chemical reaction** has occurred.

A **physical change** is one where no new substance is formed.

It is not always easy to *see* if anything new has been made and so, in order to make sure, the chemical properties of the 'new' substance should be compared with those of the starting material. Further evidence to look for is shown in Table 2.2, together with examples of each type of change.

Chemical change	Physical change
New substance formed with different chemical properties.	No new substance formed.
Heat usually produced or absorbed.	No heat change, usually.
Generally difficult to reverse.	Generally easy to reverse.
Mass of *residue* often differs from that of starting material.	Mass of *residue* usually same as that of starting material.
Examples	
Adding aluminium sulphide to water.	Adding copper(II) sulphate crystals to water.
Digesting food.	Freezing water.
Burning coal.	Using an electric fire.

Table 2.2 Characteristics and examples of chemical and physical changes

Note: if heat is produced during a reaction it is said to be **exothermic** whereas if heat is absorbed the reaction is **endothermic**.

Questions on Chapter 2

1 You have been given a rod of unknown material which is believed to be a metal. What tests would you carry out to verify this belief?

2 Without naming any substance more than once, give examples of (a) two metallic elements, (b) two solid non-metallic elements, (c) two gaseous elements, (d) two gaseous compounds, (e) two compounds which are soluble in water, (f) two compounds which are insoluble in water.

3 Distinguish between an element, a compound and a mixture. Classify each of the following as an element, compound or mixture, giving reasons for your choice: (a) sugar, (b) ink, (c) wine, (d) copper(II) sulphate crystals, (e) water, (f) rock salt, (g) copper foil.

4 Give three important differences between compounds and mixtures, illustrat-

ing your answer by reference to aluminium, sulphur and aluminium sulphide.

5   Give four ways in which a chemical change differs from a physical change. Do you think that the following are physical or chemical changes? (a) The boiling of water, (b) the magnetisation of a strip of iron, (c) the decaying of plants on a compost heap, (d) the heating of copper in air. Give reasons for your answers.

6   When a white solid, A, is strongly heated it gives off water and the residue has a smaller mass than the original material. When water is added to this residue a great deal of heat is produced and the original solid, A, is formed again. Does A undergo a physical or a chemical change when it is heated? Explain how you reached your conclusion.

7   Read the following statements:
(a) A pure sample of a solid compound can be distinguished from an impure sample by a melting point determination.
(b) A mixture of iron and sulphur can be distinguished from the compound iron(II) sulphide by treatment with dilute sulphuric acid.
(c) Ammonium chloride can be purified by sublimation.
(d) Sodium nitrate can be purified by crystallisation.

Describe, with essential practical details, how you would carry out the tests in (a) and (b), and the purifications in (c) and (d). Explain how the results of your experiments would enable you to distinguish in (a) between a pure and an impure sample, and in (b) between the mixture and the compound.   (C)

3 The atmosphere and combustion

In this chapter we investigate that common, and yet mysterious substance which surrounds us all – the air. We cannot see it, taste it, nor (usually) smell it, so how do we know it is there? What is it made of? Is it an element, a compound or a mixture? To try to find answers to these questions we will begin by looking at a common chemical reaction which takes place in air, namely combustion, and then see where our discoveries lead us.

Experiments

3.1.a To find if there is a change in mass when magnesium burns in air

PROCEDURE Coil up about 15 cm of magnesium ribbon so that it fits into a crucible, place the lid in position and find the mass of the whole apparatus. Heat the crucible and contents on a pipe-clay triangle as in Figure 2.1.a but in this case keep the lid in place. When the bottom of the crucible is red hot, lift the lid slightly for a second or so and observe what happens to the magnesium. Replace the lid so that no smoke escapes. Repeat the procedure at 5-second intervals until no further change takes place and then heat for about a minute with the lid off. Replace the lid, allow the crucible to cool and find the mass of the apparatus and contents.

Questions

1   What happened to the magnesium (a) on raising the lid, (b) on replacing the lid? What conclusion can you draw from this?
2   How did the mass of the crucible and ash compare with the mass of the crucible and magnesium? Did you expect this result?
3   If the mass increased, where do you think the extra material came from? If the mass decreased, where did the lost material go? Your answer to question 1 should give you a clue.

3.1.b The action of heat on copper – a closer look

PROCEDURE   Fold a square piece of copper foil of side 4 cm in half and then fold over a strip 0.5 cm wide round the open edges. (**Caution**: be careful to avoid cutting your fingers.) Press the copper as flat as possible so that air is excluded from the inside of the resulting 'envelope'. Now carry out Experiment 2.1.c using the copper envelope but after the final measurement of mass, open the envelope and examine it.

Questions

1 In what way did the mass of the foil change?
2 What difference was there in the appearance of the foil on the inside and on the outside of the envelope?
3 Can you offer an explanation for the results and is it in agreement with your conclusions from Experiment 3.1.a?

3.1.c To find if there is a change in volume when white phosphorus is burned in air ★ ★

This experiment must not be performed by pupils.

This experiment involves the use of white phosphorus, a poisonous element which catches fire spontaneously (i.e. of its own accord) in warm air. It is stored under water and must *never* be touched by hand.

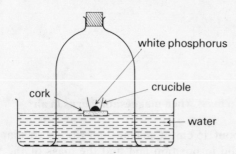

white phosphorus

crucible

cork

water

Figure 3.1.c

PROCEDURE Using tongs, place a small piece of phosphorus (a cube of side 0.5 cm) in a crucible, floating on a cork in a pneumatic trough. Cover this with a bell-jar as shown in Figure 3.1.c, ignite the phosphorus by touching it with a warm glass rod and quickly insert the stopper in the jar. Watch the water level inside the jar carefully and when the reaction ceases, leave the apparatus to cool.

Questions

1 What happened to the water level inside the jar at first? Can you explain this?
2 What do you think happened to the smoke from the burning phosphorus?
3 What happened to the water level inside the jar by the end of the experiment? Can you offer an explanation for this which agrees with those given in Experiments 3.1.a and 3.1.b?
4 Was (a) all of the air or (b) all of the phosphorus used up? (The white phosphorus may have changed to the red form during the experiment – any red solid remaining *is* phosphorus.)
5 Does the answer to question 4 indicate that air is probably made up of (a) one gas or (b) at least two gases, only one of which supports combustion?
6 Would you expect more air to be consumed if a larger piece of phosphorus were used? Explain your answer.

3.1.d Is there really a change in mass when substances burn? ★ ★

This experiment must not be performed by pupils. (See Experiment 3.1.c.)

In Experiments 3.1.a and 3.1.b you should have found that the residue in each case had a greater mass than the starting materials. Often the reverse is true – can you think of any examples where this is so? In this experiment combustion is carried out in a *sealed* flask so that the mass of the air (and of any gases that may be produced) is included in the results.

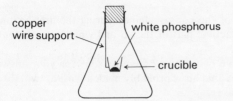

copper wire support

white phosphorus

crucible

Figure 3.1.d

PROCEDURE   Using tongs, place a small piece of phosphorus (a cube of side 0.5 cm) in a crucible in the apparatus shown in Figure 3.1.d. Make sure that the stopper is firmly in place and find the mass of the apparatus. The phosphorus will smoulder but after several days the reaction should cease. When this has happened, predict whether there will be a change in mass, explaining your reasoning, and then measure the mass to see if your prediction is correct. Finally, release the stopper, listening carefully for any noise, and find the mass once more.

Questions

1   Did the mass change while the flask was sealed? Was your prediction correct? If it was not, think again and try to explain the result obtained, saying whether you think that matter is (a) created, (b) destroyed, or (c) rearranged in some way, when phosphorus burns in air.
2   Did the mass change on removing the stopper? What do you think happened in view of the sound you heard when this was done?
3   If the magnesium in Experiment 3.1.a had been burned in a sealed flask, do you think a change in mass would have been detected?

3.1.e   To find how much of the air reacts with heated copper★★

By now you should know that air is necessary for combustion but that only part of it combines with the burning substance. In this experiment you will find the percentage by volume of 'active air' in the atmosphere.

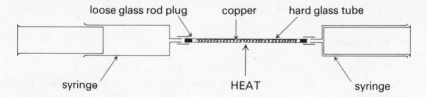

loose glass rod plug   copper   hard glass tube

syringe   HEAT   syringe

Figure 3.1.e

PROCEDURE   Set up the apparatus as shown in Figure 3.1.e, with one syringe containing 100 cm³ of air and the other containing none. Heat the copper strongly and pass air over it several times from one syringe to the other until no further change in volume takes place. Allow the tube to cool and note the volume of the residual gas. Heat the copper again, this time in a different place, and pass the residual gas over it to see if any further reaction occurs.

Questions

1 How did the appearance of the copper change during the first heating?
2 What is the name of the 'active' part of the air and what is its percentage by volume in the sample taken?
3 Was the residual gas active or inactive? Give reasons for your answer.
4 Does this experiment show that air is probably made up of (a) two, (b) at least two, or (c) more than two gases?
5 Do you think that the mass of the whole apparatus and contents changed during the experiment? Explain your answer.

3.1.f To isolate some active air ★ ★

This experiment must not be performed by pupils.

When mercury is heated in air it slowly changes to a red ash (mercury(II) oxide). Because the reaction takes a long time and mercury vapour is poisonous, it is more convenient to use a sample of ash which has already been made.

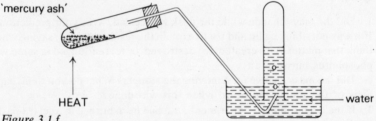

Figure 3.1.f

PROCEDURE Place about 1 g of 'mercury ash' in a Pyrex test tube and fit it with a bung and delivery tube. Heat the ash and, after allowing the first few bubbles to escape into the atmosphere, collect two test tubes full of the liberated gas over water (Figure 3.1.f). Remove the delivery tube from the water *before* you stop heating, allow the test tube to cool and examine it carefully. Place a glowing wood splint in the mouth of a tube of the gas and note the result.

Questions

1 Why were the first few bubbles of gas not collected?
2 Why was the delivery tube removed from the water before heating was stopped?
3 Do you think that the 'mercury ash' decomposed to give mercury and active air back again? Give reasons for your answer.

3.2.a To investigate the conditions for the rusting of iron ★

PROCEDURE Clean eight iron nails, about 3 cm long, by rubbing them with emery cloth and set up four boiling tubes as shown in Figure 3.2.a. Anhydrous calcium chloride absorbs water and ensures that the air in tube

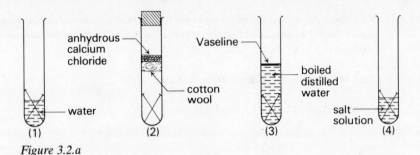

anhydrous calcium chloride

Vaseline

water

cotton wool

boiled distilled water

salt solution

(1) (2) (3) (4)

Figure 3.2.a

(2) is dry. The distilled water in tube (3) should be boiled for a few minutes to ensure that no air is dissolved in it; if Vaseline is placed on the hot water it will melt and, on cooling, re-solidify and seal the surface to exclude air. Leave the tubes for a few days, topping up the liquids in tubes (1) and (4) if necessary.

Questions

1 Which of the following is *essential* for rusting to take place: (a) air, (b) water, (c) both air and water?
2 Which nails rust the most rapidly?
3 Why should cars be washed *underneath* during winter?
4 How do we stop each of the following from rusting: (a) bicycle handlebars, (b) a bicycle frame, (c) steel dustbins, (d) food cans?

3.2.b To find which gas in the air is used up when iron rusts ★

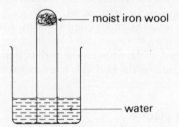

moist iron wool

water

Figure 3.2.b

PROCEDURE Push a plug of iron wool to the bottom of a test tube and add distilled water so that the level is 10 cm below the mouth of the tube. Place your thumb over the end of the tube, invert the tube and place it in a 500 cm³ beaker containing 3–4 cm depth of distilled water as shown in Figure 3.2.b. Leave the apparatus for several days until the water level in the tube ceases to change and then add water to the beaker until the levels inside and outside the tube are the same (why?). Using your thumb to prevent the escape of water, turn the tube up the right way and insert a burning splint into the residual gas. Measure the distance from the water to the mouth of the tube.

Questions

1 What fraction of the air, approximately, was used up?
2 What happened to the burning splint?

3 Which gas do you think causes wet iron to rust? Give reasons for your answer.
4 Apart from the presence of water, what is the main difference between burning and rusting?

3.3.a To find if exhaled air differs from inhaled air ★

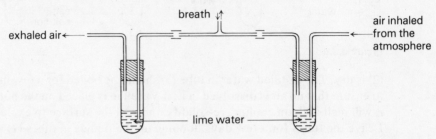

Figure 3.3.a

PROCEDURE (a) Breathe on a thermometer bulb for a few seconds and then on a mirror or other cold glass surface.
(b) Make sure that the T-piece of the apparatus shown in Figure 3.3.a is clean and then gently breathe in and out through it for about a minute. (If the lime water in either tube turns milky it shows that carbon dioxide is passing through.)

Questions

1 How does the temperature of exhaled air differ from that of the atmosphere?
2 Which contained more water, exhaled air or atmospheric air?
3 How did the carbon dioxide content of exhaled air compare with that of the atmosphere?
4 If sugar or starch is *burned* in air, heat, water vapour and carbon dioxide are produced. Refer to questions 1, 2 and 3 and explain, in simple terms, what happens inside us when we breathe.

3.3.b To examine the gas expelled from a plant ★

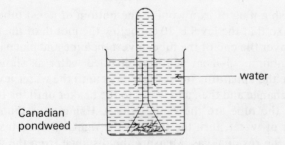

Figure 3.3.b

PROCEDURE Cover some Canadian pondweed with an inverted funnel in a beaker of water, place a test tube filled with water over the funnel (Figure 3.3.b), and leave it for a day or so in direct sunlight or near an electric light. Test the gas which collects in the tube as in Experiment 3.2.b but this time use a *glowing* splint.

Questions

1 What happened to the splint? Which gas does this indicate?
2 Do you think that this process is important? If so, why?

3.4.a Other gases in the air ★

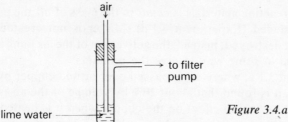

air

to filter
pump

lime water

Figure 3.4.a

PROCEDURE   (a) Fill a 500 cm³ beaker with a mixture of ice and salt and clamp it above the bench. Examine it after about 5 minutes.
(b) Draw air through lime water (Figure 3.4.a) for a few minutes and note any change.

Questions

1 What do you think was formed on the beaker in (a)? If sufficient were collected, how could you confirm this?
2 In (b) what happened to the lime water and what did this show?
3 The two substances detected by this experiment are minor constituents of the inactive part of the air. Which gas is the major constituent?
4 From the results of this experiment, do you think that air is a mixture or a compound? Explain your answer.

3.1   Combustion

When magnesium burns in air it forms an ash which has a greater mass than the original metal. Where does the extra material come from? If the magnesium is burned in a closed crucible (Experiment 3.1.a) combustion only takes place when the lid is raised and air is allowed to enter. The conclusion to be drawn from this is that when magnesium burns, it combines with the air, or something in it. It may have reacted with the crucible, which is the only other thing in contact with it, but then why should combustion not continue with the lid on? In another experiment, using copper without a crucible (Experiment 3.1.b), an increase in mass is again registered. However, where air is excluded by folding the copper, the black coating of ash does not form. Again, the conclusion is that the hot metal combines with the air. Final confirmation that air is used up in burning is obtained in Experiment 3.1.c where phosphorus is ignited over water in a bell-jar. The water level in the jar rises as the air is consumed.

Another important observation is made in this experiment, namely that even though some phosphorus is left unburned, only *part* of the air is used up. The implication is that some of the air is active and combines with burning substances while the rest is inactive.

Though the 'ash' in these experiments has a greater mass than the original solids, the reverse is frequently the case (e.g. paraffin leaves little or no residue when it burns in a domestic heater). The reason for this is that some, or all, of the products of combustion in these reactions are gases and they pass off into the atmosphere. In fact, the *increase* in mass of the products is equal to the *loss* in mass of the air. This means that the mass of *all* the reacting materials is *equal* to the mass of *all* the products, gases being included (Experiment 3.1.d). Matter is not created during burning, nor is it destroyed; instead, the active part of the air joins on to the burning material in some way to give the products.

In Experiment 3.1.e, where air is passed over heated copper by means of gas syringes, it is found that about 20% by volume of the air is active, the residual gas having no effect on the copper when it is heated for the second time. If this active air could be isolated in some way you might expect combustion in it to be more vigorous than when it is 'diluted' in the atmosphere. Experiment 3.1.f shows that this is so. It was by a similar experiment that the French chemist Antoine Lavoisier first successfully explained what happens when substances burn. Lavoisier's work was published in 1774 and it was he who named the active part of the air *oxygen*.

When a substance combines with oxygen the process is called **oxidation** and the substance is said to be **oxidised**.

3.2 The rusting of iron

Iron will only rust if both air *and* water are present, and dissolved salt is found to speed up the process (Experiment 3.2.a). Thus, when salt is used to melt ice on the roads in winter it accelerates the corrosion of cars and may involve the motorist in costly repair bills. The gas in the air which is responsible for rusting is oxygen (Experiment 3.2.b). Rusting, like combustion, is an example of *oxidation*, but in this case the temperature is much lower and the reaction occurs more slowly.

Many metals tarnish in air but the coating formed on the surface often prevents, or at least slows down, further corrosion. This is why lead on church and cathedral roofs lasts for hundreds of years. Other metals which form protective coatings when exposed to the air include zinc, copper and aluminium. Rust, unfortunately, does not prevent air and water from reaching the iron underneath it and the result is that millions of pounds are spent each year in the battle to halt corrosion, or in replacing parts which have rusted away.

3.3 Respiration and photosynthesis

When we and other animals breathe, the air which we exhale is warmer and contains more water vapour and carbon dioxide than it did when we inhaled it (Experiment 3.3.a). The burning of a food (such as sugar or starch) in air also produces water vapour, carbon dioxide and heat. Thus, it would seem that we are 'burning' food inside us to provide energy for movement and warmth. The oxidation in our bodies is taking place at a much lower temperature than that at which food burns in air, but it must be carried out very efficiently because we do not breathe out smoke or soot!

The overall change may be summed up as

$$sugar + oxygen \rightarrow carbon\ dioxide + water + energy.$$

Since combustion, rusting and respiration all involve the removal of oxygen from the atmosphere and are going on continuously, you may well be wondering why the oxygen was not all used up years ago. Experiment 3.3.b gives the answer to this problem – the correct balance of oxygen is maintained by plants during *photosynthesis*, a large tree providing enough oxygen for two humans.

> **Photosynthesis** is the process by which green plants synthesise carbohydrates from carbon dioxide and water, using sunlight as the source of energy and chlorophyll as a catalyst, oxygen being liberated into the air.

For the present it is sufficient to know that sugar is a carbohydrate and that a catalyst 'helps' a reaction to take place. The overall change may thus be written

$$carbon\ dioxide + water + energy \rightarrow sugar + oxygen.$$

This is the reverse of the equation summarising respiration.
Note: photosynthesis only takes place by day but plants also absorb oxygen and liberate carbon dioxide, just as we do, both by day and by night. However, on balance, more oxygen is liberated than is used.

Summary

When a substance burns in air it combines with oxygen, and combustion continues until all of the substance *or* all of the oxygen is used up. Other processes which involve combination with oxygen (i.e. oxidation) include the rusting of iron and respiration, but these take place at much lower temperatures. The oxygen content of the atmosphere is maintained by photosynthesis in green plants.

3.4 The composition of the air

About one-fifth of the air by volume is *oxygen*, the rest being mainly *nitrogen*. Between them these gases make up about 99% of the volume of a typical air sample. Experiment 3.4.a shows that *water vapour* and *carbon dioxide* are also present and the fact that they may be removed by physical means indicates that air is a *mixture* and not a compound. The proportion of these gases varies, depending on conditions. For example, on a humid day the water vapour content may be as high as 3 or 4%, perspiration evaporates only slowly and we feel 'sticky'; in a crowded train, the carbon dioxide concentration rises and the air becomes 'stuffy'.

In addition there are traces of the *noble gases* such as argon, neon and helium (see Table 3.4).

Gas	% by volume
Nitrogen	78.1
Oxygen	20.9
Argon	0.9
Carbon dioxide ⎫ Neon ⎬ Helium ⎪ + others ⎭	0.1

Table 3.4 The major components of dry air

Air in built-up areas will be polluted by fumes from cars, fires, factories, etc. Substances present may include dust, soot, *lead compounds* from car exhaust gases, *carbon monoxide* (16.5), *nitrogen dioxide* (17.2) and *sulphur dioxide* (18.6). Most of these are corrosive and/or poisonous.

Finally, there are usually *bacteria* and *viruses* present which may or may not be harmful.

Thus, you can see that air is a very complicated mixture indeed. The point which emerges from this is that just because a substance is *common*, this does not necessarily mean that it is *simple*, and you should bear this in mind for the future.

Questions on Chapter 3

1 Explain why magnesium gains in mass but coal loses mass when each is burned in air.

2 A chip pan has caught fire in the kitchen. Should your mother (a) pour water on it, or (b) cover it with a damp towel? Explain your answer.

3 A small child has been playing with matches and his clothes have caught fire. Explain how you would put out the flames.

4 Give the conditions necessary for iron to rust. How would you show that burning and rusting both involve the removal of oxygen from the air?

5 How could you prove that rusting is the combination of iron with *dissolved* oxygen in water and not the oxygen in the water molecules?

6 Give a simple outline of the chemistry of respiration in mammals. How is it that the percentages of oxygen and carbon dioxide in the atmosphere remain approximately constant?

7 (a) Name three gases, other than oxygen and nitrogen, which are present in the atmosphere.

(b) Describe how you would determine the percentage of oxygen in the air.

8 Argon is a minor constituent of the atmosphere and is present to the extent of about 1% in dry air. Calculate the volume of argon in your laboratory. The group

of gases to which argon belongs used to be referred to as the 'rare gases'. Do you think that this was a good name as far as argon was concerned?

9 (a) Explain what you understand by the term *photosynthesis*.

(b) Name *two* non-metallic oxides which cause pollution of the atmosphere. Indicate the sources of this pollution and mention steps being taken to minimise its effects.

(c) Compare the two processes 'combustion' as applied to fuels, and 'respiration' as applied to animals. (O & C)

4 Water and hydrogen

You probably do not think of water as an unusual substance because it is so familiar to you. However, in many ways it is surprising in its behaviour and as your course develops you will discover more and more about its unexpected properties.

Experiments

4.1.a To extract air from water★★

In the last chapter you found that there is water in the air. This experiment is designed to see if there is air in water.

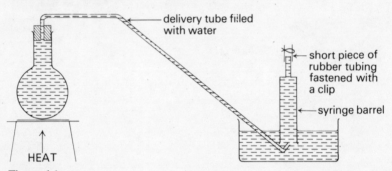

Figure 4.1.a

PROCEDURE Fill a $2\,dm^3$ round-bottomed flask to the brim with tap water and fit it with a bung and delivery tube. The delivery tube must not protrude into the flask beyond the bung and it, too, must be filled with tap water. Set up the apparatus as shown in Figure 4.1.a, making sure that there are no air bubbles anywhere, and heat the flask until the water boils (this will take a considerable time if only one bunsen is used). Carry on heating until there is no further change in the volume of gas in the syringe.

Questions

1   Does air become (a) more soluble, (b) less soluble in water as the temperature rises?
2   How did the volume of bubbles in the flask compare with the volume of gas collected in the syringe? Can you explain the difference?
3   Why is the presence of dissolved air in water important?

4.1.b To find the percentage of oxygen in dissolved air ★★

PROCEDURE Draw the boiled-out air from Experiment 4.1.a into a

100 cm^3 syringe and transfer it to the apparatus shown in Figure 3.1.e. Find the percentage of oxygen in the air sample as before.

Questions

1 What was the percentage of oxygen in the sample and how does it compare with the percentage of oxygen in atmospheric air?
2 Do you think that oxygen is (a) more soluble, (b) less soluble in water than nitrogen?

4.2.a To find what happens when copper(II) sulphate crystals are heated

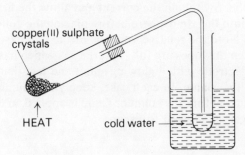

copper(II) sulphate
crystals

HEAT   cold water

Figure 4.2.a

PROCEDURE Set up the apparatus as shown in Figure 4.2.a, with the left-hand test tube about one-third-filled with dry copper(II) sulphate crystals. Heat this tube gently; if the solid goes brown/black, it is too hot. Keep the delivery tube out of the liquid which collects in the other tube (why?) and make sure that all of the solid is heated. Continue until no more liquid condenses and then find its boiling point as in Experiment 1.3.a. When the solid residue from the heated tube is cool, tip half of it into a crucible held in the palm of the hand and add half of the collected liquid. Repeat this procedure with the other half of the solid residue but this time use distilled water in place of the collected liquid. Finally, test separate samples of the collected liquid and distilled water with cobalt(II) chloride paper.

Questions

1 Why was the right-hand test tube placed in a beaker of water?
2 What colour changes did you see when the copper(II) sulphate crystals were heated?
3 What was the boiling point of the liquid? Suggest what this liquid might be.
4 What changes did you observe when (a) the liquid, (b) distilled water were added separately to the solid residue? Was your suggestion in question 3 confirmed?
5 Do you think that the effect of heat on copper(II) sulphate crystals is a chemical or a physical change? Give reasons for your answer.
6 How did the colour of the cobalt(II) chloride paper alter when you added (a) the liquid, (b) distilled water? What could the solid residue and cobalt(II) chloride paper be used to test for?

4.3.a To investigate the effect of an electric current on water ★

You have seen in Chapter 2 that compounds are often broken down by an electric current. Let us find if water may be decomposed in this way.

PROCEDURE   (a) Place distilled water in the apparatus shown in Figure 11.2.b and switch on the electricity supply. Note the reading of the ammeter and look for changes at the electrodes. Disconnect the electricity supply and dissolve some sodium fluoride (a compound consisting of the elements sodium and fluorine *only*) in the distilled water. Switch on again and repeat the observations.

(b) Transfer the solution to the U-tube of the apparatus shown in the lower half of Figure 11.2.d, making sure that the platinum electrodes are near the base of the U-tube. Switch on the current and allow the first few bubbles of gas to escape into the atmosphere before placing the collecting tubes in position (why?). Note the relative volumes of the two gases and switch off the current when the tubes are full. Test the gas which has been evolved at the positive electrode with a glowing splint and test that from the negative electrode with a flame. In each case, keep your thumb over the end of the tube until just before the splint or flame is applied so that as little as possible of the gas is allowed to escape.

Questions

1   Did the distilled water conduct electricity? How did the sodium fluoride affect the conducting properties?

2   In (b) what was the ratio of the volume of gas from the negative electrode to that of the gas from the positive electrode?

3   What happened to the glowing splint when it was placed in the gas from the positive electrode? Which gas did this indicate?

4   What happened when the gas from the negative electrode was tested with a flame? What do you think the gas was?

5   How do you know that the gases came from the water and not from the sodium fluoride?

6   Does the experiment show that water consists of (a) two, (b) at least two elements combined together?

4.3.b To find if water is formed when hydrogen burns in air ★ ★

This experiment must not be performed by pupils.

Experiment 4.3.a showed you that water can be broken down to give

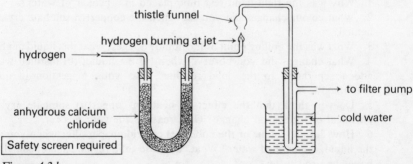

Figure 4.3.b

hydrogen and oxygen. Now we will try to synthesise water by burning hydrogen in air (i.e. combining it with oxygen).

PROCEDURE Pass hydrogen through a U-tube containing anhydrous calcium chloride to ensure that no water is present. Allow the air to be displaced, then collect a test tube full of the gas over water and ignite it well away from the apparatus. Repeat this until the sample burns *quietly*.

Set up the apparatus as shown in Figure 4.3.b, turn on the filter pump, ignite the hydrogen jet and adjust the flow of gas until the flame is about 3 cm high. After about ten minutes, turn off the hydrogen supply, disconnect the filter pump and by means of suitable tests try to identify the liquid that condenses in the tube.

Questions

1 Why was it important to test the gas before igniting the jet?
2 Why was the hydrogen dried?
3 What was the purpose of the filter pump?
4 What tests did you apply to the condensed liquid and what did they show?

4.3.c The reaction between hydrogen and oxygen★★

This experiment must not be performed by pupils.

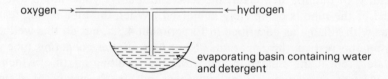

Figure 4.3.c

PROCEDURE Pass an electric current through sodium fluoride solution in a U-tube as in part (b) of Experiment 4.3.a but this time mix the gases by passing them through a T-piece. Allow the mixture of gases to bubble into water in an evaporating basin and when all the air has been displaced from the apparatus, dip the end of the delivery tube into another evaporating basin containing a solution of detergent (Figure 4.3.c). When the surface is covered with bubbles, switch off the electricity supply and replace the delivery tube in the basin of water (why?). Move the second evaporating basin at least 2 metres away from the rest of the apparatus and ignite the bubbles at arm's length.

Note: a faster flow of bubbles is obtained if dilute sulphuric acid is substituted for the sodium fluoride solution.

Questions

1 What happened on ignition? Why was the result different from that in Experiment 4.3.b?
2 What do you think was formed?

4.4.a The laboratory preparation of hydrogen

Although hydrogen can be prepared by decomposing water with an electric current, a more usual way of preparing it in the laboratory is by the action of dilute sulphuric or hydrochloric acid on zinc.

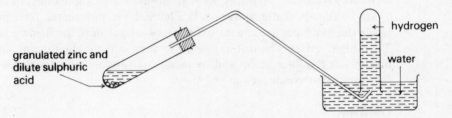

granulated zinc and dilute sulphuric acid

hydrogen

water

Figure 4.4.a

PROCEDURE   Place two or three pieces of granulated zinc in a 150×19 mm test tube, one-third-fill the tube with dilute sulphuric acid and fit a delivery tube as shown (Figure 4.4.a). Add a few crystals of copper(II) sulphate if the reaction is slow. Allow the first few bubbles of gas to escape into the atmosphere (why?) before placing the collecting tube in position. (You should fill the collecting tube with water and cover the end with your thumb, then invert the tube and remove your thumb when the end of the tube is completely immersed.) When the tube is full, test the gas with a flame as described in Experiment 4.3.a, but do this well away from the rest of the apparatus. Now *half-fill* the collecting tube with water, repeat the collection and ignite the resulting mixture, which will consist of equal volumes of hydrogen and air. By means of a similar procedure carry out the ignition test with various ratios of gas volumes. **Do not apply a flame directly to the delivery tube.**

Questions

1   Write an equation for the reaction between zinc and the acid.
2   What happened when a flame was applied to the gas and what do you think was formed? Write an equation for the reaction.
3   Which mixture gave the loudest pop, which the quietest, and which, if any, did not react?

4.4.b To compare the density of hydrogen with that of air★★

PROCEDURE   Fill a large balloon with hydrogen, secure the end tightly and place the balloon on the bench. Observe what happens. Alternatively, devise an apparatus for blowing bubbles filled with hydrogen.

Questions

1   What did you observe?
2   What conclusion can you draw concerning the density of hydrogen compared with that of air?

4.4.c To investigate the effect of hydrogen on heated copper(II) oxide ★ ★

This experiment must not be performed by pupils.

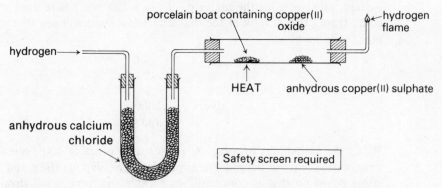

Figure 4.4.c

PROCEDURE   Fill a porcelain boat with copper(II) oxide and place it in a hard glass combustion tube. Place a small heap of anhydrous copper(II) sulphate near one end of the tube and assemble the apparatus as shown (Figure 4.4.c). After checking that a sample of the gas issuing from the jet burns quietly (see Experiment 4.3.b), ignite the hydrogen and adjust the flow so that the flame is about 3 cm high. Heat one end of the porcelain boat until the reaction starts and then remove the flame. Heat again, if necessary, to finish the reaction, and finally allow the apparatus to cool down with the hydrogen still passing.

Questions

1   Why was the hydrogen passed through anhydrous calcium chloride?
2   Did any reaction between hydrogen and copper(II) oxide occur in the cold?
3   When the reaction started, was heat given out? How could you tell?
4   What happened to the anhydrous copper(II) sulphate and what did this tell you?
5   What did the copper(II) oxide appear to change into? Write an equation for the reaction.
6   Since copper(II) oxide consists of the elements copper and oxygen only, what does this experiment tell you about the composition of water?
7   Why was it essential to let the product cool down in hydrogen?

4.1   Water, the most common liquid in the world

Water is a very widespread and important liquid indeed, covering over two-thirds of the earth's surface and occurring in the atmosphere, in rocks and in all living organisms. You may be surprised to learn that over two-thirds of your own body, by mass, is made up of it. Water is extensively used in industry for heating and cooling, for producing steam to drive turbines, and as a solvent. In addition, vast amounts are used in the home – each person in Britain, on average, uses about 160 dm³ (36 gallons) every day.

Sources of water – desalination

Have you ever wondered where our water supplies come from? In Britain, rain provides the majority of our water but where does the rain come from? If you look at the diagram below, you will see that a cyclic process is involved.

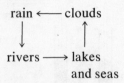

When the sun shines or a breeze passes over a sea or lake, some water evaporates, leaving behind the impurities. The vapour rises and slowly cools down so that it condenses to form many very small droplets (a cloud). If these droplets come together, they fall as rain, which eventually returns to a sea or lake so that the cycle can begin again. You have probably realised that one of the separation techniques in Chapter 1 resembles this process – which one is it?

There are many parts of the world where the rainfall is very low and it is difficult to grow crops or support animals. The resulting famine and hardship would largely be alleviated if adequate supplies of water could be provided. Even in Britain there are some areas where the demand for water has outgrown the supply, so that purification of contaminated water for re-use is necessary. (It is said that the water of the river Thames is drunk and recycled seven times before it reaches the sea!)

The sea contains a huge volume of water but it also contains many salts, making it undrinkable. The total mass of dissolved solids is vast – from one cubic metre of sea water (which you can easily visualise) you could

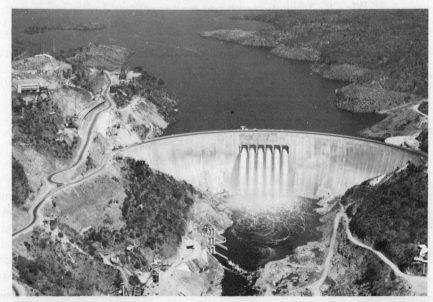

The Kariba Dam, Rhodesia

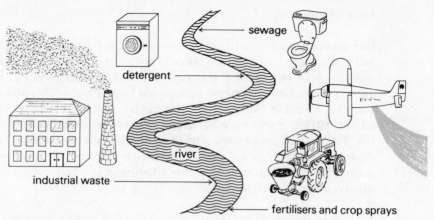

Figure 4.1 Some causes of water pollution

extract about 28 kg of sodium chloride (common salt), 7 kg of hydrated magnesium chloride, 5 kg of hydrated magnesium sulphate and 2 kg of hydrated calcium sulphate, together with smaller quantities of other salts such as potassium chloride and potassium bromide. When these salts are removed (a process known as *desalination*), the water may be used for domestic consumption and one of the most common ways of doing this is by distillation. For example, nuclear power stations produce heat to generate electricity; in dry countries some of this heat may be used to evaporate sea water instead.

Pollution of water

In recent years, water pollution has become a serious problem. The main sources of trouble are shown in Figure 4.1, and great pains are now being taken to prevent contamination of our rivers and lakes.

Water contains dissolved air, the percentage of oxygen in it being about 30%, compared with 20% in the atmosphere, because oxygen is more soluble than nitrogen (Experiments 4.1.a and 4.1.b). This oxygen is used by fish and other aquatic creatures, and also by bacteria which feed on animal and vegetable remains and keep our rivers healthy. Too much untreated sewage causes the bacteria to multiply at a tremendous rate and they may use up so much oxygen that there is insufficient left for fish to survive.

Fertilisers, insecticides, detergents and industrial waste all help to upset the delicate balance of life in the water and in extreme cases pollution is so bad that the river 'dies' – no life exists in it and it becomes a black, evil-smelling sewer. Even in small concentrations, the pollutants constitute a hazard since they can be absorbed by the fish and subsequently eaten by us.

The rivers carry waste to the sea, where it is diluted to a minute concentration and in the past it was considered that no harm could be caused to the vast oceans of the world by materials dumped in them. However, concern is now growing that marine life may be suffering from the effects of living in mankind's sink.

Domestic supplies

Our drinking water must contain none of the pollutants mentioned above and elaborate precautions are taken to ensure that they are removed. This must not lead you to think that tap water contains no dissolved solids at all – try evaporating a sample on a steam bath (Figure 1.2.f) and you will see what we mean. Although the percentage of soluble salts is very small (about 0.01%), many have beneficial effects; for example they improve the 'flat' taste of pure water, a trace of fluoride has been shown to reduce dental decay, and calcium salts are needed by the body to build bones and teeth. These calcium salts also cause hardness in water, which may be a disadvantage in certain circumstances (20.3).

Properties of water

Pure water is a colourless, odourless, tasteless liquid which freezes at 273 K (0 °C) and boils at 373 K (100 °C), provided that the air pressure is 101 325 Pa (760 mm mercury). The melting point and boiling point are considerably higher than expected, owing to the presence of hydrogen bonds; this is discussed in Chapter 12, together with the unusual fact that water *expands* on freezing whereas most liquids contract. Thus ice is less dense than water and, as you know, forms on the surface rather than at the bottom of the liquid. The density of water is a maximum at 277 K (4 °C).

Other important properties of water, which are dealt with later in the book, include its action as a solvent (12.4), its conduction of electricity (11.2), its reaction with metals (20.1), with non-metals (19.3) and with oxides (18.1).

When you have completed your studies you will realise why it has been said that water is not only the most common liquid in the world but also the most amazing one.

4.2 Tests for water

There are two chemical tests that are commonly used to show the presence of water. To understand them you need to know that many crystals are *hydrated* and contain *water of crystallisation* combined within them (12.2). If this water is driven off, the residue is said to be *anhydrous* ('without water').

White anhydrous copper(II) sulphate turns blue when water is added (Experiment 4.2.a). The reaction is reversible since when blue hydrated copper(II) sulphate is heated, water is driven off and the white anhydrous form is given.

$$CuSO_4(s) + 5H_2O(l) \rightleftharpoons CuSO_4 \cdot 5H_2O(s) + heat*$$
(white)   (blue)

A similar reaction occurs when water is added to anhydrous cobalt(II)

*The following state symbols will be used throughout in equations:
(s) = solid,   (l) = liquid,   (g) = gas,   (aq) = aqueous solution.

chloride or to paper impregnated with it. The blue anhydrous form changes to the pink hydrated one.

$$CoCl_2(s) + 6H_2O(l) \rightleftharpoons CoCl_2 \cdot 6H_2O(s) + heat$$
(blue)   (pink)

However, these reactions do not prove that a liquid *is* water; they simply show that it *contains* water. Any *aqueous* solution (i.e. solution made in water) would give positive results. Can you think how you might show that a sample of liquid was *pure* water?

4.3 The breaking and making of water

The electrolysis of water

Pure water is a very poor conductor of electricity but its conductivity is improved by dissolving a substance such as sodium fluoride in it (Experiment 4.3.a). When an electric current is passed through the resulting solution the water is decomposed and two gases are collected. The one given at the *positive* electrode relights a glowing splint, showing it to be *oxygen*, while that obtained at the *negative* electrode pops when a flame is applied. This is a test for *hydrogen*.

The synthesis of water

In Experiment 4.3.b water is synthesised by burning hydrogen in air. The hydrogen is first dried since there would be little point in proving that water had been collected if it were already present in the starting materials. The gas is then burned and, by means of a filter pump, the products of combustion are drawn through the cooled tube where a colourless liquid condenses out. This can be shown to be water by the tests mentioned in 4.2, together with measurement of boiling point and freezing point.
Note: in any experiment where hydrogen is to be burned, care must be taken to ensure that *all* air is displaced from the apparatus before a flame is brought near. Experiment 4.3.c shows what happens when a mixture of hydrogen and oxygen is ignited – a loud explosion occurs – and you can imagine the serious consequences which would result if a reaction of this sort took place within the apparatus.

What do the experiments in this section tell us? You may think at first that they prove water to be a compound made up of just hydrogen and oxygen in the ratio 2 volumes to 1 volume but if you consider carefully you will see that this is not so. In the electrolysis experiment there may have been other products which stayed in solution, while in the synthesis the hydrogen *may* have reacted with other components of the air in addition to the oxygen. Thus the only conclusion we can draw is that water contains at least two elements, hydrogen and oxygen, combined together.

However, it *is* possible to show that no other elements are present (e.g. Experiment 4.4.c) and also that a water molecule contains two atoms of hydrogen and one of oxygen. Thus its formula is H_2O.

4.4 Hydrogen

About 90% of the known universe is made up of hydrogen, although it is considerably less abundant than this on earth. Atoms of the other elements were probably made by the fusing together of hydrogen atoms in the sun and stars. This sort of reaction, which requires enormous temperatures, also takes place in the hydrogen bomb.

There is very little hydrogen in our atmosphere (why?) but it is widely found combined with other elements, as in water, North Sea Gas (methane, CH_4), petroleum and, of course, living matter.

Hydrogen was prepared by Robert Boyle in 1672 from iron nails and dilute sulphuric acid but it was not until 1766 that Henry Cavendish carried out the first careful investigation of it. The present name was provided by Antoine Lavoisier from the Greek *hydro genon* (water former).

Preparation

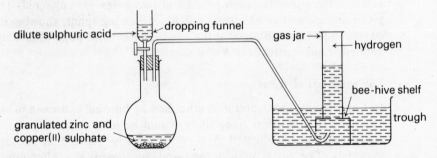

Figure 4.4 The laboratory preparation of hydrogen

Hydrogen is usually prepared in the laboratory by the action of dilute sulphuric (or hydrochloric) acid on zinc. If the reaction is too slow, the addition of a little copper(II) sulphate will speed it up (11.4). The gas is collected over water or, if preferred, in a syringe. The residue in the flask after the reaction is a solution of zinc sulphate (it will, of course, be zinc chloride if dilute hydrochloric acid is used).

$$Zn(s) + H_2SO_4(aq) \rightarrow ZnSO_4(aq) + H_2(g)$$

Figure 4.4 shows the apparatus used for the experiment. What advantages has it over the apparatus used in Experiment 4.4.a?

Test for hydrogen

A mixture of hydrogen with air or oxygen explodes when a flame is applied, forming water *only*.

$$2H_2(g) + O_2(g) \rightarrow 2H_2O(g)$$

Physical properties

Hydrogen is a colourless, odourless gas which is almost insoluble in water. It is less dense than air (Experiment 4.4.b) and, in fact, is the least dense gas known.

Chemical properties

COMBUSTION *Pure* hydrogen burns quietly in air with a faint blue flame to give water as the only product. (The equation is given under 'test for hydrogen'.)

REDUCING ACTION When dry hydrogen is passed over heated copper(II) oxide (Experiment 4.4.c) an exothermic reaction occurs and a glow spreads through the black solid. The copper(II) oxide changes to reddish brown copper and a colourless liquid is also produced. This turns anhydrous copper(II) sulphate blue and, by checking its boiling point, may be shown to be pure water.

$$CuO(s) + H_2(g) \rightarrow Cu(s) + H_2O(g)$$

The apparatus must be allowed to cool down with the hydrogen still passing so that no air can enter and oxidise the hot copper back to copper(II) oxide.

In Chapter 3 you learned that the addition of oxygen to a substance is known as oxidation but here oxygen has been *removed*. This type of reaction, which is the opposite of oxidation, is known as *reduction*. Hydrogen will *reduce* the oxide of any metal below zinc in the reactivity series (20.1).

Because hydrogen so readily combines with oxygen to form water, chemists in the past regarded it as the 'chemical opposite' of oxygen and thus they extended the definitions of oxidation and reduction as follows.

Oxidation is the addition of oxygen to, or the removal of hydrogen from a substance (see also 8.6).

Reduction is the removal of oxygen from, or the addition of hydrogen to a substance (see also 8.6).

Can you see why it is impossible for oxidation to occur without having reduction at the same time? Consider what happened to the hydrogen while it was reducing the copper(II) oxide and you should grasp the idea.

REACTION WITH OTHER ELEMENTS Hydrogen reacts with some metals and non-metals. Details will be found under the references given below.

$$2Na(s) + H_2(g) \rightarrow 2(Na^+ + H^-)(s) \quad\quad\quad (20.2)$$

$$N_2(g) + 3H_2(g) \rightleftharpoons 2NH_3(g) \quad\quad\quad (17.3)$$

$$Cl_2(g) + H_2(g) \rightarrow 2HCl(g) \quad\quad\quad (19.2)$$

Uses of hydrogen

1 In the large-scale syntheses of ammonia (17.3), hydrogen chloride (19.1) and methanol.

2 In the hardening of oils to make margarine and cooking fats (21.3).

3 In the oxy-hydrogen blowpipe and atomic hydrogen torch, used to produce high temperatures for welding, etc. The blowpipe produces

temperatures of up to 3273 K (3000 °C) by burning hydrogen in oxygen, the gases being kept separate until they reach the jet (why?).

The atomic hydrogen torch produces heat in a completely different way. Hydrogen atoms are produced by blowing the gas through an electric arc and the energy produced when they recombine to form molecules on the surface of the metal is capable of producing temperatures of the order of 3673 K (3400 °C). Since the fused metal surfaces are surrounded by hydrogen it is impossible for oxidation, which weakens the joints, to occur.

The manufacture of hydrogen

As you can imagine from the uses mentioned overleaf, hydrogen is a very important industrial chemical and hence is manufactured on a large scale. Several methods are available, e.g. it is produced as a by-product in the cracking of oils (21.6) and in the electrolysis of brine (20.2). Alternatively, the action of steam on hydrocarbons such as methane (North Sea Gas) may be used.

At a temperature of 1073 K (800 °C) steam, methane and a limited supply of oxygen react together in the presence of a nickel catalyst, to form a mixture of carbon monoxide and hydrogen, known as 'synthesis gas'.

$$2CH_4(g) + O_2(g) \rightarrow 2CO(g) + 4H_2(g) + heat$$
$$CH_4(g) + H_2O(g) \rightarrow \underbrace{CO(g) + 3H_2(g)}_{synthesis\ gas} - heat$$

The reaction with oxygen is *exothermic* while that with steam is *endothermic*. By balancing the two, the need for external heating is eliminated and the product produced more cheaply.

The synthesis gas is now treated with more steam at a temperature of 723 K (450 °C) in the presence of an iron oxide catalyst. The carbon monoxide in the mixture is oxidised to carbon dioxide and the steam reduced to give more hydrogen.

$$CO(g) + H_2O(g) \rightleftharpoons H_2(g) + CO_2(g)$$

The carbon dioxide is removed by dissolving it in water under pressure and any unchanged carbon monoxide is absorbed in a solution of copper(I) chloride in aqueous ammonia. Finally the gas is dried and, if necessary, compressed and stored.

Questions on Chapter 4

1 Describe an experiment to show that tap water contains dissolved air. How does the percentage of oxygen in this sample of air compare with that in atmospheric air and how do you account for the difference?

2 Describe the 'water cycle' in nature. Why is it important that the percentage of dissolved oxygen in river water should not fall below a certain level?

3 List the chief causes of water pollution. What precautions can be taken to keep pollution to a minimum level?

4   A blue crystalline solid, A, was heated strongly in a test tube, forming a white powder, B, and a colourless liquid, C. On adding C to B, the original substance, A, was formed. Identify A, B and C and explain the reactions. Give one other test to prove the identity of C.

5   From the following four gases – hydrogen, oxygen, nitrogen, carbon dioxide –
(a) name the two which could be *combined* together to form water;
(b) name the two which could be *mixed* to make a gas which most closely resembles air. What volume of each would be required to make $1\,dm^3$ of 'air'?
(c) Explain why the word *combined* was used in (a) but not in (b).   (SCE)

6   Oxygen and nitrogen are obtained from air by fractional distillation. Why can this method be used for this purpose yet cannot be used to obtain hydrogen and oxygen from water?   (SCE)

7   How is hydrogen manufactured on a large scale? Give three industrial uses of the gas.

8   Describe the preparation and collection of samples of (a) hydrogen from water, (b) water from hydrogen. Draw diagrams of the apparatus you would use in each case.

In (b), how would you show that the liquid obtained is pure water?   (C)

9   The compound of an element with hydrogen is called the **hydride** of that element, so that water is the hydride of oxygen.
(a) Write formulae for **one** hydride of **each** of **four** elements other than oxygen, stating whether each hydride is very soluble, sparingly soluble or insoluble in water.
(b) Describe and write equations for reactions in which (i) **one** of these hydrides burns in air or oxygen, (ii) **one** of these hydrides reduces another compound, (iii) **one** of these hydrides reacts as an acid.
(c) Give reagents, conditions and equations for reactions by which **two** of these hydrides can be conveniently prepared in the laboratory.   (C)

10   A current of dry hydrogen was passed over heated lead oxide. The lead oxide was thus changed to lead, the excess hydrogen being burnt in the air.
(a) Draw a labelled sketch of the apparatus which would be suitable for this experiment. (Do not show the source of hydrogen.)
(b) What substance other than lead is produced in this reaction?
(c) What type of change occurs when lead oxide is converted to lead?
(d) A special precaution is necessary before setting light to the excess hydrogen. What is this precaution and why is it necessary?
(e) Name a suitable drying agent for hydrogen gas.
(f) Why is it necessary to heat the lead oxide in this reaction?
(g) The instructions which might be given for this experiment would state that 'when the reaction is apparently complete, the lead should be allowed to cool *whilst the hydrogen is still passing*'. Explain why this is done.   (O & C)

11   (a) By means of a labelled diagram show how you would prepare some hydrogen, dry it, and use it to show that hydrogen can reduce a metal oxide to the metal and liquid water.
(b) From the following oxides, select two which could be reduced in this experiment: Al_2O_3, CaO, CuO, MgO, Na_2O, PbO, ZnO.
(c) For any one of these oxides that can be reduced, (i) write the equation for the reaction, and (ii) describe briefly how you would prove that the products of the reaction are the metal and water.   (L)

5   Acids, bases and salts

What picture does the word 'acid' conjure up in your mind? Probably you imagine a fuming liquid, capable of dissolving almost anything coming into contact with it. Some acids do indeed fit this description but, as you become more familiar with them, you will discover that many are far less dangerous.

Does the word 'base' mean anything to you? What about 'salt'? To most people salt is just something they use to improve the flavour of food but it means much more than this to a chemist.

Acid, base and salt are terms used by chemists to describe classes of compounds with certain characteristic properties. We shall now find out what these properties are.

Experiments

5.1.a   A simple property of acids

PROCEDURE   Place a piece of crushed acid drop on the tip of your tongue and then lick some sodium hydrogencarbonate ('bicarbonate of soda') from the *clean* palm of your hand. Repeat the procedure using a few drops of lemon juice in place of the acid drop, and magnesium hydroxide (found in Milk of Magnesia) in place of the sodium hydrogen-carbonate.
Do not taste any other substance in the laboratory.

Questions

1   Lemons, vinegar, rhubarb and acid drops all contain acids. How would you describe their taste?
2   What are bicarbonate of soda and Milk of Magnesia usually used for?
3   What was the effect of these substances on the taste of the acid drop and the lemon juice?

5.1.b   To make an indicator

Sodium hydrogencarbonate and magnesium hydroxide are examples of *alkalis* and when the sour taste of an acid is destroyed by an alkali the acid is said to be *neutralised*. Some substances give different colours in acids and alkalis. They are called *indicators* and many occur naturally in plants.

PROCEDURE   Prepare solutions of coloured materials extracted from plants. Suitable substances include the juice of blackberries, blackcurrants and elderberries, or the water in which red cabbage or beetroot has been boiled.

Place a few drops of the coloured extract in a test tube and add 1 cm depth of lemon juice or a solution of citric acid in water. Note the colour of the solution and then add small quantities of sodium hydrogencarbonate, stirring after each addition, until a little solid remains undissolved in the bottom of the tube. Make a note of the final colour. Repeat the experiment with other home-made coloured solutions and also use commercial indicators such as litmus, methyl orange and phenolphthalein. Record your results in a table under the following headings.

Coloured substance	Colour in acid	Colour in alkali

Carry out a similar experiment with one of the coloured materials but use sugar in place of the alkali.

Questions

1 Sugar is used to get rid of the sour taste in food but is it an alkali? Give a reason for your answer.
2 Why is it better to use indicators rather than the sense of taste when testing for acids and alkalis?

5.1.c To use an indicator to detect acids and alkalis

PROCEDURE   Place 1 cm depth of ammonia solution in a test tube and add a few drops of litmus solution. Note the colour given and decide whether the ammonia solution is an acid or an alkali. Repeat the experiment using solutions of sulphur dioxide, calcium hydroxide (slaked lime), carbon dioxide, white vinegar and sodium carbonate in place of the ammonia solution. List the substances in a table under the headings 'Acid' and 'Alkali'.

Question

1 Why do you think farmers add lime to the soil?

5.1.d Degree of acidity and the pH scale

So far we have used indicators simply to *distinguish* between acids and alkalis. If we wish to know *how* acidic or alkaline a substance is we must use a mixture of indicators called a **universal indicator**. This has a range of colours corresponding to the degree of acidity but since it is inconvenient to describe acidity in terms of colour we use a scale of numbers called the **pH scale**. The different colours of the universal indicator correspond to different pH values, usually ranging from 0 to 14.

PROCEDURE   Place 1 cm depth of distilled water in a *clean* test tube and add 1–2 drops of universal indicator. Alternatively, remove a drop of the water on a glass rod and spot it on a piece of universal indicator paper (i.e. paper which has been soaked in the indicator). Record the pH value and repeat the experiment using firstly dilute hydrochloric acid and then sodium hydroxide solution in place of the water. Now carry out the same procedure with solutions of substances found in the home, such as salt, sugar, white vinegar, bleach, toothpaste, etc. Draw Table 5.1.d and insert your results as instructed in the following questions.

	pH value	Colour	Substance
Increasing ↑ acidity			
Neutral			Pure water
Increasing ↓ alkalinity			

Table 5.1.d The pH values of common substances

Questions

1 Pure water is neutral, i.e. neither acidic nor alkaline. Write its pH value in the appropriate place in the second column of the table.
2 From your results, do acids have a pH which is less than, or more than the neutral value?
3 Sodium hydroxide is an alkali. Is the pH of its solution less than, or more than the neutral value?
4 Fill in the numbers 0–14 in the 'pH value' column of the table. Your answers to questions 2 and 3 should tell you whether to start with 0 or 14 at the top.
5 Complete the remaining columns in the table.

5.2.a To find how dilute acids react with metals

PROCEDURE Add a 5 cm strip of magnesium ribbon to a test tube containing about $5 cm^3$ of dilute sulphuric acid (i.e. sufficient to one-quarter-fill the tube). Collect a tube-full of the resulting gas over water and test it with a flame as described in Experiment 4.3.a. Repeat the experiment using dilute hydrochloric, dilute nitric and dilute citric acids in place of the sulphuric acid. If there is no reaction when a flame is applied, sniff the gas *cautiously* and look at it carefully against a white background to see if there is any colour. For each acid repeat the procedure with a piece of granulated zinc and then a piece of copper foil. If the evolution of gas is only slow with zinc, try adding a little copper(II) sulphate to speed it up.

Questions

1 Which gas was given in most cases? What happened when you applied a flame to it?
2 Which *metal* never gave this gas?

3 Which *acid* never gave this gas? Do you know which gas it *did* give?
4 Which acid seemed the least reactive?

5.2.b To find how dilute acids react with carbonates and hydrogen-carbonates

PROCEDURE Add a spatula measure of one of the solids in the list below to a test tube containing about 5 cm³ of one of the acids and pass the resulting gas through lime water as in Experiment 16.4.a. Repeat the experiment until each solid has been tried with each acid, using fresh lime water each time.

Solids: sodium carbonate, sodium hydrogencarbonate, potassium hydrogencarbonate, zinc carbonate.

Dilute acids: hydrochloric, nitric, sulphuric, citric.

Questions

1 What did you see in each case?
2 Which gas was given off and how could you tell what it was?
3 Did any combination of acid and solid not give the gas?
4 Which acid seemed the least reactive?

5.2.c To investigate the reaction of acids with bases

Most metallic oxides and hydroxides are bases. You will learn more about them later in the chapter.

PROCEDURE Place about 5 cm³ of dilute nitric acid in a boiling tube, find its pH by means of universal indicator paper and then add a spatula measure of zinc oxide to it. Heat the tube, shaking it continuously, until the liquid boils. If all the solid dissolves, add more, boiling after each addition, until excess remains in the bottom of the tube. Filter, and test the filtrate with universal indicator paper. Repeat the experiment using dilute hydrochloric, sulphuric and citric acids in place of the dilute nitric acid and then try each acid with copper(II) oxide in place of the zinc oxide. Record your results as in Table 5.2.c.

	Hydrochloric acid	Nitric acid	Sulphuric acid	Citric acid
pH of acid alone				
pH after adding zinc oxide				
pH after adding copper(II) oxide				

Table 5.2.c

Questions

1 Were any of the final solutions acidic?
2 Do you think that new substances have been formed during the reactions? Give a reason for your answer.

5.2.d To find if water affects the properties of acids

So far, you have used *solutions* of acids in water. Although some pure acids such as sulphuric or nitric acids are too dangerous for you to handle at this stage, those chosen for this experiment will enable you to make a discovery which may surprise you.

PROCEDURE (a) Test *anhydrous* ethanoic acid, solid citric acid and solid tartaric acid with *dry* universal indicator paper and then with a piece which has been moistened with distilled water.
(b) Add a spatula measure of solid citric acid to a piece of magnesium ribbon in a test tube and stir with a glass rod. Look for signs of a reaction, then add distilled water and collect a tube-full of the resulting gas over water. Decide what the gas is most likely to be and apply a suitable test.
(c) Add a spatula measure of solid tartaric acid to one of solid sodium hydrogencarbonate in a test tube and stir with a glass rod. Look for signs of a reaction, then add distilled water and apply a suitable test to the gas which is evolved. You should prepare for the test *before* you add the distilled water (why?).

Questions

1 Which gas was evolved in (b) and how did you identify it?
2 Which gas was evolved in (c) and how did you identify it?
3 Were the characteristic properties of these acids shown when no water was present?
4 Mixture (c), with a few other compounds added, is often found in the home – can you think where?

5.3.a To investigate the action of bases with ammonium salts

PROCEDURE Add a spatula measure of calcium hydroxide to one of ammonium chloride in a test tube and stir with a glass rod. Hold the tube in front of you and, with your hand, waft air across the top of it towards your nose. Sniff *cautiously* and if no smell is detected gradually bring the tube nearer. You should use this technique whenever testing an unknown gas for smell so that you avoid the possibility of inhaling large quantities of choking fumes. Warm the tube gently if you still smell nothing and then try again. Finally, hold a piece of universal indicator paper, moistened with distilled water, in the mouth of the tube. Repeat the experiment with copper(II) oxide and ammonium sulphate.

Questions

1 What was the gaseous product in each reaction?
2 How would you describe its smell?
3 Was it acidic or alkaline?

5.3.b   Some properties of alkalis

Alkalis are bases which are soluble in water. You already know how they affect indicators (Experiment 5.1.b); two more properties are investigated here.

PROCEDURE   (a) Pour 5 cm³ of sodium hydroxide solution into a test tube, place your thumb over the end and shake the tube so that the solution comes into contact with your skin. Rub your thumb and forefinger together and then wash off the liquid *immediately*.
(b) Add a little aluminium powder to the solution and warm gently until a reaction begins. Collect a tube-full of the gas over water in the usual way and test it with a flame. Repeat the experiment using magnesium ribbon in place of the aluminium powder.

Questions

1   How would you describe the feel of the sodium hydroxide solution?
2   Which gas was given with the aluminium? How could you tell?
3   Was any gas given with the magnesium ribbon?

5.3.c   To find if water affects the properties of alkalis

PROCEDURE   Test *dry* calcium hydroxide with *dry* universal indicator paper and then with a piece which has been moistened with distilled water.

Question

1   Is water necessary for the alkali to affect the indicator? Compare your result with those of Experiment 5.2.d.

5.4.a   Salts – an introduction

PROCEDURE   (a) Add a 15 cm length of magnesium ribbon to 5 cm³ of dilute sulphuric acid in a boiling tube. When the evolution of hydrogen slows down, heat the tube to speed up the reaction. If all the metal dissolves add further 1 cm lengths, heating each time, until excess remains in the bottom of the tube (i.e. all the acid has been used up). Boil for about 30 seconds to ensure that the reaction is complete and then filter into another boiling tube. Evaporate the filtrate until not more than one-third of it remains and then pour it into a watch glass. When it is cool, examine the resulting crystals with a hand lens.
(b) Repeat this procedure with small quantities of magnesium oxide in place of the magnesium ribbon. There will, of course, be no evolution of hydrogen in this case.

Questions

1   Did the crystals appear to be the same shape in each experiment?
2   In (a) you started with dilute sulphuric acid (i.e. sulphuric acid dissolved in water) and magnesium. Consider what was given off during the experiment and

then try to decide what the crystals consisted of. (It may help you in your explanation to know that the part of an acid left after the hydrogen has been removed is called an *acid radical.*)

3 Do you think that the crystals obtained in (b) were the same as those in (a)? If they were the same, did they contain hydrogen?

4 Since hydrogen was not given off in (b) what could have happened to it? (Hint: magnesium oxide is a compound of magnesium and oxygen only, and no oxygen was given off.)

5.4.b To prepare some salts by double decomposition

Read about the main methods of salt preparation in section 5.4 before carrying out this, and the next two experiments.

PROCEDURE Add about $5\,cm^3$ of copper(II) sulphate solution to an equal volume of sodium carbonate solution and stir with a glass rod. Filter, rinse the precipitate with distilled water and then dry it between sheets of filter paper. Repeat the experiment using solutions of sodium chloride and silver nitrate, then hydrochloric acid and lead(II) nitrate. In the last case, do not filter but place the mixture in a boiling tube and heat, *with shaking*, until it boils. If the precipitate does not dissolve add small quantities of distilled water, boiling after each addition, until it does. Leave the solution until completely cool and then filter it.

Note: Several of the solutions are very poisonous; therefore, *wash your hands* before leaving the laboratory.

Questions

1 In each case, name the precipitate and give a word equation or a chemical equation for the reaction.

2 Which of the precipitates seemed unstable? What happened to it?

3 In what way did the appearance of the product in the final experiment differ from that of the rest?

4 Name three more salts which could be prepared by this method and name the starting materials in each case. (Refer to Table 5.4.2, page 77, for solubilities.)

5.4.c To prepare a salt by the titration method

PROCEDURE Place $30\,cm^3$ of sodium hydroxide solution in a conical flask and note the temperature. Fill a burette with dilute hydrochloric acid, run $20\,cm^3$ into the flask and note the temperature again. Stir carefully with a glass rod, withdraw a drop of solution and spot it on a piece of universal indicator paper. If the solution is still alkaline (and it should be), add $1\,cm^3$ portions of acid, stirring and spotting after each addition until the solution is neutral or *just* acidic. Transfer the liquid to an evaporating basin and evaporate to dryness as in Experiment 1.2.f. Examine the residue with a hand lens.

Questions

1 Which salt was formed? Give a word equation or a chemical equation for the reaction.

2   Why is it better to spot the solution on universal indicator paper rather than to add indicator to the flask?

3   What evidence was there that a reaction took place when the acid was added to the alkali?

4   Why was the solution evaporated to dryness instead of being crystallised?

5   Name three more salts which could be prepared by this method and name the starting materials in each case. (Refer to Table 5.4.2, page 77, for solubilities.)

5.4.d   To prepare a salt by the insoluble base method

PROCEDURE   Place 30 cm^3 of dilute sulphuric acid in a 100 cm^3 beaker. Heat, but do not boil, and add copper(II) oxide in small portions, stirring after each addition until excess solid remains undissolved in the bottom of the beaker. Filter and obtain crystals in the usual way (see Experiment 1.2.c).

Questions

1   What was the salt? Write a word equation or a chemical equation for the reaction.

2   Name three more salts which could be prepared by this method and name the starting materials in each case. (Refer to Table 5.4.2, page 77, for solubilities.)

5.4.e   To prepare two salts from one pair of reagents

Salts are prepared by replacing the hydrogen of an acid directly or indirectly by a metal. Sulphuric acid, H_2SO_4, has *two* replaceable hydrogen atoms in each molecule and in this experiment you will make two sodium salts, one where all the hydrogen has been replaced by the metal and one where only half has been replaced.

PROCEDURE   Use sodium hydroxide solution and dilute sulphuric acid and proceed as in Experiment 5.4.c, but this time crystallise the salt instead of evaporating to dryness (see Experiment 1.2.c). Then repeat the whole experiment using the same quantity of acid but only *half* as much sodium hydroxide solution as was required in the first case. Compare the shapes of the crystals obtained by the two methods.

Questions

1   Were the crystals from each experiment the same shape? Describe and sketch them.

2   Name the first salt prepared and give an equation for the reaction.

3   Suggest a name for the second salt, given that its formula is $NaHSO_4$, and write an equation for the reaction in which it was formed.

5.4.f   To compare the properties of sodium sulphate with those of sodium hydrogensulphate

Since half of the replaceable hydrogen in sulphuric acid is still present in sodium hydrogensulphate you might expect the salt to show acidic properties.

PROCEDURE   Test solutions of sodium sulphate and sodium hydrogen-sulphate with universal indicator paper and then devise experiments to compare the reactions, if any, of the two salts with magnesium and with sodium carbonate.

Questions

1   What was the pH of each solution and what did it show?
2   Describe the experiment with the magnesium, including any tests you carried out on the products, and write an equation for the reaction.
3   Repeat question 2 for the sodium carbonate.
4   What are your conclusions about the nature of sodium sulphate and of sodium hydrogensulphate?

5.5.a   The effect of water on acids and alkalis in terms of ions ★

This experiment and the remaining ones in the chapter are best left until after you have studied the mole, ions and electrolysis (Chapters 9 and 11).

PROCEDURE   (a) Dip a thermometer into water and then, with a drop of water still on the bulb, lower it into a gas jar of hydrogen chloride. Record the temperature just before the thermometer enters the gas and look for a change in the reading. Carry out the experiment again, using methylbenzene in place of the water.
(b) Repeat (a) with water and ammonia and then with trichloromethane and ammonia.
(c) Carry out the procedure described in part (a) of Experiment 11.2.b but use only the following solutions: 1 M hydrochloric acid, a solution of dry hydrogen chloride in methylbenzene, 1 M aqueous ammonia solution, and a solution of dry ammonia in trichloromethane. Rinse and dry the electrodes between each experiment.

Questions

1   In which case(s) did the thermometer indicate that a reaction was taking place between the gas and the solvent?
2   Which solutions seemed to contain ions, those in water or those in the other solvents?
3   Write down the formulae of the common acids and consider possible ways in which they might form ions. You should find that all the acids have a characteristic ion – what is it? (Remember that you have only produced a *theory*. Do you have any experimental evidence to back it up?)
4   Suggest a characteristic ion for alkalis by using the same approach as in question 3.
5   Write equations for the formation of *ions* from reactions between gas and solvent *molecules*, bearing in mind your answers to the previous questions.

5.5.b   Acidic and alkaline properties in terms of ions ★

PROCEDURE   Use the same solutions as in part (c) of the previous experiment for the following tests.

(a) Test each solution with dry litmus paper and then with a piece which has been moistened with distilled water.

(b) To a sample of each hydrogen chloride solution in a test tube, add a 5 cm strip of magnesium ribbon. Collect any gas evolved over water and identify it.

(c) To a sample of each hydrogen chloride solution in a test tube, add a spatula measure of sodium carbonate. Identify any gas evolved.

Questions

1   Since the characteristic ions in acids and alkalis must be responsible for the various colours shown by indicators, explain the colour changes (or lack of them) when the solutions were tested with dry and with moist litmus paper.

2   Did any reaction take place between the magnesium or the sodium carbonate and the non-aqueous solutions? Why do you think this was?

3   Write *ionic* equations for any reactions which did occur.

5.5.c   The strengths of acids and alkalis – an introduction

PROCEDURE   (a) Use universal indicator paper to find the approximate pH values of 1 M hydrochloric acid, 1 M ethanoic acid, 1 M aqueous sodium hydroxide solution and 1 M aqueous ammonia solution.

(b) To 5 cm³ of each acid in separate test tubes add *simultaneously* a 3 cm strip of magnesium ribbon. Shake the tubes gently and note the relative times taken for the pieces of magnesium to dissolve completely.

(c) To 5 cm³ of each alkali in separate test tubes add a little aluminium powder. Warm both tubes together and note which solution gives the more vigorous reaction.

Questions

1   Which acid appeared to be the 'weaker' according to the universal indicator paper? Was this borne out by the result in (b)?

2   Which alkali appeared to be the 'weaker' according to the universal indicator paper? Was this borne out by the result in (c)?

3   Why was it important for the solutions to be of the same concentration for this experiment?

5.5.d   The strengths of acids and alkalis in terms of ions★

PROCEDURE   Carry out the procedure described in part (a) of Experiment 11.2.b, but use 1 M hydrochloric acid, 1 M ethanoic acid, 1 M aqueous sodium hydroxide solution and 1 M aqueous ammonia solution.

Questions

1   All the beakers contained the same number of moles of solute. What can you say about the relative numbers of *ions* present?

2   What connection is there between the strengths of acids and alkalis and the number of ions formed in aqueous solution?

PROCEDURE   Place 5 cm³ of distilled water in a test tube and test it with universal indicator paper. Dissolve a spatula measure of iron(III) chloride in the water and test again with the universal indicator paper. Add a 5 cm strip of magnesium ribbon, collect any gas evolved over water and identify it. Repeat with solutions of aluminium chloride and sodium chloride.

Questions

1   Which solutions, if any, were (a) acidic, (b) neutral, (c) alkaline? How could you tell?
2   Which characteristic ions must have been present to give the gas you collected and what must they have come from?
3   Compare the results obtained using (a) iron(III) chloride, (b) sodium chloride. Since they both contain chloride ions, which ion must have been responsible for the acidic nature of the solution?
4   Bearing in mind your answer to question 2, write an ionic equation for the reaction between the metal ion you chose in question 3 and water. (If you look carefully at a solution of the chosen salt which has been standing for some time you should see something which gives you a further clue.)
5   You probably know that sodium carbonate solution is alkaline. Compare this with the result obtained for sodium chloride solution and, using similar arguments to those in questions 2 and 3, deduce an ionic equation which accounts for this fact.

5.1   An introduction to acids, alkalis and indicators

You must have come into contact with acids long before you set foot inside a chemistry laboratory. For example, vinegar, lemon juice, rhubarb and unripe fruit are familiar substances and they all contain acids. One thing which they obviously have in common is their sour taste, and this is a property shared by all acids.

If you have ever been unfortunate enough to suffer from 'acid indigestion' you may well have been given bicarbonate of soda (sodium hydrogencarbonate) or Milk of Magnesia (magnesium hydroxide in water) to relieve the pain. These 'antacids' are examples of alkalis and presumably they destroy or 'neutralise' the acids in some way. They certainly get rid of the sour taste of an acid, as is shown in Experiment 5.1.a. (You should realise that it is *too much* acid in the stomach which causes indigestion – hydrochloric acid is always present and plays an important part in breaking down food.)

Taste is not only unreliable in detecting acids and alkalis but is obviously far too dangerous to use as a test for unknown substances. Materials which have different colours in acids and alkalis are more reliable and safer (Experiments 5.1.b and 5.1.c). These are called *indicators* and one of the most common ones used in the laboratory is litmus, an extract of lichens. It is red in acids, blue in alkalis and purple in neutral solutions, i.e. solutions which are neither acidic nor alkaline.

Things which taste more sour than others are presumably more acidic. To measure the degree of acidity or alkalinity of a solution, a scale of numbers called the pH scale has been chosen, its usual range being from 0 to 14. Solutions with a pH value of 7 are neutral, those with a pH value of less than 7 are acidic, and those with a pH value greater than 7 are alkaline. A *universal indicator* is a mixture of indicators giving different colours corresponding to different pH values.

Many of the materials found in everyday life are acidic or alkaline. Table 5.1 shows the approximate pH values of a few of these materials.

	pH value	Substance
Increasing acidity	0 1 2 3 4 5 6	Vinegar Lemon juice
Neutral	7	Pure water
Increasing alkalinity	8 9 10 11 12 13 14	Bleach Scouring powder

Table 5.1 The pH scale

5.2   Acids

The acids most commonly found in the laboratory are hydrochloric acid (HCl), nitric acid (HNO_3), sulphuric acid (H_2SO_4) and ethanoic (acetic) acid (CH_3COOH). The first three are examples of 'mineral' acids while the last one is an 'organic' acid. These terms refer to the original sources of the compounds. In this chapter we are mainly concerned with *dilute* acids, i.e. ones which have been dissolved in relatively large volumes of water.

Properties of dilute acids

Taste   Dilute acids taste sour (5.1).

pH   Dilute acids have a pH value of less than 7 and turn blue litmus paper red (5.1).

With metals   Most dilute acids effervesce (fizz) to give hydrogen when magnesium, zinc or iron is added (see Experiment 5.2.a). For this reason Sir Humphry Davy, in 1816, defined acids as 'hydrogen producers'.

Exceptions: (a) Copper does *not* give hydrogen with any acid. The reason for this is discussed in the section on the electrochemical series (11.1). (b) Instead of producing hydrogen, dilute nitric acid gives one or more of the oxides of nitrogen with most metals (17.2). Magnesium *will* liberate hydrogen from very dilute solutions of it.

With carbonates and hydrogencarbonates Dilute acids effervesce to give carbon dioxide when a carbonate or hydrogencarbonate is added (see Experiment 5.2.b).

Exceptions: Dilute sulphuric acid does *not* effervesce with marble chips owing to the formation of an insoluble layer of calcium sulphate on the surface of the chips.

With bases Dilute acids are neutralised by bases, i.e. the characteristic properties of the two classes of compound cancel out. This is considered in more detail in 5.4.

With water Experiment 5.2.d illustrates the fact that *anhydrous* acids do not give the characteristic reactions listed above. Thus, the term 'acid' really applies only to *aqueous solutions* of the compounds. The explanation for this is best left until later (5.5).

5.3 Bases and alkalis

Bases

Bases are usually metallic oxides or hydroxides (see also *Bases*, 5.5).

A base may be considered to be the 'chemical opposite' of an acid since the two react together to give a 'neutral' substance which has the properties of neither.

Properties of bases

With ammonium salts When a base is mixed with an ammonium salt and warmed, ammonia is evolved (Experiment 5.3.a). This may be detected by its smell and its effect on damp universal indicator paper (ammonia is the only *common* alkaline gas).

With acids As already mentioned, acids and bases neutralise one another (see also 5.4).

Alkalis

Alkalis are bases which are soluble in water (see also page 81).

Many metallic oxides and hydroxides are *not* soluble in water and thus are not alkalis. This means that, though all alkalis are bases, not all bases are alkalis. (Similarly, all dogs are four-legged animals but not all four-legged animals are dogs.)

The common laboratory alkalis are potassium hydroxide (KOH), sodium hydroxide (NaOH), aqueous ammonia solution (often called 'ammonium hydroxide') and calcium hydroxide ($Ca(OH)_2$).

Properties of alkalis

Taste Aqueous solutions of alkalis taste bitter. Do *not* check this for yourself.

Feel   Aqueous solutions of alkalis feel soapy (Experiment 5.3.b). They react with the skin.

pH   Aqueous solutions of alkalis have a pH value of more than 7 and turn red litmus blue (5.1).

With metals   Aqueous solutions of alkalis do not usually give hydrogen with metals but they do react with a few, such as aluminium and zinc, to give the gas (Experiment 5.3.b).

With ammonium salts   Since all alkalis are bases, ammonia is given on warming with ammonium salts.

With acids   Neutralisation occurs (5.4).

With water   Water is necessary for the characteristic properties of alkalis to be shown. (This is explained in 5.5.)

5.4   Salts

What is a salt?

Let us consider what happens when magnesium reacts with dilute sulphuric acid (Experiment 5.4.a). Hydrogen is given off and a clear solution is left. This presumably contains the magnesium and the acid radical, i.e. the part of the acid left after the hydrogen has been removed. It also contains the water which was originally used to dilute the acid. If this water is evaporated off, crystals are obtained, and these must consist of the magnesium combined with the acid radical. In other words, the metal has replaced the hydrogen in the acid to form a new compound. This compound is called a *salt*.

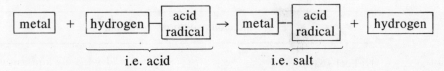

If the experiment is carried out with magnesium oxide in place of the magnesium the *same* salt is formed and so the metal again replaces the hydrogen of the acid. However, in this case, there is no evolution of gas. What, then, happens to the hydrogen of the acid? It must form something which stays in the solution remaining at the end of the reaction. Since magnesium oxide consists of magnesium and oxygen only, the most probable explanation is that the hydrogen of the acid combines with the oxygen of the base to form water. This does in fact happen and, in general, if an acid and a base neutralise one another the products of the reaction are a salt and water *only*.

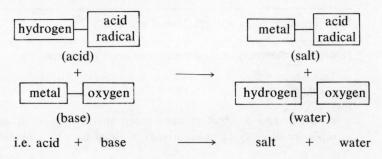

When the *metal itself* reacts with the acid the replacement of hydrogen is said to be *direct*; if a *compound of the metal* is used the replacement is *indirect*, because the metal has been combined with something else before being placed in the acid.

Dilute acids

---

1 Taste sour.
2 pH value less than 7, litmus turns red.
3 With active metals (e.g. magnesium, zinc, iron),

$$\text{acid} \quad + \text{metal} \rightarrow \quad \text{salt} \quad + \text{hydrogen}$$
$$2HCl(aq) + Mg(s) \rightarrow MgCl_2(aq) + \quad H_2(g)$$

4 With bases,

$$\text{acid} \quad + \text{ base } \rightarrow \quad \text{salt} \quad + \text{water}.$$
$$2HNO_3(aq) + CuO(s) \rightarrow Cu(NO_3)_2(aq) + H_2O(l)$$

5 With carbonates,

$$\text{acid} \quad + \text{carbonate} \rightarrow \quad \text{salt} \quad + \text{ water} + \text{carbon dioxide}.$$
$$H_2SO_4(aq) + ZnCO_3(s) \rightarrow ZnSO_4(aq) + H_2O(l) + \quad CO_2(g)$$

Note: dilute nitric acid does not give hydrogen with metals.

Aqueous solutions of alkalis

---

1 Feel soapy.
2 pH value greater than 7, litmus turns blue.
3 With a few metals (e.g. aluminium, zinc) potassium and sodium hydroxides give hydrogen and complex compounds.
4 With acids,

$$\text{alkali} \quad + \quad \text{acid} \quad \rightarrow \quad \text{salt} \quad + \text{ water}.$$
$$2KOH(aq) + H_2SO_4(aq) \rightarrow K_2SO_4(aq) + 2H_2O(l)$$

5 With ammonium salts,

$$\text{alkali} \quad + \text{ammonium salt} \rightarrow \quad \text{salt} \quad + \text{ water} + \text{ammonia}.$$
$$NaOH(aq) + \quad NH_4Cl(s) \quad \rightarrow NaCl(aq) + H_2O(l) + \quad NH_3(g)$$

Note: insoluble bases give reactions 4 and 5 only.

Table 5.4.1. Summary of the properties of acids and bases

Definitions

A **salt** is the product formed when the hydrogen of an acid is replaced directly or indirectly by a metal (or ammonium radical).

An **acid** is a compound containing hydrogen which can be replaced directly or indirectly by a metal (see also pages 81 and 82).

Neutralisation is the reaction between an acid and a base in such quantities that the products are a salt and water *only* (see also page 82).

The properties of acids and bases are summarised in Table 5.4.1.

Names of salts

Salts are named from the 'parent' acids which have had their hydrogen atoms replaced by a metal or ammonium radical.

Hydrochloric acid gives chlorides   Nitric acid gives nitrates
Sulphuric acid gives sulphates   Carbonic acid gives carbonates
Ethanoic (acetic) acid gives ethanoates (acetates)

Methods of preparing salts

There are three main methods of preparing salts and you need to know whether or not a particular salt is soluble in water before you can decide which method to choose. Table 5.4.2 shows the solubilities of the common bases and salts and you should try to remember it.

Bases	Oxides	All *insoluble*. Potassium, sodium and calcium oxides react violently with water to form a solution of the corresponding hydroxide. Ammonium oxide does not exist.
	Hydroxides	All *insoluble* except potassium, sodium and 'ammonium' hydroxides. Calcium hydroxide is slightly soluble.
Salts	Carbonates	All *insoluble* except potassium, sodium and ammonium carbonates.
	Chlorides	All *soluble* except silver and lead chlorides. Lead chloride is soluble in hot water.
	Nitrates	All *soluble*.
	Sulphates	All *soluble* except barium and lead sulphates. Calcium sulphate is only slightly soluble.

Note: all common potassium, sodium and ammonium salts are soluble.

Table 5.4.2. Solubilities of the common bases and salts in water

Detailed instructions for the preparation of salts are given in Experiments 5.4.b, 5.4.c and 5.4.d. The following summaries give the main points of the methods.

1 DOUBLE DECOMPOSITION METHOD (for *insoluble* salts and hydroxides)

In this method the starting materials 'swap chemical partners' to form the insoluble product. Solutions containing the required components are mixed and the precipitate filtered off, washed with distilled water and dried.

e.g. $Zn(NO_3)_2(aq) + Na_2CO_3(aq) \rightarrow ZnCO_3(s) + 2NaNO_3(aq)$

zinc	+	sodium	→	zinc	+	sodium
nitrate		carbonate		carbonate		nitrate
(solution)		(solution)		(precipitate)		(remains in solution)

$CuSO_4(aq) + 2KOH(aq) \rightarrow Cu(OH)_2(s) + K_2SO_4(aq)$

copper(II)	+	potassium	→	copper(II)	+	potassium
sulphate		hydroxide		hydroxide		sulphate
(solution)		(solution)		(precipitate)		(remains in solution)

Remember that *solutions* of the starting materials are used; therefore both of them must be *soluble* in water. Also, make sure that the second product of the reaction is soluble so that a mixture of two precipitates is not obtained. Such a mixture would probably be impossible to separate.

2 THE TITRATION METHOD (for *potassium*, *sodium* and *ammonium* salts)

Dilute acid is run from a burette into a solution of alkali until it is neutral. Crystals are obtained in the usual way.

e.g. $HNO_3(aq) + KOH(aq) \rightarrow KNO_3(aq) + H_2O(l)$

| nitric acid | + | potassium | → | potassium | + | water |
| | | hydroxide | | nitrate | | |

| acid | + | base | → | salt | + | water |

Since solutions of potassium, sodium and ammonium *carbonates* are alkaline they may be used instead of the hydroxides.

3 THE INSOLUBLE BASE METHOD (for *soluble* salts, except those in 2)

The required dilute acid is warmed and the insoluble base added until no more will dissolve. The excess is filtered off and the filtrate evaporated and crystallised.

e.g. $H_2SO_4(aq) + MgO(s) \rightarrow MgSO_4(aq) + H_2O(l)$

| sulphuric | + | magnesium | → | magnesium | + | water |
| acid | | oxide | | sulphate | | |

| acid | + | base | → | salt | + | water |

Insoluble carbonates or in some cases the metals themselves (e.g. magnesium, zinc, iron) may be used in place of the insoluble base. The procedure is exactly the same.

4 DIRECT COMBINATION (for a few *anhydrous* salts)

Some anhydrous salts (e.g. iron(III) chloride) react with water and so cannot be prepared by 'wet' methods. In these cases, it is sometimes possible to directly combine the elements which make up the salt by heating them together.

e.g. $2Fe(s) + 3Cl_2(g) \rightarrow 2FeCl_3(s)$ (see Experiment 19.2.b)

The method may also be used for salts which do not react with water. It is usually restricted to those which consist of two elements.

The information given in Figure 5.4 should help you to decide which method to choose in order to prepare a particular salt.

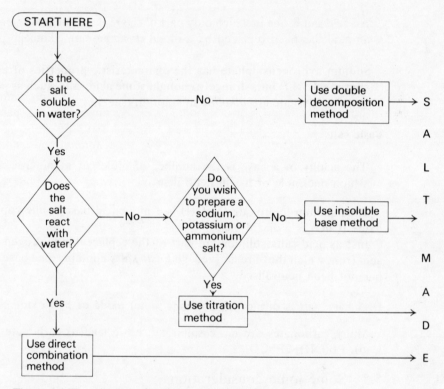

Figure 5.4 Flow diagram for the preparation of a salt

Normal salts and acid salts

The **basicity** of an acid is the number of hydrogen atoms in one molecule of it which are replaceable by a metal.

It follows that hydrochloric acid (HCl) is *mono*basic, sulphuric acid (H_2SO_4) is *di*basic and phosphoric(V) acid (H_3PO_4) is *tri*basic. Ethanoic acid (CH_3COOH) has four hydrogen atoms in each molecule but only the last one in the formula is replaceable by a metal. Thus it has a basicity of *one*.

> A **normal salt** is one in which all of the replaceable hydrogen of an acid has been replaced by a metal or ammonium radical.

If an acid has more than one replaceable hydrogen atom in each molecule it is possible for salts of it to be formed in which only one of these atoms (or perhaps two, in the case of a tribasic acid) has been replaced by a metal.

e.g. $NaOH(aq) + H_2SO_4(aq) \rightarrow NaHSO_4(aq) + H_2O(l)$ (Experiment 5.4.e)
sodium
hydrogensulphate

Salts of this type, which themselves contain replaceable hydrogen, fit the simple definition of an acid given earlier, and thus are called *acid salts*.

> An **acid salt** is one in which only part of the replaceable hydrogen of an acid has been replaced by a metal or ammonium radical.

Sodium hydrogensulphate has the characteristic properties of an acid (Experiment 5.4.f) but, strangely enough, some acid salts, such as sodium hydrogencarbonate are *alkaline* in solution (see 5.6).

Basic salts

> The **acidity** of a base is the number of moles of monobasic acid which one mole of it will neutralise.

Thus sodium hydroxide (NaOH) is a monoacid base, while copper(II) oxide (CuO) is a diacid base.

Just as acid salts still contain part of the replaceable hydrogen of the acid from which they are derived, so *basic salts* contain some base which has not been neutralised.

> A **basic salt** is one which contains metal oxide or hydroxide.

Many carbonates are basic salts, e.g. basic lead(II) carbonate (white lead), $Pb(OH)_2 \cdot 2PbCO_3$.

5.5 Some ionic considerations

The effect of water on the properties of acids and alkalis

A solution of hydrogen chloride in an organic solvent such as methylbenzene has very different properties from an aqueous solution of the gas. Only the solution in water behaves as an acid and, since heat is given out when this solution is made (Experiment 5.5.a), it would seem that a reaction of some kind takes place between the *covalent* molecules of hydrogen chloride and water. As the aqueous solution conducts electricity the products of this reaction must be *ions*.

Since all acids contain hydrogen and all give hydrogen at the cathode when their aqueous solutions are electrolysed, the most likely explanation is that all dilute acids contain hydrated hydrogen ions, $H^+(aq)$. The hydrated hydrogen ion is often written as if each ion were joined to *one* water molecule, and is then referred to as the *oxonium* ion, H_3O^+. Thus when hydrogen chloride dissolves in water the reaction is given by

$$HCl(g) + \text{water} \rightarrow H^+(aq) + Cl^-(aq)$$

or $\qquad HCl(g) + H_2O(l) \rightarrow H_3O^+(aq) + Cl^-(aq).$

As these hydrated hydrogen ions are the only particles which acids have in common, they must be responsible for the characteristic properties mentioned in 5.2. Thus we may redefine an acid as follows.

An **acid** is a substance giving hydrated hydrogen ions in aqueous solution (see also page 82).

Similar reasoning may be applied to the dissolving of ammonia in trichloromethane and in water, leading to the conclusion that the *covalent* molecules of gas react with those of water to form *ions*. Since most alkalis are metallic hydroxides, made up of positive metal ions and negative hydroxide ions, we can assume that the characteristic ion in alkaline solutions is the hydrated hydroxide ion, $OH^-(aq)$.

$$(Na^+ + OH^-)(s) + \text{water} \rightarrow Na^+(aq) + OH^-(aq)$$

and for ammonia,

$$NH_3(g) + H_2O(l) \rightleftharpoons NH_4^+(aq) + OH^-(aq)$$

A new definition of an alkali is thus:

An **alkali** is a substance giving hydrated hydroxide ions on dissolving in water.

The reason why water is necessary for acidic and alkaline properties to be shown is now clear – without it the characteristic ions are not present.

Reactions of acids and alkalis

If hydrated hydrogen ions alone are responsible for the characteristic properties of dilute acids it should be possible to write ionic equations for the standard reactions mentioned earlier in the chapter in which the ions of the acid radicals, being spectator ions (9.5), do not appear. For example:

1 *Acid + active metal*

$$2H^+(aq) + Mg(s) \rightarrow Mg^{2+}(aq) + H_2(g)$$

or $\qquad 2H_3O^+(aq) + Mg(s) \rightarrow Mg^{2+}(aq) + H_2(g) + 2H_2O(l)$

2 *Acid + alkali* (neutralisation)

$$H^+(aq) + OH^-(aq) \rightarrow H_2O(l)$$

or $\qquad H_3O^+(aq) + OH^-(aq) \rightarrow 2H_2O(l)$

3   *Acid + carbonate*

$$2H^+(aq) + CO_3^{2-}(aq) \rightarrow H_2O(l) + CO_2(g)$$

or $$2H_3O^+(aq) + CO_3^{2-}(aq) \rightarrow 3H_2O(l) + CO_2(g)$$

Similarly, the reaction of an alkali with aluminium is given by

$$2Al(s) + 6H_2O(l) + 2OH^-(aq) \rightarrow 2[Al(OH)_4]^-(aq) + 3H_2(g).$$

Bases

As bases are metallic oxides or hydroxides, containing positive metal ions together with negative oxide or hydroxide ions, they too may be redefined.

Bases are substances containing oxide (O^{2-}) or hydroxide (OH^-) ions (see also below).

Ionic equations for the two main reactions of insoluble bases are thus:

1   *Base + acid* (neutralisation)

$$(M^{2+} + O^{2-})(s) + 2H^+(aq) \rightarrow H_2O(l) + M^{2+}(aq)$$

or $$(M^{2+} + 2OH^-)(s) + 2H^+(aq) \rightarrow 2H_2O(l) + M^{2+}(aq)$$

2   *Base + ammonium salt*

$$O^{2-}(s) + 2NH_4^+(s) \rightarrow H_2O(l) + 2NH_3(g)$$

or $$OH^-(s) + NH_4^+(s) \rightarrow H_2O(l) + NH_3(g)$$

From the ionic equations for neutralisation given above, it follows that:

Neutralisation is the reaction between the hydrogen ions of an acid and the hydroxide or oxide ions of a base to form water, a salt being formed at the same time.

Strengths of acids and alkalis

Strong acids (those with low pH values) and strong alkalis (those with high pH values) conduct electricity well in aqueous solution and are therefore strong electrolytes. Weak acids and alkalis (those with pH values nearer 7) are weak electrolytes. You will find a discussion of these terms in 11.2.

The Brønsted–Lowry theory of acids and bases

In 1923 a Danish chemist, J. N. Brønsted, and an Englishman, T. M. Lowry, independently put forward the following new definitions for acids and bases.

An **acid** is a proton donor and a **base** is a proton acceptor.

These definitions include all the familiar acids and bases but extend the range.

e.g.
$$HCl(g) + H_2O(l) \rightleftharpoons H_3O^+(aq) + Cl^-(aq) \qquad \text{(i)}$$

In the forward reaction the water molecule accepts a proton from the hydrogen chloride molecule and therefore is a *base*. In the back reaction the oxonium ion is the acid and the chloride ion the base.

Thus when an acid donates a proton it becomes a base and each acid–base pair of this type is referred to as a *conjugate acid–base pair*.

e.g.
$$\underset{\text{acid}}{HCl} \xrightleftharpoons[\text{proton accepted}]{\text{proton donated}} \underset{\text{base}}{Cl^-}$$

Clearly, the stronger the acid, the weaker is its conjugate base (i.e. the greater the tendency of the acid to donate protons, the smaller the tendency of the conjugate base to accept them and reverse the process). Thus hydrogen chloride is a strong acid in water but the aqueous chloride ion is a weak base.

The formation of conjugate acid–base pairs when ammonia reacts with water may be shown in a similar way.

$$\underset{\text{base 1}}{NH_3(g)} + \underset{\text{acid 2}}{H_2O(l)} \rightleftharpoons \underset{\text{acid 1}}{NH_4^+(aq)} + \underset{\text{base 2}}{OH^-(aq)} \qquad \text{(ii)}$$

Notice that the water acts as a *base* in (i) and as an *acid* in (ii).

Substances which may act as both acids and bases are said to be **amphoteric**.

Make sure that you can sort out the conjugate acid–base pairs in the next section. The first one is rather tricky but the others are straightforward.

5.6   Hydrolysis of salts

Hydrolysis is the reaction of a compound with water such that the hydroxyl group of the latter remains intact.

Iron(III) chloride solution is acidic (Experiment 5.6.a) because of hydrolysis. Hydrated hydrogen ions are formed from the water as shown:

$$FeCl_3(aq) + 3H_2O(l) \rightleftharpoons Fe(OH)_3(s) + 3HCl(aq)$$

i.e. $\quad Fe^{3+}(aq) + 3H_2O(l) \rightleftharpoons (Fe^{3+} + 3OH^-)(s) + 3H^+(aq)$

or $\quad Fe^{3+}(aq) + 6H_2O(l) \rightleftharpoons (Fe^{3+} + 3OH^-)(s) + 3H_3O^+(aq)$

An aqueous solution of sodium carbonate is alkaline, again because of hydrolysis, and once more the water provides the characteristic ion.

$$Na_2CO_3(aq) + 2H_2O(l) \rightleftharpoons H_2CO_3(aq) + 2NaOH(aq)$$

i.e. $\quad CO_3^{2-}(aq) + 2H_2O(l) \rightleftharpoons H_2CO_3(aq) + 2OH^-(aq)$

Hydrogencarbonates are *acid* salts (5.4) and yet their solutions are *alkaline*. This apparent contradiction is explained by the following hydrolysis equation.

$$NaHCO_3(aq) + H_2O(l) \rightleftharpoons H_2CO_3(aq) + NaOH(aq)$$

i.e. $\qquad HCO_3^-(aq) + H_2O(l) \rightleftharpoons H_2CO_3(aq) + OH^-(aq)$

Note: the existence of H_2CO_3 molecules is doubtful. 'Carbonic acid' is probably just an aqueous solution of carbon dioxide (see page 301).

Questions on Chapter 5

1 Here are three definitions of an acid. Write a comment showing how far each is satisfactory and indicate which of these definitions is best.
(a) Corrosive liquid which turns litmus red.
(b) Compound which contains hydrogen.
(c) Compound which yields hydrogen ions (or protons) to other substances.
(SCE)
2 Give two chemical properties in each case (apart from their action on indicators) which are typical of (i) a dilute acid, (ii) a dilute aqueous alkali. Illustrate your answer by specific examples. (L)
3 In a cellar you find four stoppered bottles. Their labels, which have dropped off, read dilute hydrochloric acid, sodium carbonate solution, calcium hydroxide solution (lime water) and distilled water. A packet of blue litmus paper is lying nearby along with a rack of test tubes. Without leaving the cellar to obtain any other materials, how would you go about labelling the bottles correctly? Explain clearly your tests and results and give balanced equations where applicable.
(SCE)
4 Describe concisely how you would prepare crystals of zinc sulphate starting from zinc. Write the equation for the reaction and give the formula for the crystals. (O)
5 You are supplied with a solution of sodium hydroxide, crystalline lead nitrate, iron filings, dilute hydrochloric acid and a source of chlorine.
Describe how you would prepare samples of the anhydrous chlorides of sodium, iron(III) and lead(II). (O & C)
6 You are provided with the following substances: water, calcium carbonate, anhydrous sodium sulphate, dilute nitric acid.
Using a source of heat and the usual laboratory apparatus, but no other chemical substance, describe fully how you could prepare (i) a sample of calcium sulphate (calcium sulphate is almost insoluble in water), (ii) a solution of calcium hydrogen-carbonate (calcium bicarbonate), (iii) crystals of hydrated sodium sulphate ($Na_2SO_4 \cdot 10H_2O$). (C)
7 The following is an outline of one method for the preparation of crystals of zinc sulphate, $ZnSO_4 \cdot 7H_2O$.
'An excess of zinc carbonate is added to 100 cm^3 of 2.0 M sulphuric acid in a beaker and the mixture is warmed until no further reaction takes place. The mixture is filtered and the filtrate evaporated until the volume is about 25 cm^3. This liquid is left to cool and the crystals are separated, dried and weighed.'
(a) Write an equation for the reaction.
(b) Why is an excess of zinc carbonate used?
(c) How can you tell when the reaction has stopped?
(d) Why is it necessary to filter after the reaction has stopped?

(e) Why is the filtrate not evaporated to dryness?

(f) What is the relative molecular mass of $ZnSO_4 \cdot 7H_2O$?

(g) What maximum mass of zinc sulphate crystals could be formed from 100 cm^3 of 2.0 M sulphuric acid?

(h) It was found, in an actual preparation, that the mass of crystals obtained was much less than the maximum mass calculated in (g). Explain this result.

(i) Would a similar method be suitable for the preparation of lead(II) sulphate? Give reasons for your answer. (C)

8 Comment on the following statement. 'An acid salt is one which dissolves in water to give a solution which has a pH of less than 7.'

9 Write ionic equations for the reactions between

(a) solutions of copper(II) sulphate and sodium carbonate,

(b) dilute sulphuric acid and zinc,

(c) calcium hydroxide solution and dilute nitric acid,

(d) dilute hydrochloric acid and sodium carbonate solution.

10 'Concentrated ethanoic acid is a strong acid; dilute sulphuric acid is a weak one.' Is this true? Explain your answer.

11 What do you understand by the terms *acid* and *alkali*? Explain, using the ionic theory, the reaction which occurs between acid and alkali.

Describe briefly how you would prepare sodium hydrogensulphate solution from sodium hydroxide solution. Give an equation for the reaction and say what you would expect the main impurity to be. (SUJB)

12 Explain why a solution of hydrogen chloride in water turns blue litmus red, whereas a solution of dry hydrogen chloride in methylbenzene has no effect on dry blue litmus paper. (SUJB)

6 Atoms and molecules

In your course so far you have carried out experiments and made observations but you probably do not know what is taking place *within* the materials which are reacting together. In the same way, someone may be able to drive a car but until he understands its construction he will not know how it works. If you are going to find out more about what is happening in a reaction you will need to learn something about the nature of matter (i.e. the 'stuff' of which the world is composed).

No doubt you know already that scientists consider that matter consists of minute particles called atoms, but what evidence is there that atoms actually exist? We cannot *prove* their existence to you but you should find that the results of the following experiments are best explained by the idea that matter *does* consist of many tiny particles.

Experiments

6.1.a. What happens when a solid dissolves in a liquid?★

PROCEDURE   Place a crystal of ammonium dichromate(VI) in 50 cm³ of water in a 100 cm³ beaker. Observe the mixture after 15 minutes and again after several hours.

Questions

1   What did you observe?
2   Which explanation is best: (a) the ammonium dichromate(VI) crystal stretched to fill the liquid, or (b) the ammonium dichromate(VI) crystal and water are made up of tiny particles which mixed with one another?

6.1.b What happens when a solution crystallises?

PROCEDURE   *Gently* warm a microscope slide over a bunsen flame. Place a drop of hot, saturated ammonium dichromate(VI) solution on the slide and observe it closely through a microscope.

Questions

1   What did you observe?
2   Which explanation is best: (a) the ammonium dichromate(VI) which was filling the drop shrank to form crystals, or (b) tiny particles of ammonium dichromate(VI) left the solution and built up in layers to form crystals?

6.1.c How does the volume change when salt dissolves in water? ★

PROCEDURE   Place 400 g of salt in a 1 dm³ volumetric flask. Add sufficient water to cover the solid, swirl the flask to remove air bubbles and then top up to the mark with more water. Stopper the flask, shake vigorously to dissolve the salt, and observe the level of the solution.

Question

1   A possible explanation of the result is that water particles have spaces between them and the salt particles enter these when the solid dissolves. Does this explanation satisfy you?

6.1.d   The cleavage of crystals ★ ★

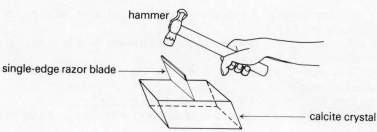

Figure 6.1.d

PROCEDURE   (a) Place the sharp edge of a razor blade parallel to one of the faces of a calcite crystal (Figure 6.1.d) and hit the blade with a hammer.
(b) Repeat, using some other orientation of the blade.

Questions

1   How did the shape of the crystals obtained in (a) compare with that of the original crystal? How did this differ from the shape in (b)?
2   Does the idea that a crystal is composed of layers of particles arranged in a regular pattern explain these results? Give a reason for your answer.

6.1.e   Is hydrogen more dense than air? ★ ★

PROCEDURE   Place a jar of hydrogen above one of air and ensure a good seal using Vaseline. After a few minutes test the gas in both jars with a flame.

Questions

1   What, if anything, did you observe when the flame was applied to (a) the top jar, (b) the bottom jar? Did you expect these results?
2   Which statement best explains the result: (a) hydrogen is more dense than air and slowly flows down into the bottom gas jar, pushing air up into the other one, or (b) hydrogen and air are composed of tiny particles which intermingle?

6.2.a   How far can we divide matter? ★ ★

The previous experiments suggest that matter does consist of tiny particles, but just how small are they? How many of them would you expect to find in 1 g of substance – a hundred, a thousand, a million, or even more?

PROCEDURE   Make a solution of potassium manganate(VII) containing $1 \, g \, dm^{-3}$, and place $20 \, cm^3$ of it in a $1 \, dm^3$ volumetric flask, top up to the mark with water and shake well. Repeat the dilution procedure and, by means of a dropping pipette, place 1 drop of the solution on a watch glass. Add one drop at a time to a $10 \, cm^3$ measuring cylinder and count the number of drops per cm^3.

Questions

1   (a) How many g of potassium manganate(VII) were there in $1 \, cm^3$ of the original solution?
(b) How many g of potassium manganate(VII) were there in $1 \, cm^3$ of the second solution?
(c) How many g of potassium manganate(VII) were there in $1 \, cm^3$ of the third solution?
(d) How many drops of solution were there in $1 \, cm^3$?
(e) How many g of potassium manganate(VII) were there in 1 drop of the third solution?
2   Was the colour of the potassium manganate(VII) visible in the third solution? Could the solution be diluted further?
3   Do you think that one drop of the third solution contained only one particle of potassium manganate(VII) or could it be divided further?
4   Suppose that one drop of the third solution *did* contain only one particle of potassium manganate(VII). What would be its mass?

6.2.b   To find the size of an oil molecule ★

The last experiment gives you some idea of the mass of a particle of matter but it does not tell you very much about its size. In this experiment you will find the diameter of an oil particle (molecule) by using a solution of $0.1 \, cm^3$ of oil (oleic acid) in $1 \, dm^3$ of petroleum ether. When this solution is dropped on to a water surface, it spreads out and the petroleum ether evaporates, leaving a layer of oil one molecule thick.

PROCEDURE   (a) Half-fill a plastic tray with water and sprinkle the surface with powdered talc. Using a dropping pipette, allow 1–2 drops of the oil solution to fall into the centre of the tray and observe what happens.
(b) Make a loop from a piece of cotton about 30 cm long, cut off the loose ends and grease the cotton with Vaseline. Place fresh water in another tray and float the loop in it. Add drops of the oil solution to the water surface inside the loop until the loop is taut, circular in shape and does not dent when touched. Measure the diameter of the oil film. Finally, find the number of drops of solution per cm^3 as in Experiment 6.2.a.

Questions

1 What did you see when the oil was added to the water surface covered with talc? Can you explain your observations?

2 Can you explain why the loop became taut?

3 What was the diameter of the oil film in the loop? Calculate the area (A) of the film in cm^2, given that $A = \pi r^2$, where $\pi = 22/7$ and r = radius of loop.

4 There was $0.1 \, cm^3$ of oil in $1 \, dm^3$ of solution.

(a) How many cm^3 of oil were there in $1 \, cm^3$ of solution?

(b) How many cm^3 of oil were there in 1 drop of solution?

(c) How many drops of solution were added to the loop?

(d) How many cm^3 of oil were added to the loop?

5 The volume (V) of the film is equal to the volume of oil added to the loop. It is also equal to the area (A) multiplied by the thickness (T); i.e., $V = AT$, therefore $T = V/A$. Calculate the thickness of the film in cm from this equation.

6 Assuming that the film is one molecule thick, and that molecules are tiny spheres, what is the diameter of an oil molecule?

6.3.a Looking at smoke ★★

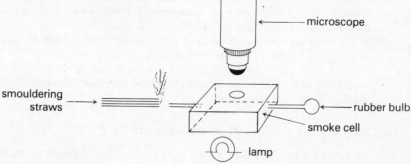

Figure 6.3.a

PROCEDURE Place the smoke cell on the microscope stage and arrange the lamp so that the side of the cell is illuminated (Figure 6.3.a). Ignite a bunch of straws, blow out the flame so that the straws are smouldering and then, by means of the bulb, draw smoke into the cell. Finally, focus the microscope on the smoke particles.

Questions

1 Smoke is made up of tiny particles of carbon. What other particles are present in the cell? Are they bigger, or smaller than those of smoke?

2 What did you see when you looked through the microscope? What do you think causes these effects?

6.3.b Do all gas particles move at the same rate? ★★

The explanation you arrived at in Experiment 6.3.a should have included the idea that gas molecules are continually moving in a random manner. Do they all move at the same rate?

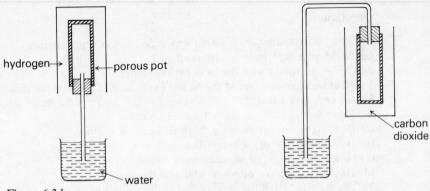

Figure 6.3.b

PROCEDURE (a) Set up the apparatus as shown in Figure 6.3.b, placing a gas jar of hydrogen over the porous pot. Note what happens at the end of the tube, then remove the gas jar and again observe the results.
(b) Repeat the experiment using carbon dioxide in place of the hydrogen but modify the apparatus as shown.

Questions

1   The porous pot has many very small tunnels through it. Which molecules seemed to move faster through these tunnels, those of hydrogen or those of air? How did you reach this conclusion?
2   How did the speed of the carbon dioxide molecules through the pot compare with that of the air molecules? What led you to this conclusion?
3   Air is more dense than hydrogen but less dense than carbon dioxide. Does there seem to be any connection between the rate of movement through the pot and the density of the gas concerned?

6.3.c   Do gases and liquids mix at the same rate? ★ ★

PROCEDURE (a) Place a few drops of bromine (**caution**: highly corrosive liquid, poisonous vapour) in a gas jar and cover immediately with a cover slip.
(b) Place about 2 cm depth of tetrachloromethane in a test tube and carefully add an equal volume of an aqueous solution of bromine. Place a piece of white paper behind each container and observe them both for a few minutes.

Questions

1   Did the bromine mix with (a) the air, (b) the tetrachloromethane? Were the rates of mixing the same?
2   What conclusions can you draw about the relative rates of movement of the particles of gases and liquids?
3   How quickly would you expect the particles of solids to mix?

6.3.d   The compressibility of solids, liquids and gases ★

PROCEDURE Place your finger over the end of a gas syringe filled with air and try to push the piston in. Repeat, using a syringe filled with water.

Questions

1 What do you think would happen if the experiment were repeated with a syringe filled with ice?
2 In which of the three states of matter, solid, liquid and gas, do you think that the particles are (a) closely packed, (b) sparsely distributed? Explain your answer.
3 Can you explain the difference in rigidities of solids, liquids and gases in terms of the movement of the particles *relative to one another*?

6.3.e   The effect of temperature change on the speed of molecules★★

PROCEDURE   Stand a large treacle tin, with the lid firmly in place, on a tripod behind a safety screen. Heat the tin by means of a bunsen burner and keep well clear.

Questions

1 What happened to the pressure of the gas inside the tin when it was heated?
2 The pressure exerted by the gas is due to gas molecules colliding with the walls of the tin. Does the result of the experiment indicate that the molecules move (a) faster, or (b) slower as the temperature is raised?

6.3.f   Change of state – what happens to the particles? ★

PROCEDURE   Place several ice cubes in a beaker of water, stir well and note the temperature of the mixture. Warm the beaker and continue stirring and reading the temperature until the ice melts. Then heat strongly and boil the water for a short time.

Questions

1 Did the volume change very much when the ice melted? How did the volume of water which evaporated compare with the volume of steam produced? Do your answers agree with the conclusions drawn in Experiment 6.3.d about the packing of particles?
2 Did the temperature change when (a) the ice was melting, (b) the water was boiling? Since heat energy was being supplied all the time, what do you think it was being used for? (Hint: think about the changes which were occurring in the packing of the particles.)

6.4.a   To compare the masses of reactants and products in a chemical reaction

The results of the previous experiments in this chapter can all be explained if you assume that matter consists of particles. The experiments in this section enable you to find out more about the particles themselves.

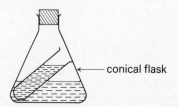

Figure 6.4.a

PROCEDURE   Select any two solutions found in your laboratory which you think will react to form a precipitate by double decomposition. *Check with your teacher that your choice is satisfactory before going any further.* Place one of the solutions in a 100 cm³ conical flask and the other in a 75 × 12 mm test tube and arrange the apparatus as shown in Figure 6.4.a, making sure that the two liquids do not come into contact with one another. Find the mass of the apparatus and contents. Gently tilt the flask so that the solutions mix and react, then find the mass again. Compare your results with those of other members of the class.

Questions

1   How did the mass of the reactants compare with that of the products?
2   Do you think that particles of matter were (a) created, (b) destroyed, (c) rearranged in some way during the reactions?

6.4.b   To see if all samples of a compound are identical

In this experiment members of the class analyse samples of copper(II) oxide from different bottles to see if the percentage of copper in each sample is the same.

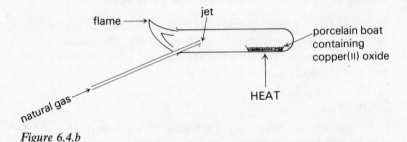

Figure 6.4.b

PROCEDURE   Find the mass of a porcelain boat, fill it with a *dry* sample of analytical grade copper(II) oxide and then find the mass of the boat and contents. (It is important to heat the copper(II) oxide just before the experiment and allow it to cool down in a desiccator to ensure that no water is present.) Push the boat into a 150 × 19 mm test tube clamped horizontally, taking care not to spill any of the solid. Connect a glass tube to the gas supply by means of rubber tubing, turn on the supply and ignite the gas issuing from the tube. (**Caution:** do not turn the gas full on.) Adjust the flow so that the flame is about 3 cm high and lodge the glass tubing in the mouth of the test tube as shown in Figure 6.4.b. Heat the boat until all the copper(II) oxide appears to have been changed to copper, allow it to cool down *with the gas still passing* and then find the mass of the boat plus copper. Repeat the heating, cooling and weighing until the mass is constant.

Questions

1   Why was it essential to keep passing the gas until the copper had cooled down?
2   Why was the last part of the experiment repeated until the mass was constant?

3 What is the name given to the type of chemical change undergone by the copper(II) oxide?

4 Set out your results as shown.

Mass of porcelain boat   = a g
Mass of porcelain boat + copper(II) oxide = b g
Mass of porcelain boat + copper = c g
∴ Mass of copper(II) oxide = (b − a) g
∴ Mass of copper = (c − a) g
∴ Mass of oxygen = (b − a) − (c − a) g

$$\therefore \text{Percentage of copper in copper(II) oxide} = \frac{\text{mass of copper}}{\text{mass of copper(II) oxide}} \times 100$$

∴ Percentage of oxygen in copper(II) oxide = ?

5 Compare the class results. Allowing for experimental error, does the composition of copper(II) oxide vary from sample to sample or is it constant?

6 Which of the following explanations best fits the results? (a) The masses of copper and oxygen atoms and the ratio in which they combine are fixed. (b) The masses of copper and oxygen atoms and the ratio in which they combine are variable.

6.4.c   Analysis of two oxides of lead ★ ★

This experiment must not be performed by pupils.

If two elements combine to form more than one compound, is there any relationship between their ratios by mass in each substance?

PROCEDURE   The procedure for this experiment is the same as for Experiment 6.4.b. Two solids, lead(II) oxide and lead(IV) oxide are analysed and hydrogen is used instead of natural gas (North Sea Gas). (**Caution:** all air must be expelled from the apparatus before igniting the gas – see Experiment 4.3.b for details.)

Questions

1 What was the product in each case? What sort of reaction have the oxides undergone?

2 Set out your results as in the previous experiment and calculate the mass of (a) lead, (b) oxygen in each oxide.

3 How many g of oxygen were combined with 1 g of lead in (a) lead(II) oxide, (b) lead(IV) oxide? Is there a simple ratio between the two figures?

4 What can you say about the *relative* numbers of oxygen atoms combining with one atom of lead in each oxide?

6.1   Evidence for the existence of atoms

You probably think that the idea that matter consists of atoms is a modern one. In fact, the philosophers of Ancient Greece first put forward this theory. These men did not carry out experiments but reached their conclusions by *thinking* about the nature of matter. They decided that if a piece of material were divided a sufficient number of times, a fundamental particle would be obtained which could not be divided further. (The word atom comes from the Greek '*atomos*', meaning 'indivisible'.) Their theories remained untested for over 2000 years and even now we cannot *prove* that their ideas were correct because we cannot see atoms, even if we use a microscope.

However, if we assume that matter is *not* continuous (i.e. *not* capable of being divided up indefinitely) and does consist of tiny particles – either atoms or groups of atoms called *molecules* – then the results of many experiments may be explained. Thus, the dissolving of a solid in a liquid (Experiments 6.1.a and 6.1.c) or the mixing of two gases (Experiment 6.1.e) may be interpreted in terms of the intermingling of the particles of the various substances. The fact that the volume of salt solution formed in Experiment 6.1.c is less than that of the salt plus water before dissolving indicates that there are spaces between the water particles which those of salt can enter when it dissolves.

Crystals

Crystals of a particular compound all have more or less the same shape. This can be easily understood if we assume that every crystal consists of layers of particles packed together in a regular pattern. This idea also explains, as shown in Figure 6.1.1, why cleaving a crystal parallel to one of the faces gives a clean cut whereas an attempt to cleave it in some other way gives a jagged cut (Experiment 6.1.d).

Figure 6.1.1 Cleaving a crystal

Crystals usually have flat faces. This is readily explained by considering the following example. Imagine 25 polystyrene balls arranged so that they form a square, 5×5. Sixteen polystyrene balls may be arranged on top of these, forming a square, 4×4, and the whole process repeated until a pyramid is obtained (Figure 6.1.2). What can you say about the sides of the pyramid?

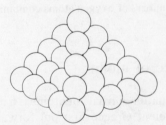

Figure 6.1.2

Seeing atoms

As we have said, the results of the experiments in this chapter can best be explained in terms of the particle theory of matter and hence they provide evidence to support it. There is no evidence *against* the theory but definite proof of its validity escapes us because atoms are too small for us to see in the normal way. However, in recent years, scientists have developed a device known as the *field-ion microscope*. This gives pictures which should convince most people that atoms do exist. The photograph below shows the field-ion micrograph of the element tungsten in which we can 'see' its atoms.

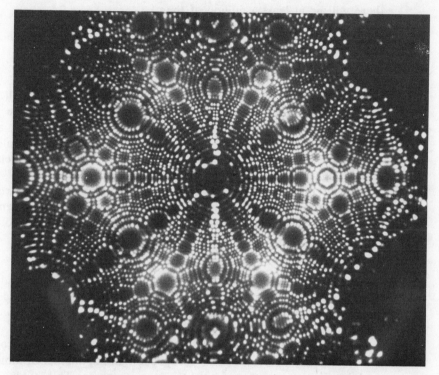

Field-ion micrograph of tungsten – each spot represents a tungsten atom

6.2 How big are atoms?

The masses of atoms

In Experiment 6.2.a, a solution of potassium manganate(VII) is diluted until the colour is pale pink. If you assume that a single drop contains only one particle of potassium manganate(VII), then you should be able to calculate its mass to be about 10^{-8} g, i.e. $1/100\,000\,000$ g. In fact, a drop of solution contains many millions of the particles which are responsible for the pink colour. Each one has a mass of about 10^{-22} g and consists of one manganese atom and four oxygen atoms, so you can see that the mass of the individual atoms must be very small indeed!

The sizes of atoms

In Experiment 6.2.b, it is found that the diameter of an oleic acid molecule is about 10^{-7} cm. Again, this is a very small figure but, since each of these molecules contains a total of fifty-four atoms (carbon, hydrogen and oxygen), the diameter of an individual atom must be smaller still. Not all atoms are the same size, but the average diameter is actually about 10^{-8} cm (i.e. 0.1 nm). The following example may help you to understand what these figures really mean.

Sharpen your pencil and mark a small dot on a piece of paper. How many carbon atoms do you think there are stretching across the middle of the dot? The answer depends on how well you sharpened your pencil but there will be at least a million of them!

6.3 The kinetic theory

Brownian motion

Some of the most convincing evidence in favour of the particle theory of matter is provided by the smoke cell in Experiment 6.3.a. The smoke particles reflect the light and bright specks are seen moving around continually in a random manner. The effect is due to air molecules, themselves too small to be seen, colliding with the smoke particles and causing them to bounce off in different directions.

This sort of motion was first observed by the botanist, Robert Brown, in 1827, when he looked through a microscope at pollen grains suspended in water. Brown could not explain the movement of the grains – can you?

Diffusion

Experiment 6.1.e shows that gases have a tendency to spread in all directions and this too is explained by the idea that gases are made up of *moving* particles. The process of spreading is known as *diffusion* but it does not occur at the same rate for all gases.

In Experiment 6.3.b, hydrogen molecules pass into the porous pot faster than the molecules of air escape, so that the number of molecules, and hence the pressure, inside the pot increases. This explains why bubbles come from the bottom of the tube. When the jar is removed the hydrogen inside the pot diffuses out faster than the air diffuses in. Hence the pressure inside decreases and the liquid moves up the tube. Eventually equilibrium is reached when all the hydrogen has escaped and the rate at which air molecules pass into the porous pot is equal to the rate at which others leave it. The pressure is then constant and equal to atmospheric pressure so that the level of water in the tube is the same as that in the beaker.

The arguments are reversed when the pot is surrounded by carbon dioxide. The carbon dioxide diffuses into the porous pot more slowly than the air diffuses out, hence the pressure inside the pot decreases and the water moves up the tube. In general, then, the denser the gas the slower the diffusion process.

Diffusion is not restricted to gases but takes place in liquids as well. Thus, when a solid dissolves in water (Experiment 6.1.a), the water molecules remove successive layers of particles from the crystal. These diffuse to produce a solution of uniform concentration. However, Experiment 6.3.c shows that diffusion occurs much more slowly in liquids than in gases and the obvious conclusion to be drawn from this is that molecules of liquids move more slowly than those of gases. Can you think of another possible reason for the slowness of diffusion in liquids? You should get a clue to the answer by considering why it is easier to walk along a school corridor in the middle of a lesson rather than just after the bell has rung.

Diffusion does not *seem* to take place in solids; therefore you could conclude that the particles do not move. However, if clean, flat metal surfaces are placed together for some time in a vacuum they weld together, so there must be *some* movement. This might be useful for, say, welding together the components of a space station, but it could cause problems with moving parts.

The states of matter

The differences in the physical properties of the three states of matter, solid, liquid and gas, may be explained by considering the packing and relative movement of the particles within them.

A solid is difficult to compress (Experiment 6.3.d) and this suggests that its particles are close together. It is also rigid and has a definite shape, which implies that the particles are held in absolutely fixed positions by forces of attraction acting between them. However, since solids diffuse into one another it follows that the particles *must* move and in order to explain this we say that they *vibrate* and *rotate* about fixed points.

A liquid is also difficult to compress but, unlike a solid, it will flow. Its particles are considered to be close together, with attractive forces acting between them, but free to move relative to one another in a random manner.

As a gas is easy to compress, its particles must be relatively far apart and since it diffuses throughout its containing vessel the particles must be free to move quite independently.

When the temperature of a solid is raised, the particles vibrate more and more violently (i.e. their kinetic energy increases) until they are moving to such an extent that they can no longer be held in an ordered arrangement by the forces of attraction. When this happens the solid melts. The fact that there is a fairly small change in volume (Experiment 6.3.f) is an indication that the spacing between the particles is similar in both the solid and the liquid states.

Raising the temperature of a liquid increases the speed of movement of the particles until their kinetic energy is sufficient to overcome the forces of attraction between them. At this point the liquid boils. You should realise that the particles do not *all* have the same energy. As you know, water evaporates slowly at room temperature, so some molecules must be moving fast enough to escape from the liquid well below its boiling point. The volume change on converting a liquid to a gas is larger than that on melting a solid (for example, 1 cm^3 of water at $100 \,^\circ\text{C}$ forms 1700 cm^3 of vapour at the same temperature) and this supports the idea that the particles in a gas are relatively far apart.

On cooling, the reverse changes occur. The particles of the gas gradually slow down as the temperature falls until the forces of attraction are able to condense them together and form a liquid. Cooling the liquid causes further loss of kinetic energy until eventually the particles settle into a solid crystal where they vibrate and rotate about fixed points.

When a solid melts or a liquid boils, heat energy is supplied and yet the temperature remains constant. This energy must be used up in overcoming the forces holding the particles together.

For most substances the heat required to convert a certain mass of solid to liquid at the melting point is less than that required to convert the same mass of liquid to gas at the boiling point. From this we can conclude that the attractive forces in a solid are greater than those in a liquid and much greater than those in a gas. Also, since these forces fall off as the distance between the particles increases, we have further evidence that the particles in a solid are closer together than those in a liquid and much closer than those in a gas.

The theory we have been discussing is concerned with the idea that matter is made up of *moving* particles. It is called the **kinetic theory**, from the Greek 'kinesis', meaning 'motion'.

Summary

Table 6.3 summarises some differences in the structure and properties of solids, liquids and gases.

	Solid	Liquid	Gas
Compressibility	Difficult to compress	Difficult to compress	Easy to compress
Molecular packing	Close	Close	Sparse
Shape	Fixed	Variable	Variable
Molecular movement	Vibrate and rotate about fixed points	Free to move	Free to move
Attractive forces	Relatively high	Relatively high	Very small
Diffusion	Very slow	Slow	Rapid

Table 6.3

6.4 The chemical laws and Dalton's atomic theory

The law of conservation of mass

In Experiment 6.4.a it is found that, within the limits of experimental error, the mass of a flask and contents before a reaction is the same as that at the end. After performing many experiments and always obtaining the same result, Antoine Lavoisier put forward the *law of conservation of mass* which may be stated as follows.

Matter can neither be created nor destroyed during the course of a chemical reaction.

In Experiment 3.1.a, magnesium is burned in air and it is found that the mass of the residue is greater than the mass of the magnesium. Is this a contradiction of the law? You will realise, of course, that it is *not*. If the mass of the magnesium *and the oxygen* with which it combines is compared with that of the magnesium oxide formed, then the law is obeyed. Can you think of any other examples which *seem* to contradict the law?

You may be interested to learn that according to modern thinking the law *is* incorrect. Einstein's theory of relativity regards mass and energy as being convertible from one to the other, so that when energy is given out in a reaction there must be a decrease in mass – the mass has been converted to energy. Similarly when energy is absorbed during a reaction the mass must increase because the energy has been converted to mass. However, this mass change is so small as to be undetectable. For example, it has been calculated that the heat evolved by the combustion of 4000 kg of phosphorus would result in a loss of mass of 0.01 g!

The law of constant composition

Different samples of copper(II) oxide are reduced to copper in Experiment 6.4.b and all are found to contain the same percentage by mass of copper. As a result of a large number of experiments of this type, the Frenchman Joseph Proust (in 1797) stated the *law of constant composition*.

All pure samples of the same chemical compound contain the same elements combined in the same proportions by mass.

The law of multiple proportions

When orange lead(II) oxide and brown lead(IV) oxide are reduced to lead, as in Experiment 6.4.c, it is found that there is twice as much oxygen combined with 1 g of lead in lead(IV) oxide as there is in lead(II) oxide. These results are typical of those covered by the *law of multiple proportions*, put forward by John Dalton.

If two elements, A and B, combine to form more than one compound, then the different masses of A which combine separately with a fixed mass of B are in a simple ratio.

Dalton's atomic theory

Although the particle theory of matter had been suggested from the time of the Ancient Greeks, it was not until 1808 that John Dalton, a

Manchester lecturer, proposed a *quantitative* theory concerning the masses of atoms and the ratios in which they combine. His theory explained the chemical laws, themselves deduced from experimental results. Its main points were as follows.

1 Matter is composed of minute particles called atoms.
2 Atoms are indivisible and cannot be destroyed.
3 Atoms of a particular element all have the same properties and the same mass.
4 Atoms of different elements have different properties and different masses.
5 When elements combine to form compounds, the atoms combine in simple whole numbers.

The law of conservation of mass is explained by points 1 and 2. If atoms are indestructible then a chemical reaction must simply involve their rearrangement to give the products.

The law of constant composition is explained by points 3, 4 and 5. In copper(II) oxide the copper and oxygen atoms are combined in a simple fixed ratio and, since atoms of any one element always have the same mass, it follows that the percentage by mass of copper and oxygen will always be the same.

Dalton *deduced* the law of multiple proportions from his atomic theory. Although he was a very poor experimenter, others carried out accurate experiments which verified the law and thus gave further support to his ideas about atoms. The explanation of the results obtained in Experiment 6.4.c is simply that there are twice as many atoms of oxygen combined with one atom of lead in lead(IV) oxide as there are in lead(II) oxide.

Shortcomings of Dalton's atomic theory

Since Dalton's time a number of facts have been discovered which contradict his theory. For example, atoms are now considered to be made up of smaller particles, and the existence of isotopes means that atoms of a particular element do not all have the same mass (see 6.5). In addition, molecules of plastics are known to consist of huge numbers of atoms joined together and so Dalton's final point must also be invalid.

However, you should not regard Dalton's atomic theory as worthless. It provided a sound working basis for chemists throughout the major part of the nineteenth century and even today you would have great difficulty in proving it wrong in your own laboratory. For the most part, matter *seems* to behave as if it were made up as Dalton suggested.

How can we define atoms and molecules?

So far we have used the terms 'atom' and 'molecule' without saying precisely what they mean. Any definition which we produce must take into account the fact that atoms are not indivisible as Dalton suggested.

An **atom** is the smallest particle of an element which can exist *and still retain the ordinary chemical properties of that element.*

This definition is valid because the smaller particles from which atoms are built up do *not* have the chemical properties of the element concerned.

Many elements in their natural state consist not of separate atoms but of groups of atoms joined together, e.g. oxygen, O_2; phosphorus, P_4; sulphur, S_8. In addition, atoms of *different* elements may join together in groups to form particles of a compound. These groups of atoms are called *molecules*.

A **molecule** is the smallest particle of an element or compound which can exist *naturally* and still retain the ordinary chemical properties of that element or compound.

Symbols

Dalton produced a series of symbols for the elements known to him so that their chemical reactions could be written down in shorthand form. Some of these symbols are shown in the photograph below. A few of Dalton's 'elements' were not elements at all – can you spot any?

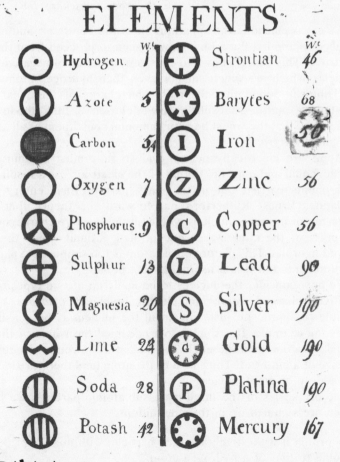

Dalton's elements

The system of symbols now used was developed by Jöns J. Berzelius in 1813. In it the initial letter, or the initial plus one other letter, of an element is used to represent *one atom* of it. Thus C is the symbol for carbon and Ca the symbol for calcium. If you look at the list on pages 466-7. you will see that some of the symbols do not fit this pattern. The reason is that in a few cases the Latin names were used. Thus Fe, the symbol for iron, comes from the word **ferrum**, Pb (lead) from **plumbum**, and Cu (copper) from **cuprum**.

6.5 Atomic structure

Dalton's atomic theory suggests that an atom is the smallest particle of an element that can possibly be obtained, but since this idea was put forward experiments have been carried out which show otherwise.

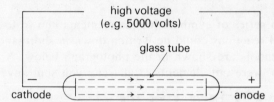

Figure 6.5 A gas
discharge tube

If a gas is placed in the apparatus shown in Figure 6.5 and its pressure reduced to about 1 Pa, a blue glow is seen to proceed from the negative electrode. This can be shown to consist of a stream of negatively charged particles which are much smaller even than hydrogen atoms. In 1897, J.J.Thomson found that the mass and charge of these particles was constant, no matter what substance was chosen for the cathode or the gas. This means that they must be a fundamental constituent of all atoms; they are called *electrons.*

Atoms are electrically neutral and so they must contain a positive charge, equal and opposite to that of the electrons. As a result of certain experiments carried out in the early part of this century, the New Zealander Ernest Rutherford put forward the theory that the atom consisted of a minute nucleus where all the positive charge and most of the mass of the atom were concentrated, around which the electrons moved in orbits. This picture of the atom is too simple to explain modern developments but it is a useful starting point.

We now consider the nucleus to be made up of two types of particle; positively charged *protons* and electrically neutral *neutrons.* You should be able to appreciate just how small the nucleus is by considering the following example. If a single atom occupied the whole of the space in your laboratory then the nucleus, situated at its centre, would be about the size of a pin-head! Thus most of the atom (and therefore most of you) is made up of empty space.

The properties of the three main sub-atomic particles we have mentioned are summed up in their definitions.

A **proton** is a positively charged particle, with mass approximately equal to that of a hydrogen atom.

A **neutron** is a neutral particle, with mass approximately equal to that of a hydrogen atom.

An **electron** is a negatively charged particle, its charge being equal but opposite to that of a proton, and its mass approximately 1/1840 of that of a proton.

Since atoms are electrically neutral the numbers of protons and electrons in an atom must be equal.

Atomic number and mass number

You will find in Chapter 7 that it is the number and arrangement of electrons in an atom (and hence the number of protons) which determine its chemical properties.

The number of protons in the nucleus of an atom is its **atomic number** and gives its position in the periodic table.

The total number of protons and neutrons in the nucleus of an atom is its **mass number**.

We can include information about the atomic number and mass number of an atom when we write down its symbol. Thus $^{12}_{6}C$ represents an atom of carbon with an atomic number of 6 and a mass number of 12; its nucleus contains 6 protons and 6 neutrons. How many protons and neutrons are there in the nucleus of an atom of $^{27}_{13}Al$?

Arrangement of electrons

The electrons surround the nucleus and are found at varying distances from it in groups known as energy levels, or 'shells'. Each shell contains electrons with similar energies and can contain up to a certain maximum number. The first shell, nearest the nucleus, can contain up to two electrons, the second up to eight and the third up to eighteen.

The movement of electrons within the shells is complicated and difficult to explain at this level and for this reason we usually represent the shells by concentric rings drawn round the nucleus. However, you must remember that the electrons are *not* travelling in circular orbits and also that the drawings are not to scale.

The simplest atom of all is that of hydrogen, consisting of one proton and one electron.

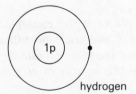

hydrogen

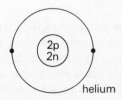

helium

A helium atom has an atomic number of 2 and a mass number of 4 and therefore is made up of 2 protons, 2 neutrons and 2 electrons. The first shell can only contain up to two electrons and if any more have to be accommodated, as in $_3^7Li$ (lithium) for example, they have to go into the second shell which is further away from the nucleus. The second shell can

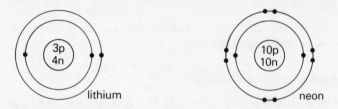

lithium

neon

contain up to eight electrons and so this will not be full until we get to neon, atomic number 10, electronic arrangement 2,8. The next element, sodium, with atomic number 11, has an electron in the third shell, i.e. 2,8,1.

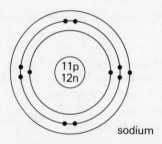

sodium

You will find in Chapter 8 that there is a special stability associated with having eight electrons in a shell. Thus in a potassium atom, atomic number 19, there is an electron in the fourth shell, even though the third is not full. The electronic arrangement is 2,8,8,1 and *not* 2,8,9.

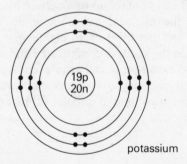

potassium

Isotopes

What determines the identity of an atom? We know, for example, that all oxygen atoms have an atomic number of 8 and since no other atoms behave exactly like those of oxygen or have the same atomic number, it would seem reasonable to assume that it is the number of protons in the nucleus of an atom which identifies it.

In 1913, J. J. Thomson reported that he had detected *two* types of neon atoms which seemed to differ in mass. His assistant, Francis Aston, managed to separate these two forms of neon and went on to show that other elements also had atoms with different masses. The explanation for this is that atoms of the *same element* may have *different numbers of neutrons* in their nuclei. Their chemical properties are identical, because they all have the same number of protons and electrons.

Atoms of the same element with different numbers of neutrons in their nuclei are called **isotopes**.

e.g. Chlorine is made up of two isotopes, $^{35}_{17}Cl$ and $^{37}_{17}Cl$.

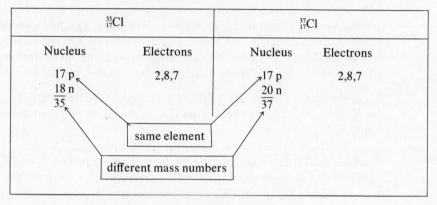

$^{35}_{17}Cl$		$^{37}_{17}Cl$	
Nucleus	Electrons	Nucleus	Electrons
17 p	2,8,7	17 p	2,8,7
18 n		20 n	
35		37	

same element

different mass numbers

Isotopes are difficult to detect and separate, which is why Dalton did not suspect their existence. Their name is derived from the Greek 'isos topos', meaning 'same place' (in the periodic table).

Relative atomic mass – a more accurate version

The simple definition of relative atomic mass given in Chapter 9 is obviously unsatisfactory since it assumes that all atoms of an element have the same mass. A more accurate definition is based on the mass of a $^{12}_{6}C$ atom.

The **relative atomic mass** of an element is the number of times that the average mass of one of its atoms is greater than one-twelfth of the mass of an atom of $^{12}_{6}C$.

The 'average mass' mentioned has to take account of the relative abundances of the various isotopes. For example, about three-quarters of all chlorine atoms are $^{35}_{17}Cl$ and about one-quarter are $^{37}_{17}Cl$. The relative atomic mass is thus

$$(\tfrac{3}{4} \times 35) + (\tfrac{1}{4} \times 37) = 35.5$$

The instrument used to measure the relative masses and abundances of isotopes is called a *mass spectrometer*. (Its mode of action is too complicated to be discussed here.)

Relative molecular mass

This too must be defined more accurately in terms of $^{12}_{6}C$.

> The **relative molecular mass** of an element or compound is the number of times that the average mass of one of its molecules is greater than one twelfth of the mass of an atom of $^{12}_{6}C$.

6.6   Radioactivity

In 1896 the French chemist Henri Becquerel discovered that uranium salts gave off radiation which passed through wrapping paper around a photographic plate and blackened the plate itself. The phenomenon was investigated by Pierre and Marie Curie who named it *radioactivity*.

The radiation is due to the spontaneous splitting up of the nuclei of certain atoms and is of three distinct types.

1   *α-particles*   These are positively charged particles and were shown by Rutherford to be helium nuclei (i.e. made up of two protons and two neutrons).

2   *β-particles*   These are electrons. You may well wonder how an electron can be emitted from the *nucleus* of an atom. This is explained by saying that a neutron splits up to give a proton and an electron, the proton being retained by the nucleus and the electron ejected.

3   *γ-rays*   When either an α- or a β-particle is emitted, the remaining nucleus usually has an excess amount of energy. This extra energy is given out in the form of γ-rays, which are like X-rays but with even shorter wavelength.

All atoms having an atomic number greater than 83 (bismuth) are radioactive. In addition, many of the lighter elements have radioactive isotopes but these only occur to a minute extent.

When the nucleus of a radioactive atom gives off an α- or a β-particle an atom of a *new element* is formed. If this new atom is stable the disintegration process stops, but if it is radioactive then the splitting-up process continues. A typical radioactive nucleus is that of a uranium-238 atom which emits an α-particle to give a thorium-234 nucleus.

$$^{238}_{92}U \rightarrow \, ^{234}_{90}Th + \, ^{4}_{2}He$$

This new nucleus is unstable and will emit a β-particle to form a protactinium-234 nucleus.

$$^{234}_{90}Th \rightarrow \, ^{234}_{91}Pa + e^-$$

Further disintegration occurs until finally a stable $^{206}_{82}Pb$ atom is left.

Notice that the emission of an α-particle gives a nucleus of atomic number 2 less, and mass number 4 less, than the parent nucleus. When a β-particle is emitted the new nucleus (daughter nucleus) has an atomic number 1 more than that of the parent nucleus, while its mass number is the same. Make sure that you understand these processes.

Half-life

The rate at which nuclei disintegrate varies enormously from one element to the next. Since radioactive decay is an exponential process (Figure 6.6) we can never measure the time taken for *all* of the material to disintegrate, but we can find the *half-life*.

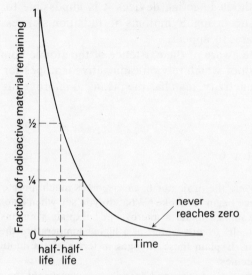

Figure 6.6 Radioactive decay curve

The **half-life** of an isotope is the time taken for the mass of radioactive material to be reduced to half its initial value.

This can vary from millions of years as in thorium-232 to less than a microsecond as in polonium-212 and is the same no matter how much material is present at the start. Thus 1000 kg of carbon-14 (half-life 5700 years) takes 5700 years to decay to 500 kg and 1 g of it takes exactly the same time to decay to 0.5 g. It then takes a further 5700 years for the 0.5 g to become 0.25 g and so on.

The uses of radioactive isotopes

Radioactive isotopes have wide and varied uses, a few of which are listed below.

1   Nuclear power stations generate electricity from the energy produced when isotopes like uranium-235 decay.

2   In living animals and plants the percentage of the radioactive isotope $^{14}_{6}C$ remains constant because it is continually being replaced. In dead tissue this does not happen. Hence by comparing the percentage of carbon-14 in a dead sample with that in living matter the age of the sample can be found.

3   If radioactive compounds are injected into a patient the functioning of glands and the flow of blood around the body may be checked by means of a Geiger counter or some other detecting device.

4   Cancer may be cured in some cases by killing the cancerous cells by exposing them to radiation.

The dangers of radiation

Radiation kills all living cells. The most penetrating type is γ-radiation, but all radioactive sources must be handled with care. They are generally stored and transported in lead containers which must be thick enough to prevent harmful radiation from escaping. Perhaps the most dangerous aspect of all is that without detecting devices it is impossible to tell if exposure to radiation is occurring. Symptoms of 'radiation sickness' take some time, possibly years, to appear.

You will, of course, be aware of the existence of the atomic bomb and the hydrogen bomb, both of which rely on radioactive isotopes for their operation. Thus, in radioactivity, man has the means to improve his lot or to destroy the world.

Questions on Chapter 6

1   Why is it that in a cinema the projector beam becomes much more obvious when people in the audience begin to smoke? What do you see when smoke is lit by a beam of light and examined under a microscope? Explain your answer.

2   When ice is heated it melts to give water; at a higher temperature the water boils to give water vapour. Explain these changes in terms of the motion and spacing of the water molecules.

3   State the law of conservation of mass and describe an experiment to verify it. When magnesium is heated in air, it increases in mass; does this fact contradict the law? Explain your answer.

4   State the law of constant composition and describe briefly how you would prove it. How does Dalton's atomic theory account for this law?

Three samples of copper(II) oxide weighed 8.0 g, 7.3 g and 11.6 g respectively and were reduced by hydrogen to give 6.4 g, 5.8 g and 9.3 g of copper. Show that these results are in agreement with the law of constant composition.

5   Draw diagrams to represent the structures of the following atoms: $^{14}_{7}N$, $^{19}_{9}F$, $^{32}_{16}S$, $^{40}_{20}Ca$.

6   (a) Explain briefly the differences between the three types of radiation given off by radioactive elements. In what way are such elements (i) useful, (ii) dangerous?

(b) What is X in each of the following equations?

   (i) $^{238}_{92}U \rightarrow {}^{234}_{90}Th + X$

   (ii) $^{32}_{15}P \rightarrow X + \beta$

7   (a) Give an account of Dalton's ideas about atoms and the way in which they combine. Explain briefly why it has been necessary to modify these ideas.

(b) Describe and explain an experiment which provides evidence for the view that a gas consists of very small particles in random motion.   (C)

8   In an experiment to find the exact mass of copper contained in a sample of copper oxide, some of the oxide was weighed and heated in a stream of hydrogen. The instructions for carrying out the experiment listed the following precautions. (i) Allow the hydrogen to pass through the apparatus for some time before lighting it at the other end. (ii) At the end of the experiment allow the copper to cool before stopping the flow of hydrogen.

(a) Suggest a reason for each of these precautions.

(b) Copper oxide can also be reduced by heating it with carbon. Explain whether this would be a suitable method for this particular experiment.

(c) Write the formulae for copper(I) oxide and copper(II) oxide.

(d) The following results were obtained in this experiment.

$$\text{Mass of copper oxide at the start} = 35.75 \text{ g}$$

$$\text{Mass of copper remaining} \quad = 31.75 \text{ g}$$

Use these results to find which of the two oxides of copper was used in the experiment. (SCE)

9 On reduction, 12.00 g of a metallic oxide A gave initially 11.20 g of a lower oxide B and ultimately 10.40 g of the metal X.

(a) Calculate the mass of X which is combined with 16 g of oxygen in each oxide.

(b) State the chemical law illustrated by these results.

(c) The relative atomic mass of X is 208. Deduce the simplest formula of the oxide A. (AEB)

10 Atom A has a mass number of 239 and an atomic number of 93. Atom B has a mass number of 239 and an atomic number of 94.

(a) How many protons has atom A?

(b) How many neutrons has atom B?

(c) Are atoms A and B isotopes of the same element? Explain.

(d) Explain whether a geologist would or would not be likely to locate compounds containing these atoms in the course of his work. (SCE)

11 From the following table, write down the values of (a), (b), . . . (l).

Element	Atomic number	Mass number	Number of protons	Number of neutrons	Electronic configuration
P	13	27	(a)	(b)	(c)
Q	(d)	(e)	(f)	71	2,8,18,18,5
R	(g)	40	20	(h)	(i)
S	17	(j)	(k)	18	(l)

(SCE)

12 $^{24}_{12}Mg$ and $^{26}_{12}Mg$ are symbols for two of the isotopes of magnesium. Compare atoms of these isotopes with respect to (a) the compositions of their nuclei, (b) their electronic configurations. (W)

7 The periodic table

You have seen earlier that your studies are simplified if substances can be placed in classes and general patterns of behaviour established. The periodic table, which you will find printed in full on pages 464–5, is an extremely useful classification and helps us not only to understand the properties of a familiar element but also to predict how an unfamiliar one should behave. Since it covers just about every aspect of chemistry, it is probably best if at first you simply see how it is built up; you can then refer back to it at appropriate points in your course.

Experiments

7.1.a To find a basis for classification

In the last century chemists found that relative atomic mass gave a good basis for classifying elements, but is it the best one?

PROCEDURE Using the table printed on pages 466–7, write down the elements with atomic numbers 1–20 (a) in order of relative atomic masses, and (b) in order of atomic numbers.

Questions

1   If you have studied the reactivity series (20.1) you will know that the elements lithium, sodium and potassium have similar properties. How many elements come between them in list (a) and in list (b)?
2   Helium, neon and argon also have similar properties. How many elements come between them in list (a) and in list (b)?
3   Magnesium and calcium show certain similarities. How many elements separate them in the two lists?
4   Which list gives the most consistent pattern, (a) or (b)?

7.2.a To find a reason for the pattern

PROCEDURE   Add the electronic arrangement of the elements to list (b) in the last experiment, remembering that the maximum numbers of electrons found in each shell *for these elements* is 2,8,8. (The third shell is not actually full when it contains eight electrons but this need not worry you here.)

Questions

1   What have lithium, sodium and potassium atoms in common, as far as the arrangement of their electrons is concerned?

2   What is common to the electronic arrangement of magnesium and calcium atoms?

3   Have helium, neon and argon atoms anything in common?

4   Does there seem to be a connection between the chemical properties of an element and the arrangement of the electrons in its atoms?

7.3.a   Do the physical properties of the elements follow a pattern?

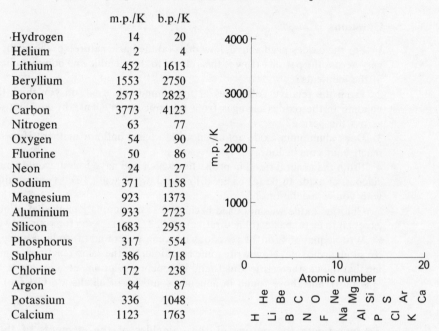

	m.p./K	b.p./K
Hydrogen	14	20
Helium	2	4
Lithium	452	1613
Beryllium	1553	2750
Boron	2573	2823
Carbon	3773	4123
Nitrogen	63	77
Oxygen	54	90
Fluorine	50	86
Neon	24	27
Sodium	371	1158
Magnesium	923	1373
Aluminium	933	2723
Silicon	1683	2953
Phosphorus	317	554
Sulphur	386	718
Chlorine	172	238
Argon	84	87
Potassium	336	1048
Calcium	1123	1763

Figure 7.3.a The melting points and boiling points of the first twenty elements

PROCEDURE   Draw a histogram of the melting points of the first twenty elements against their atomic numbers. Mark the axes as shown in Figure 7.3.a. Repeat the exercise using the boiling points of the elements.

Questions

1   Is there a repeating pattern (i.e. a periodic variation) and is it perfect?

2   Study the histograms in Figure 7.3 (page 118). Do you think that the periodic table would be useful in predicting the physical properties of a newly discovered element, provided that its atomic number were known?

7.4.a   To investigate the nature of the oxides of the elements of the third period ★ ★

PROCEDURE   (a) Add a spatula measure of sodium oxide to 5 cm³ of distilled water in a test tube. Shake well and test the liquid with universal indicator paper. Repeat the experiment using magnesium oxide, *freshly prepared* aluminium oxide (see Experiment 20.4.a), silicon(IV) oxide and phosphorus(V) oxide (**care**: violent reaction) in place of the sodium oxide.

Bubble some sulphur dioxide into 5 cm³ of distilled water in a test tube and find the pH of the solution.

(b) Add a spatula measure of freshly prepared aluminium oxide to 5 cm³ of dilute hydrochloric acid in a boiling tube. Heat until the liquid boils and note whether any solid has dissolved.

(c) Repeat (b) using 2 M sodium hydroxide solution in place of the dilute hydrochloric acid.

Questions

1 For the oxides used above, how does (a) the basic nature, (b) the acidic nature vary across the period? How is this related to the metallic and non-metallic nature of the elements?

2 From the relative volatilities of the compounds used, do you think that the bonding in the oxides changes from (a) ionic to covalent, (b) covalent to ionic across the period?

3 Does aluminium oxide act as an acid, a base, both, or neither? How did you reach your conclusion?

4 From the general trend in properties which you have found, would you *expect* silicon(IV) oxide to be (a) basic, (b) neutral or (c) acidic? Did your results agree with your expectations?

5 Chlorine oxide was not used because it is dangerously explosive. Would you expect it to be (a) basic, (b) neutral, (c) acidic? Give a reason for your answer.

6 Write equations for the reactions between water and (a) sodium oxide, Na_2O, (b) sulphur dioxide. Now write ionic equations for the same reactions, remembering that the characteristic ions found in aqueous solutions of acids are $H^+(aq)$ (or $H_3O^+(aq)$), and those found in aqueous solutions of alkalis are $OH^-(aq)$.

7.4.b To investigate the nature of the chlorides of the elements of the third period ★ ★

PROCEDURE (a) Heat a spatula measure of sodium chloride in a test tube. Repeat the experiment, using anhydrous aluminium chloride, and then phosphorus pentachloride, in place of the sodium chloride.

(b) Place 5 cm³ of distilled water in a test tube and measure its temperature. Add a spatula measure of sodium chloride, stir gently with the thermometer and note the temperature change, if any. Finally, test the solution with universal indicator paper.

Repeat the experiment, using anhydrous aluminium chloride, and then phosphorus pentachloride, in place of the sodium chloride.

Questions

1 From the relative volatilities of the compounds used, do you think that the bonding in the chlorides changes from (a) ionic to covalent, (b) covalent to ionic across the period? Explain your answer.

2 Which chlorides *react* with water: ionic or covalent ones? Which compound do you think is *always* formed if a chloride reacts with water? (The universal indicator test should give you a clue.)

3 Write an equation for one of the reactions, bearing in mind your answer to question 2.

7.1 Early attempts at classification of elements

So far, the only attempt which we have made to classify elements has been to divide them into metals and non-metals. In a similar way we could divide a crowd of people into males and females but obviously there would be many differences between the individuals in each group. If we wanted to sort the people into smaller groups whose members shared similar characteristics, we would ask them to stand together in *families*. In the same way, as chemists discovered more and more substances, they realised that 'families' of elements with similar properties existed. The problem as far as elements were concerned was to find a reason for these similarities.

The first step in solving the problem was taken by the German Johann Döbereiner, in 1829. He noticed that, where groups of three similar elements occurred, the relative atomic mass of the middle element came approximately half-way between those of the other two. One example of such a group, which Döbereiner called a *triad*, is found in the metals calcium, strontium and barium. These have relative atomic masses 40, 88 and 137 respectively. Look at the list on pages 466–7 and see if you can find any others.

Although you should have no difficulty in finding two or three triads you will soon realise that for many elements the idea does not work. However, spurred on by Döbereiner's discovery, other scientists looked for further connections between relative atomic mass and chemical properties.

In 1864, the Englishman John Newlands found that if the elements were arranged in order of increasing relative atomic masses then a pattern appeared. Starting at any given element, the eighth one from it was, as he put it, 'a kind of repetition of the first'. Because of the similarity of this repetition to the musical scale he called it the *law of octaves*. The first twenty-one elements known to Newlands are listed below in rows of seven, so that similar elements come in the same place in each row.

H	Li	Be	B	C	N	O
F	Na	Mg	Al	Si	P	S
Cl	K	Ca	Cr	Ti	Mn	Fe

You will see that the law of octaves is not very reliable. For example, phosphorus and manganese are not similar, nor are sulphur and iron, and the pattern breaks down in many places if the list is extended.

The first periodic table

The final break-through in classification of elements according to relative atomic mass was made by Dmitri Mendeléeff in Russia, in 1869. He arranged the elements in order of increasing relative atomic masses in a similar way to Newlands *but*, in order to obtain a better repetition, he left gaps for elements which he assumed must exist but had not yet been discovered. He found that many properties then followed a repeating pattern, or *periodic variation* (see Figure 7.3, page 118).

Table 7.1 shows a slightly modernised form of Mendeléeff's *periodic* table and you can see that it is made up of seven horizontal series, or *periods*, arranged so that similar elements fall into vertical columns, or *groups*. For example, all the alkali metals are in group I A and all the halogens in group VII B. Not only did Mendeléeff have to leave gaps for undiscovered elements, he also had to correct some relative atomic masses which had been determined erroneously, to make the elements fit into the right groups.

Group	I		II		III		IV		V		VI		VII		VIII
Sub-group	A	B	A	B	A	B	A	B	A	B	A	B	A	B	
1st period	H														
2nd period	Li		Be			B		C		N		O		F	
3rd period	Na		Mg			Al		Si		P		S		Cl	
4th period	K	Cu	Ca	Zn	—	—	Ti	*	V	As	Cr	Se	Mn	Br	Fe Co Ni
5th period	Rb	Ag	Sr	Cd	Y	In	Zr	Sn	Nb	Sb	Mo	Te	—	I	Ru Rh Pd
6th period	Cs	Au	Ba	Hg	La	Tl	—	Pb	Ta	Bi	W	—	—	—	Os Ir Pt
7th period	—		—		—		Th		—		U				

Table 7.1 Mendeléeff's periodic table (modernised form)

Perhaps the most striking evidence of the importance of the periodic table is found in Mendeléeff's prediction of the properties of missing elements such as germanium, which comes in the gap marked with an asterisk in group IV. He made his predictions by considering the properties of the neighbouring elements, and the following comparison shows you just how close he was.

	Properties of 'eka-silicon', Es (predicted by Mendeléeff in 1871).	Properties of germanium, Ge (discovered by Winkler in 1886).
Relative atomic mass	72	72.6
Density	5.5 g cm^{-3}	5.47 g cm^{-3}
Appearance	Dirty grey metal.	Lustrous grey metal.
With air	Will give a white powder, EsO_2, on heating.	Gives a white powder, GeO_2, on heating.
With water	Will decompose steam with difficulty.	Only decomposes steam on heating strongly.
With acids	Slight reaction only.	No reaction with HCl, or dilute H_2SO_4. Attacked by hot, concentrated H_2SO_4.
With alkalis	Will react more readily than with acids.	Attacked by hot, concentrated aqueous alkalis.
Properties of oxide, MO_2	Refractory, specific gravity 4.7, less basic than TiO_2 and SnO_2.	Refractory, specific gravity 4.7, feebly amphoteric, i.e. feebly basic and acidic.
Properties of chloride, MCl_4	Liquid, b.p. less than 373 K (100 °C), specific gravity 1.9 at 273 K (0 °C).	Liquid, b.p. 359 K (86 °C), specific gravity 1.89 at 291 K (18 °C).

Some criticisms of the early periodic classification

In spite of the success of the periodic table in grouping together elements with similar properties, correcting the values of relative atomic masses and predicting the existence of unknown elements, it was not a perfect arrangement. For example, some elements had to be listed in the *wrong* order according to their relative atomic masses so that they could be fitted into the correct group with respect to their properties. Thus, though iodine closely resembles chlorine and bromine, its relative atomic mass would put it in a different group from them. (Consult the modern form of the table on pages 464–5 and find out which one.) The noble gases were not known when Mendeléeff produced his table but their discovery has led to an instance of this 'wrong' order occurring within the first twenty elements. Look at the modern form of the table again and see if you can find the 'error'.

There are several other difficulties which we shall not discuss here, but the most important criticism of classification according to relative atomic mass is that it does not explain *why* the elements in a given group are similar to one another.

A new basis for classification

As a result of studies concerning the X-rays given out when different elements were bombarded by electrons, Henry Moseley, an Englishman, discovered that some fundamental property of the atom increased by one unit for successive elements in the periodic table. He suggested that one proton (and therefore one electron) was added to the atom on going from one element to the next. Nowadays we refer to the number of protons in an atom as its atomic number (6.5), and when elements are arranged in order of atomic numbers the problems of 'wrong' order in the periodic table disappear – iodine and tellurium take their correct places, as do argon and potassium.

Moseley published his findings in 1914 but soon afterwards, like millions of other young men, he was to die in the First World War. However, his work had provided a new basis for the classification of elements and had paved the way for an explanation of the periodic table.

Element	Atomic number	Number of neutrons	Mass number	Electronic arrangement
Hydrogen	1	0	1	1
Helium	2	2	4	2
Lithium	3	4	7	2,1
Beryllium	4	5	9	2,2
Boron	5	6	11	2,3
Carbon	6	6	12	2,4
Nitrogen	7	7	14	2,5
Oxygen	8	8	16	2,6
Fluorine	9	10	19	2,7
Neon	10	10	20	2,8
Sodium	11	12	23	2,8,1
Magnesium	12	12	24	2,8,2
Aluminium	13	14	27	2,8,3
Silicon	14	14	28	2,8,4
Phosphorus	15	16	31	2,8,5
Sulphur	16	16	32	2,8,6
Chlorine	17	18 or 20	35 or 37	2,8,7
Argon	18	22	40	2,8,8
Potassium	19	20	39	2,8,8,1
Calcium	20	20	40	2,8,8,2

Table 7.2 The atomic structures of the most common isotopes of the first twenty elements

7.2 Electronic arrangement and the periodic table

The atomic structures of the most common isotopes of the first twenty elements, in order of atomic numbers, are shown in Table 7.2.

You can see that the number of electrons in the outermost shells of the atoms follows a *periodic* variation, just as the properties of the elements do, and also that atoms of similar elements, such as sodium and potassium, have the *same* numbers of electrons in their outermost shells. At last we have found a property of the *atoms* which follows the same pattern as the properties of the *elements*. Thus, if we can connect the behaviour of an atom with the arrangement of its outermost electrons, the reason for the existence of the periodic table will be clear.

Consider what happens when atoms come into contact with one another. Which parts will interact? It seems likely that it will be the outer regions and so it is reasonable to suppose that the outermost electrons in an atom *do* govern its behaviour.

If we arrange the first twenty elements in rows so that similar elements fall into vertical columns then we arrive at the first three periods of the modern periodic table. Hydrogen and helium have been placed apart from the others for reasons which will be discussed shortly.

				H	He		
Li	Be	B	C	N	O	F	Ne
Na	Mg	Al	Si	P	S	Cl	Ar
K	Ca						

If you refer to pages 464–5 you will see that the table becomes more complicated when element 21 is reached. A centre block has to be added because of the way in which the electron shells fill up but a discussion of this is beyond the scope of this book.

For the outer blocks or *main groups* the number of electrons in the outermost shell of the atom is the same as the group number, except for the noble gases (e.g. atoms of the alkali metals in group I all have one electron in the outermost shell while those of the halogens in group VII all have seven). Can you find a connection between the group number and the oxidation number of an element? (This topic is dealt with in Chapter 8.)

The first period (hydrogen and helium) remains a problem. Although the properties of helium show that it is a noble gas, its atoms do not possess a stable octet of electrons and so it has been separated from the others. The position of hydrogen is even more difficult to decide. It has one electron in its atom and can form a singly charged positive ion like the alkali metals in group I. On the other hand, like the halogens in group VII, it only needs to gain one electron to attain the electronic structure of the nearest noble gas, helium. In its reactions it shows some of the properties of each group and yet does not really fit into either of them. Thus, together with helium, it is isolated from the rest of the periodic table.

7.3 Periodic variation in physical properties of the elements

If you draw the melting point and boiling point histograms according to the instructions in Experiment 7.3.a you should find that a periodic variation exists. Figure 7.3 shows that periodicity also occurs in other physical properties. The pattern is not perfect but elements in a particular

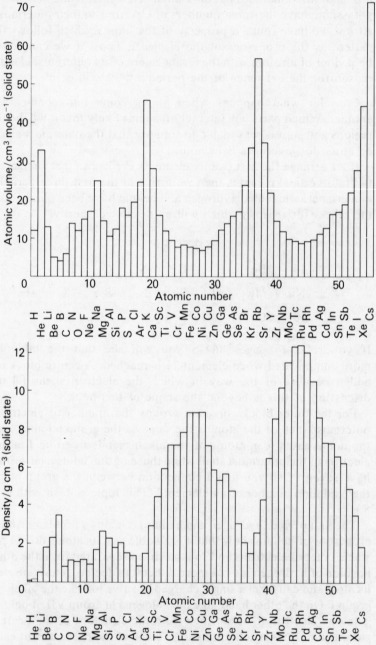

Figure 7.3 Histograms showing how atomic volume and density vary across the periodic table

group usually come at more or less the same position on each 'hill'. Find each alkali metal and halogen to confirm this.

You will see that the centre block of *transition metals* (atomic numbers 21 to 30 and 39 to 48) makes the pattern more complicated after the third period but that periodicity still exists. Cover up this block and see if the variation for the main group elements is the same as in the earlier periods.

7.4 Trends in properties across the third period (sodium to argon)

The detailed properties and reactions of these elements are dealt with elsewhere in the book. What is important here is to find the way in which these properties vary. You have already seen that the pattern of variation in physical properties is not repeated exactly in each period and this applies to chemical properties too. However, it is extremely useful to know what *general* trends are found.

Note: the comments made in the remainder of this chapter apply to the *main group elements* only. For simplicity the transition metals are *excluded*.

The elements

The most obvious variation to be seen in going across a period is that the elements change from metals to non-metals. Somewhere in between there are elements which have some of the properties of both types and these are referred to as *semi-metals* or *metalloids*. The point of change is not sharp but is indicated by the 'stairway' which starts between beryllium and boron on the periodic table chart on pages 464–5.

The reactivity of the elements is connected with the number of electrons which must be lost or gained to attain the noble gas octet (8.1). Thus sodium, which only needs to lose one electron per atom, is more reactive than magnesium, which needs to lose two. Similarly, chlorine is more reactive than sulphur. Argon, of course, is chemically inert.

Another property which depends on the number of outer electrons in the atom is the oxidation number (8.6). The most commonly-found oxidation numbers going across this period are $+1, +2, +3, +4, -3, -2, -1, 0$. Can you explain this sequence?

The hydrides

The bonding in the hydrides changes from ionic to covalent, going across the period. Metallic hydrides are unusual in that those of the more electropositive metals contain the hydride ion, H^-, where hydrogen is behaving like a halogen. They react with water to form hydrogen and the corresponding alkali.

e.g. $$NaH(s) + H_2O(l) \rightarrow NaOH(aq) + H_2(g)$$

or $$H^-(s) + H_2O(l) \rightarrow OH^-(aq) + H_2(g)$$

As the bonding becomes more covalent, so the acidic nature of the hydrides increases. Thus silicon hydride is insoluble in water and neutral,

Element	Sodium	Magnesium	Aluminium	Silicon	Phosphorus	Sulphur	Chlorine	Argon
Appearance	Silvery metal	Silvery metal	Silvery metal	Black solid	Yellow solid	Yellow solid	Greenish gas	Colourless gas
Electronic structure	2,8,1	2,8,2	2,8,3	2,8,4	2,8,5	2,8,6	2,8,7	2,8,8
Most common oxidation number	+1	+2	+3	+4	−3	−2	−1	0
Atomic radius/nm	0.157	0.136	0.125	0.117	0.110	0.104	0.099	0.19
Melting point/K (°C)	371 (98)	923 (650)	933 (660)	1683 (1410)	317 (44)	386 (113)	172 (−101)	84 (−189)
Bonding and structure	←——— Giant metallic lattice ———→			Giant atomic lattice (covalent)	←——— Covalent molecules ———→			Free atoms

Hydride	NaH	MgH_2	$(AlH_3)_n$	SiH_4	PH_3	H_2S	HCl
Melting point/K (°C)	←——— Decompose before melting ———→			88 (−185)	139 (−134)	188 (−85)	158 (−115)
Bonding and structure	←—— Giant ionic lattice ——→		←——————— Covalent molecules ———————→				

Hydride	NaH	MgH₂	(AlH₃)ₙ	SiH₄	PH₃	H₂S	HCl
With water	Gives H₂ and NaOH (strong alkali)	Gives H₂ and Mg(OH)₂ (weak alkali)	Gives H₂ and Al(OH)₃ (amphoteric)	Insoluble	Slightly soluble. Neutral solution	Weakly acidic solution	Strongly acidic solution

Oxide	Na_2O	MgO	Al_2O_3	SiO_2	P_4O_{10}	SO_3	Cl_2O_7
Melting point/K (°C)	1466 (1193)	3348 (3075)	2318 (2045)	2001 (1728)	836 (563)	303 (30)	182 (−91)
Bonding and structure	←——— Giant ionic lattice ———→			Giant atomic lattice (covalent)	←——— Covalent molecules ———→		
Nature	←— Decreasingly basic —→		Amphoteric	←——— Increasingly acidic ———→			

Chloride	NaCl	$MgCl_2$	$AlCl_3$	$SiCl_4$	PCl_3	SCl_2	–
Melting point/K (°C)	1074 (801)	987 (714)	sublimes	205 (−68)	182 (−91)	195 (−78)	–
Bonding and structure	←— Giant ionic lattice —→		←——— Covalent molecules ———→				
With water	←— Dissolve —→		←——— Hydrolysed ———→				

Table 7.4

phosphine (PH_3) is slightly soluble and neutral, hydrogen sulphide solution is weakly acidic and hydrogen chloride solution strongly so.

e.g. $\qquad\qquad HCl(g) + \text{water} \rightarrow H^+(aq) + Cl^-(aq)$

or $\qquad\qquad HCl(g) + H_2O(l) \rightarrow H_3O^+(aq) + Cl^-(aq)$

The oxides

You probably know already that most metallic oxides are basic and that they are ionic compounds. Those of the more electropositive metals react with water to form alkaline solutions.

e.g. $\qquad\qquad Na_2O(s) + H_2O(l) \rightarrow 2NaOH(aq)$

or $\qquad\qquad O^{2-}(s) + H_2O(l) \rightarrow 2OH^-(aq)$

As with the hydrides, the bonding changes from ionic to covalent going across the period and the acidity of the compounds increases. (This acidity trend only applies if, for an element with several oxides, the one having the greatest proportion of oxygen is chosen, i.e. P_4O_{10}, SO_3, Cl_2O_7.) Experiment 7.4.a shows that aluminium oxide is amphoteric, that is, both acidic and basic, and you probably deduced from your results that silicon(IV) oxide is neutral. However, it is *acidic*, in accordance with the general trend and gives rise to salts (silicates). The reason for its apparent neutrality is that it is insoluble in water. Sulphur dioxide, like most non-metallic oxides, is acidic and it dissolves in water to give sulphurous acid.

$$SO_2(g) + H_2O(l) \rightleftharpoons H_2SO_3(aq)$$

or $\qquad\qquad SO_2(g) + H_2O(l) \rightleftharpoons 2H^+(aq) + SO_3^{2-}(aq)$

The chlorides

Again the bonding changes from ionic to covalent across the period. The ionic chlorides simply dissolve in water but the covalent ones *react* to give an acidic solution because of hydrolysis (5.6).

e.g. $\qquad\qquad AlCl_3(s) + 3H_2O(l) \rightarrow Al(OH)_3(s) + 3HCl(aq)$

$\qquad\qquad\qquad PCl_3(s) + 3H_2O(l) \rightarrow H_3PO_3(aq) + 3HCl(aq)$

Summary

The various changes in properties which have been discussed are summarised in Table 7.4. Remember that the pattern will not be repeated *exactly* in other periods but that the same general trends will be observed for the main group elements. Also the dividing line between ionic and covalent bonding, like that between metals and non-metals, is not sharp. Thus the compounds near to the line may show some properties of both types of bond.

7.5 Trends in properties going down groups

It is interesting to study the variation in properties found going down the groups in the periodic table. The changes are generally less marked than those seen in going across a period, since all the members of a particular group have atoms with the same number of electrons in their outermost shells. These points are discussed in detail in sections 19.3, 20.2 and 20.3 but the general conclusions which emerge are that on going down any group (1) metallic nature *increases*, (2) reactivity of a metal *increases* and (3) reactivity of a non-metal *decreases*.

For the main groups (I–VII), the results of this section and the previous one are summed up in Figure 7.5.

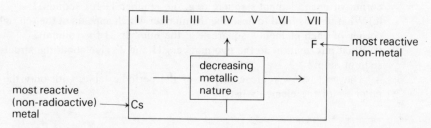

Figure 7.5

A word of warning

The periodic table is not a magic arrangement giving perfectly regular patterns of behaviour. It must be stressed that the trends observed are *general* and that there are exceptions to them. Also the centre block of transition metals has been excluded from our discussions in order that the arguments developed should be as straightforward as possible. Nevertheless, the periodic table *is* extremely useful in simplifying the study of chemistry and will often be referred to in later chapters.

Questions on Chapter 7

1 From the following list of elements: bromine, helium, hydrogen, mercury, neon, phosphorus, potassium, silicon, sodium, name:
(a) one which is stored under oil in the school laboratory
(b) one which must be stored under water in the school laboratory
(c) one which forms an acidic oxide
(d) one which forms a basic oxide
(e) a metal which is liquid at room temperature
(f) a noble gas which is used in meteorological balloons.   (SCE)
2 The atomic number of lithium, Li, is 3.
(a) On what grounds do you infer that lithium is a metallic element?
(b) Name two elements which you would expect to resemble lithium chemically.
(c) Which of the three elements, lithium and the two referred to in (b), would you expect to be the least chemically reactive? Give a reason.
(d) State any two properties that you would expect lithium chloride to possess.

(W)

3   The following table gives some information about the elements in the period sodium to argon in the periodic table.

11 23 Na	12 24 Mg	13 27 Al	14 28 Si	15 31 P	16 32 S	17 35.5 Cl	18 40 Ar

(a) Which of the elements in this period are metals?

(b) Which of the elements in this period are gaseous at room temperature and pressure?

(c) Two of the elements in this period burn in air to form oxides which dissolve in water to produce acidic solutions. Name these two elements.

(d) What term is used to describe the number which appears in the top left-hand corner of each element's square (e.g. the number 11 for sodium)?

(e) What term is used to describe the number which appears in the top right-hand corner of each element's square (e.g. the number 23 for sodium)?

(f) What information do the two numbers 11 and 23 give about the structure of a sodium atom?   (L)

4   The grid below represents part of the periodic table, with only the atomic numbers of the elements listed.

1							2
3	4	5	6	7	8	9	10
11	12	13	14	15	16	17	18

Write down the atomic number of

(a) one noble gas shown,

(b) the most reactive metallic element shown,

(c) the most reactive non-metallic element shown,

(d) a solid, non-metallic element which forms an oxide of the type EO_2 (where E is a symbol for a suitable element),

(e) a metallic element which forms an oxide of the type E_2O_3,

(f) a solid element which forms a chloride of the type ECl_3,

(g) an element which forms an ion of the type E^{2-}, and

(h) an element which forms an ion of the type E^{2+}.   (L)

5   For each of the pairs of elements (a) sodium and potassium, (b) chlorine and bromine, give *three* properties that illustrate their chemical similarity.

   Suggest an electronic explanation for this similarity.   (O & C)

6   'Non-metallic elements of valency n usually form with hydrogen, compounds of formula XH_n. These compounds, called hydrides, are generally covalent and gaseous. They vary from being very soluble to insoluble in water and may be acidic, basic or neutral in character.'

   Illustrate these statements by reference to the simple hydrides of chlorine, sulphur, nitrogen and carbon.   (C)

   Further questions which require an understanding of the periodic table will be found following Chapters 8 and 20.

8 Bonding and structure

By now you should be familiar with the idea that matter is made up of atoms, and atoms themselves consist of smaller particles, protons, neutrons and electrons. Since the nucleus of every atom is surrounded by a cloud of negatively charged electrons, and like charges repel one another, how do atoms join together to form compounds?

Experiments

8.1.a The electrolysis of copper(II) chloride solution ★

PROCEDURE   Set up the circuit shown in Figure 2.1.b (page 30) but place the electrodes in about 50 cm³ of 1 M copper(II) chloride solution in the beaker. Allow the current to flow for about 5 minutes and, if any gas is evolved, hold moist indicator paper above the surface of the solution where the bubbles are bursting. Sniff the gas *very carefully* (as described in Experiment 5.3.a, page 66). Switch off and examine the carbon rods.

Questions

1   Which gas was given off and how did you identify it?
2   Which electrode did the gas come from? For particles to be attracted to this electrode what charge must they carry, positive or negative?
3   Which substance do you think was deposited on the other electrode? What can you say about the charge on the particles which were attracted to this electrode?
4   Does copper(II) chloride solution seem to contain (a) neutral molecules of copper(II) chloride or (b) oppositely charged particles of copper and chlorine?
5   What is the formula of copper(II) chloride? Since a solution of copper(II) chloride is electrically neutral, what can you deduce about the *relative sizes* of the charges on the individual particles of copper and chlorine in it? How are these values linked with the oxidation numbers (or the valencies) of the two elements?

8.2.a The effect of an electric current on tetrachloromethane ★

PROCEDURE   Set up the apparatus as in the previous experiment but use 50 cm³ of tetrachloromethane in place of the copper(II) chloride solution. (**Caution**: avoid breathing tetrachloromethane vapour.)

Questions

1   Did the tetrachloromethane conduct an electric current? How did you reach this conclusion?
2   Does this result indicate that tetrachloromethane consists of ions?

3 Tetrachloromethane has the formula CCl_4 and if it *does* consist of ions it must be made up of $C^{4+} + 4Cl^-$. Can you suggest why these ions might *not* form?

8.3.a To investigate some of the differences between ionic and covalent compounds

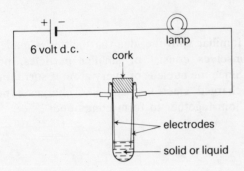

Figure 8.3.a

PROCEDURE Use samples of sodium chloride, lead(II) iodide, potassium carbonate, paraffin wax, naphthalene and tetrachloromethane for each of the following experiments.

(a) Find whether each of the solids has a high or low melting point by heating a little of it in an ignition tube. Do not heat for longer than about two minutes and stop heating immediately if the solid melts.

(b) Fill the rounded part of a test tube with one of the compounds, add 5 cm³ of distilled water and shake the tube to see if the compound dissolves. Repeat this experiment with the other compounds and keep the solutions for part (d).

(c) Repeat (b) using methylbenzene in place of the water.

(d) Place a little of one of the solid compounds in a 12 × 75 mm test tube and insert two electrodes as shown in Figure 8.3.a. (These may conveniently be made by straightening out all but the centre 'U' of a paper clip.) Hold the electrodes in position by means of a cork making sure that they do not touch. Complete the circuit; note whether the lamp glows and then heat the solid until it *just* melts. Repeat the experiment using the other compounds and any solutions obtained in (b) and (c). Do *not* heat any of the liquids.

Questions

1 Which of the compounds are (a) ionic, (b) covalent?

2 Which type of compounds had high melting points and which had low ones? Can you explain these results in terms of the forces of attraction between the particles?

3 Which substances were soluble in water: ionic or covalent ones?

4 Which substances were soluble in the organic solvent methylbenzene?

5 Which *solids*, if any, conducted electricity?

6 Which of the *liquids* conducted electricity?

7 Can you explain the results of questions 5 and 6?

8 Write down a list of the typical properties of (a) ionic and (b) covalent compounds.

To investigate the structures of some solids

The states of matter were discussed in Chapter 6 and no doubt you will remember that for a solid to melt or a liquid to boil the forces of attraction between the particles have to be overcome. Hence, the ease with which a solid may be turned into a liquid or vapour should give some clue as to the strength of the forces holding the particles together.

PROCEDURE   Place a crystal of iodine in a test tube and heat it gently. If no change is apparent, heat it more strongly. Repeat the experiment using separate samples of potassium chloride, graphite and silicon(IV) oxide in place of the iodine.

Questions

1 What did you observe in each case?
2 In which solid(s) would you say that the forces of attraction between the particles are (a) strong, (b) weak?
3 In Experiment 8.3.a you should have found that ionic compounds have high melting points owing to the strong forces of attraction acting between the oppositely charged ions. Do you think that *all* solids which have high melting points are ionic? Give a reason for your answer.

8.1   The ionic (electrovalent) bond

When atoms combine during a chemical reaction, only their outermost parts come into contact. Thus it seems likely that the forces which hold them together involve *electrons* in some way.

The noble gases, which are found in group 0 in the periodic table, are known to be extremely unreactive, so much so that helium, neon and argon form no compounds at all. Presumably this is because there is something particularly stable about the arrangement of electrons in their atoms. These arrangements are shown in Table 8.1 and you can see that, with the exception of helium, all the atoms have an outermost shell of eight electrons.

Element	Atomic number	Electronic arrangement
Helium	2	2
Neon	10	2,**8**
Argon	18	2,8,**8**
Krypton	36	2,8,18,**8**
Xenon	54	2,8,18,18,**8**

Table 8.1 The electronic structures of the noble gases

Since there seems to be a special stability associated with an octet of electrons in the outermost shell, we might expect there to be a tendency for other atoms to lose or gain electrons in order to achieve this stability for themselves. The question is, can they do this? If we assume that they *can* and we then treat chemical reactions simply as processes where electrons are redistributed between different atoms to give each one the electronic structure of the nearest noble gas, then we are able to explain a great deal about chemical bonds and the properties of elements and compounds.

The formation of potassium chloride

Consider the electronic arrangements in potassium and chlorine atoms. Each potassium atom (2,8,8,1) has one electron in its outermost shell while each chlorine atom (2,8,7) has seven. You can see (Figure 8.1.1) that if the outer electron is transferred from an atom of potassium to one of chlorine then both will have an octet. This is what happens when the two elements react together to form potassium chloride.

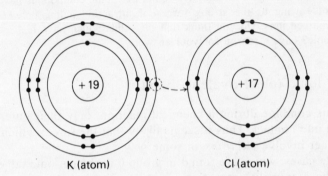

K (atom) Cl (atom) *Figure 8.1.1*

When the potassium atom has lost its electron it is left with a net positive charge of one unit, since it still has nineteen protons (i.e. nineteen positive charges) but now has only eighteen electrons (i.e. eighteen negative charges). Similarly, the chlorine atom acquires unit negative charge since it contains the original seventeen protons but now has eighteen electrons. Charged particles which are formed in this way are called *ions*. Their symbols are the same as those of the atoms from which they are derived but with the charge added, i.e. K^+ and Cl^-.

An **ion** is an electrically charged particle formed from an atom or group of atoms by the loss or gain of one or more electrons.

The oppositely charged potassium and chloride ions will attract one another (Figure 8.1.2) and this force of attraction is known as an *ionic* (or *electrovalent*) *bond*. The formation of such a bond always involves *transfer of electrons*.

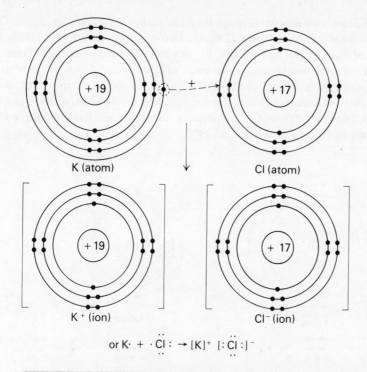

K (atom)   Cl (atom)

K⁺ (ion)   Cl⁻ (ion)

or K· + ·C̈l: → [K]⁺ [:C̈l:]⁻

Figure 8.1.2

An **ionic (electrovalent) bond** is a bond formed by the transfer of one or more electrons from one atom to another.*

The charge on an ion is equal to its **oxidation number** (8.6).

The number of electrons in the outermost shell of an atom of a main group element is equal to its group number. Thus, for metals outside the centre block of the periodic table, the number of electrons lost per atom is equal to the group number; for non-metals, the number of electrons gained by each atom is equal to (8 minus group number).

You may have noticed that both of the ions in potassium chloride have the same *electronic* structure (that of argon). However, their different charges and *nuclear* structures give them *very* different properties. In addition you must remember that ions do not behave in the same way as the corresponding atoms. For example, when potassium chloride is added to water it simply dissolves but when potassium metal is placed in water, a violent reaction takes place. Can you explain why the potassium ion is stable but the atom is not?

Calcium oxide

The bonding in calcium oxide can be explained in a similar manner to that in potassium chloride. Calcium atoms have the electronic arrangement,

*The number of electrons lost or gained by an atom in forming an ion is often called its *valency*.

2,8,8,2; i.e. two electrons more than the stable argon configuration. On the other hand, oxygen atoms (2,6) are two electrons short of a stable octet. Hence when calcium burns in oxygen to form calcium oxide, two electrons are transferred from each calcium atom to an oxygen atom to give calcium ions and oxide ions (Figure 8.1.3). Since the calcium atom has lost two electrons the ion formed has a net positive charge of two units and is written as Ca^{2+}. Similarly, the oxide ion has a negative charge of two units and is written as O^{2-}.

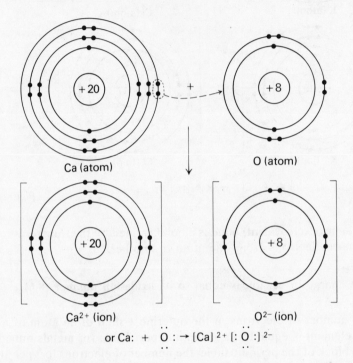

or Ca: $+$ $\overset{..}{\underset{..}{O}}$: $\rightarrow$ [Ca] $^{2+}$ [: $\overset{..}{\underset{..}{O}}$:] $^{2-}$

Figure 8.1.3

Magnesium chloride

So far we have only considered the combination of atoms which need to lose or gain the same numbers of electrons to reach the electronic structure of the noble gas nearest to them in the periodic table. What happens if different numbers of electrons are involved? To answer this question we will consider the change which occurs when magnesium reacts with chlorine to form magnesium chloride.

The magnesium atom (2,8,2) must lose *two* electrons in order to attain a stable octet, but each atom of chlorine (2,8,7) can only accept *one*. Hence the magnesium atom must give its two electrons to two chlorine atoms as illustrated in Figure 8.1.4.

Evidence for the existence of ions

When copper(II) chloride solution is electrolysed (Experiment 8.1.a), copper is deposited on the negatively charged electrode, and chlorine gas

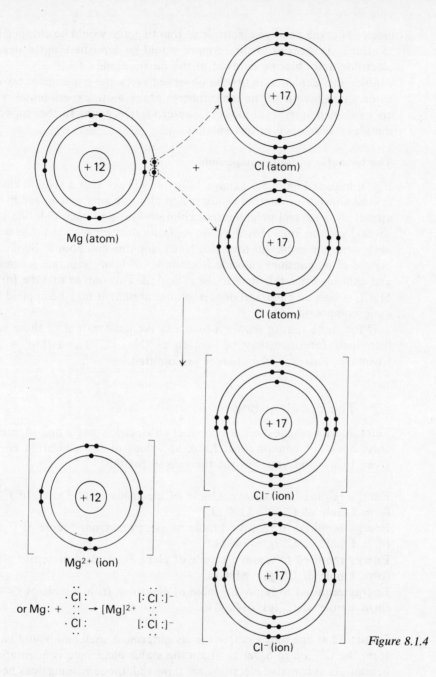

$$\text{or Mg:} + \begin{matrix} \cdot \ddot{C}l : \\ \vdots \\ \cdot \ddot{C}l : \end{matrix} \rightarrow [Mg]^{2+} \begin{matrix} \ddot{} \\ [: \ddot{C}l :]^- \\ \ddot{} \\ [: \ddot{C}l :]^- \\ \ddot{} \end{matrix}$$

Figure 8.1.4

is liberated at the positively charged electrode. Let us see if these results fit in with our ideas about ions.

When copper combines with chlorine to form copper(II) chloride we would expect atoms of copper, being metallic, to lose electrons and form positive ions. These electrons would be transferred to chlorine atoms, giving negative ions. In copper(II) chloride solution, the positive copper ions could be given electrons at the negative electrode and thus be changed back to atoms. Similarly, the negative chloride ions could give up electrons at the positive electrode so that they too would be changed back to atoms. As a result of this, copper would be deposited on the negative electrode and chlorine evolved at the positive one.

Since this explanation fits the observed facts, the experiment provides evidence in favour of the ionic theory. Many more experimental results are explained in terms of ions in Chapter 11, thus giving further support to the ideas developed in this section.

The formulae of ionic compounds

If you look at Figure 8.5.1 (page 140) you will see that a sodium chloride crystal consists of alternate sodium and chloride ions. Each sodium ion is attracted to several neighbouring chloride ions and each chloride ion to several sodium ions. However, *no molecules are present* and thus we can only write an *empirical formula*, NaCl, for this compound. Similarly, a crystal of magnesium chloride is made up of many separate magnesium and chloride ions, this time in the ratio 1:2. This means that the formula $MgCl_2$ is also an empirical one. A similar argument may be applied to all ionic compounds.

If for some reason we wish to stress the ionic nature of these solids, then their formulae may be written as $(Na^+ + Cl^-)$ and $(Mg^{2+} + 2Cl^-)$. Normally, however, the charges are omitted.

8.2   The covalent bond

Tetrachloromethane does not conduct an electric current and so, presumably, does not contain ions. Look at values of the ionisation energies given below and you will see the reason for this.

Energy required to remove 1 mole of electrons from 1 mole of C(g) to form 1 mole of $C^+(g) = 1100\,kJ$.
Energy required to remove 1 mole of electrons from 1 mole of $C^+(g)$ to form 1 mole of $C^{2+}(g) = 3400\,kJ$.
Energy required to remove 1 mole of electrons from 1 mole of $C^{2+}(g)$ to form 1 mole of $C^{3+}(g) = 8100\,kJ$.
Energy required to remove 1 mole of electrons from 1 mole of $C^{3+}(g)$ to form 1 mole of $C^{4+}(g) = 14\,200\,kJ$.

Carbon (2,4) has four electrons in its outermost shell and would have to form the C^{4+} ion in order to attain the stable electronic configuration of helium. As successive electrons are removed, the remaining ones become more firmly held by the positive charge on the nucleus. So much energy is required to form the C^{4+} ion that the carbon and chlorine atoms combine by an easier method – they *share* pairs of electrons. Each atom contributes one electron to a shared pair and in this way they all achieve stable octets (Figure 8.2.1). The atoms are held together by the attraction of the two nuclei for the shared pair of electrons. This type of bonding is known as *covalent bonding*, and since fixed numbers of atoms are joined directly to one another, the particles formed are molecules.

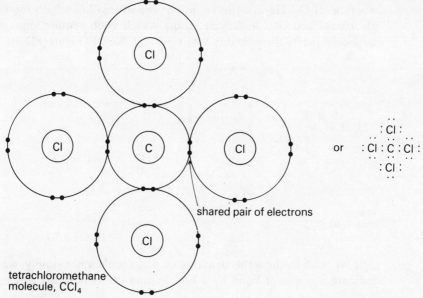

or
$$: \overset{..}{\underset{..}{Cl}} :$$
$$: \overset{..}{\underset{..}{Cl}} : \overset{..}{\underset{..}{C}} : \overset{..}{\underset{..}{Cl}} :$$
$$: \overset{..}{\underset{..}{Cl}} :$$

shared pair of electrons

tetrachloromethane
molecule, CCl_4

Figure 8.2.1

(Since the electrons of the inner shells are not involved in forming bonds they have been omitted from the diagram.)

A **covalent bond** is one formed by the sharing of a pair of electrons by two atoms.*

A few more examples should make the principles of covalent bonding clear.

AMMONIA, NH_3 A nitrogen atom (2,5) is three electrons short of its stable octet while a hydrogen atom is one electron short of the helium configuration. Thus *three* hydrogen atoms form covalent bonds with *one* nitrogen atom (Figure 8.2.2).

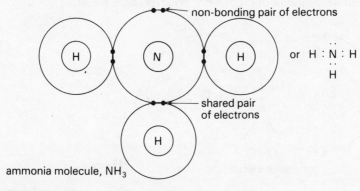

non-bonding pair of electrons

or H $:$ N $:$ H
         H

shared pair
of electrons

ammonia molecule, NH_3

Figure 8.2.2

*The number of electron pairs shared by an atom is often called its *valency*.

WATER, H_2O Here we have an oxygen atom (2,6) which requires two electrons, and two hydrogen atoms which each require one. A water molecule therefore contains two covalent bonds (Figure 8.2.3).

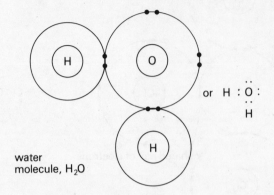

or   H : O :
           H

water
molecule, H_2O

Figure 8.2.3

If we wish to show the structure of a molecule more simply, we usually indicate a covalent bond by a line joining the two atoms concerned. e.g.

$$Cl—C—Cl \quad H—N—H \quad H—O$$

with Cl above and below the C, and Cl below; H below the N; and H below the O.

Molecules of gaseous elements

You are probably aware that most of the elements which are gases at room temperature are made up of molecules containing two atoms, e.g. Cl_2, N_2, O_2. You should also know that these elements occur towards the right hand side of the periodic table and therefore their atoms need to *gain* electrons to reach the electronic configuration of the nearest noble gas. They do this by forming covalent bonds.

CHLORINE (2,8,7)   Each atom needs one electron and thus a Cl_2 molecule contains one covalent bond (Figure 8.2.4).

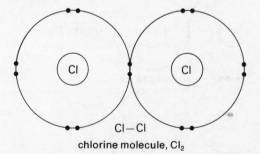

Cl—Cl

chlorine molecule, Cl_2

Figure 8.2.4

OXYGEN (2,6)   In this case each atom requires two electrons and so in the O_2 molecule *two* electron pairs must be shared, forming a *double* covalent bond (Figure 8.2.5).

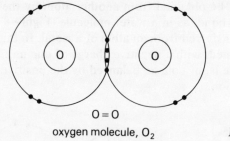

O = O

oxygen molecule, O_2

Figure 8.2.5

NITROGEN (2,5) Reasoning similar to that given for oxygen brings us to the conclusion that a nitrogen molecule, N_2, contains a *triple* covalent bond, i.e. three shared pairs of electrons (Figure 8.2.6).

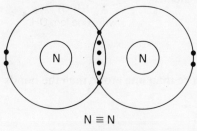

N ≡ N

nitrogen molecule, N_2 · *Figure 8.2.6*

If you have followed the arguments so far you should have no difficulty in working out the structures of other common covalent molecules. Try hydrogen (H_2), hydrogen chloride (HCl) and carbon dioxide (CO_2) as examples. Can you explain why hydrogen chloride molecules have the formula HCl and not H_2Cl or HCl_2?

The bonding in radicals

In Chapter 9 a radical is defined as a group of atoms which occurs in compounds but which cannot exist on its own. Consider, as an example, the hydroxyl radical, OH. If, as seems likely, it is formed by the sharing of a pair of electrons between an oxygen atom and a hydrogen atom, the oxygen atom will still be one electron short of an octet (Figure 8.2.7). Such an unstable arrangement is unlikely to last long and since most radicals may be shown to be similarly short of electrons, this explains why they have no separate existence of their own.

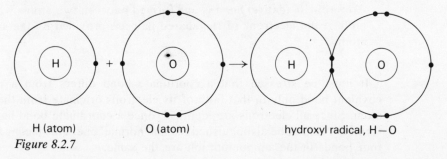

H (atom)   O (atom)   hydroxyl radical, H — O

Figure 8.2.7

The extra electron may be obtained from another atom by the formation of a second covalent bond, as in a water molecule (Figure 8.2.3), or alternatively it may be transferred from an atom of a metal. In this case a *hydroxide ion* will be formed, with a negative charge of one unit (Figure 8.2.8). The negative charge is, of course, balanced by the positive charge on the resultant metallic ion.

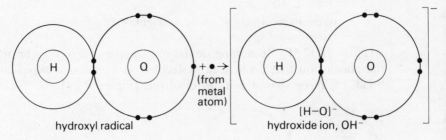

Figure 8.2.8

In general, if an *ion* contains more than one atom, then the bonds within it are *covalent*.

Coordinate bonding

If you look at the electronic configuration of the nitrogen atom in an ammonia molecule (Figure 8.2.2) you will see that one of the electron pairs in the outer shell is not involved in bonding. It is known as a *non-bonding pair* or *lone pair*.

It is possible for a lone pair of electrons to be shared with an atom or ion which is *two* electrons short of the electronic structure of the nearest noble gas, the resulting bond being known as a *coordinate* or *dative bond*. For example, a hydrogen ion, H^+, can become bonded to an ammonia molecule to form an ammonium ion as shown.

$$H^+ \leftarrow : \underset{\underset{H}{|}}{\overset{\overset{H}{|}}{N}}-H \longrightarrow \left[\underset{\underset{H}{|}}{\overset{\overset{H}{|}}{H-N}}-H \right]^+$$

ammonium ion, NH_4^+

A **coordinate (dative) bond** is one formed between two atoms where both of the electrons of the shared pair are provided by the same atom.

It must be stressed that a coordinate bond differs from a normal covalent bond only in that both of its electrons originate from the same atom. Since all electrons are identical, once a coordinate bond has been formed it cannot be distinguished from a 'normal' one. This means that all four bonds in the ammonium ion are the same.

A summary of the main types of bonding

1   Bonds are formed when electrons are redistributed among atoms to give each one the stable electronic configuration of the nearest noble gas in the periodic table. Usually this results in the formation of an octet of electrons.

2   An *ionic* bond is formed when one or more electrons is *transferred* from an atom of an element on the left-hand side of the periodic table (i.e. a metal) to an atom of one on the right-hand side (i.e. a non-metal). Ionic compounds are made up of ions.

3   A *covalent* bond is formed when a pair of electrons is *shared* between two atoms which both need to gain electrons (i.e. non-metals). Covalent compounds are made up of molecules.

8.3   Some differences between ionic and covalent compounds

The main differences between ionic and covalent compounds are brought out in Experiments 8.1.a, 8.2.a and 8.3.a. They are summarised in Table 8.3.

Property	Ionic compounds	Covalent compounds
State	Usually crystalline solids	Often liquids or gases
m.p.	High	Generally low
Solubility in water	Generally soluble	Generally insoluble
Solubility in organic solvents	Generally insoluble	Generally soluble
Conduction of electricity	Conduct electricity when molten or when in solution, but not when solid	Do not conduct electricity when solid, liquid or in solution (unless ions are formed)

Table 8.3 Differences between ionic and covalent compounds

Ionic compounds have high melting points because a large amount of energy has to be provided to overcome the strong forces of electrostatic attraction holding the oppositely charged ions together. On the other hand the forces of attraction between neighbouring molecules in covalent compounds are weak and so the melting points are generally low. (There must be *some* attraction between the molecules, otherwise no solid covalent compounds would exist – see 8.5, *molecular lattices*.)

We must emphasise that *covalent bonds are not weak*. It is the attractive forces *between* the molecules which are weak, the covalent bonds *within* the molecules being strong. Make sure that you understand this point.

The ways in which water and other solvents act on ionic and covalent compounds are more difficult to explain. They are discussed in detail in Chapter 12.

Ionic compounds do not conduct an electric current when in the solid state because the ions are only able to vibrate about fixed positions in the crystal lattice and cannot migrate towards the electrodes. When the solid is melted or dissolved in water the ions are free to move and hence can carry an electric current. Covalent compounds cannot conduct an electric current because they do not contain ions. However, some of them react with water to *form* ions and the resulting solutions do, of course, conduct electricity. Two common compounds which behave in this way are hydrogen chloride and ammonia.

8.4 The shapes of covalent molecules

Any molecule consisting of only two atoms must be linear, i.e. the atoms must be in a straight line. We can explain the shapes of other covalent molecules by assuming that the electron pairs (both bonding and lone pairs) of the outermost shell of the central atom repel one another and take up positions in which they are as far apart as possible. In the determination of molecular shape, a double bond behaves in the same way as a single one.

The first step to take in predicting the shape of a covalent molecule is to work out its electronic structure. Then the total number of bonds and lone pairs around the central atom must be found (remembering that a double bond counts as *one*) and the correct arrangement decided upon. The shapes which you are most likely to encounter at this level are shown in Figure 8.4.

Examples

METHANE, CH_4 The carbon atom has four bonding pairs of electrons and mutual repulsion causes these to take up a *tetrahedral* distribution around it. Thus the molecule consists of four hydrogen atoms at the corners of a regular tetrahedron with the carbon atom at its centre.

$$H \overset{\overset{\ddot{}}{:}}{:} \overset{\ddot{}}{C} : H \qquad \qquad \text{angle } H\text{–}C\text{–}H = 109° 28'$$

AMMONIA, NH_3 In this molecule the nitrogen atom has four pairs of electrons in its outer shell, three bonding pairs and one lone pair (see Figure 8.2.2). The arrangement of these pairs is again tetrahedral but the shape of the molecule (i.e. the distribution of its *atoms*) is *pyramidal*, with the nitrogen atom at the top. A lone pair exerts a greater repulsive effect than a bonding pair and pushes the hydrogen atoms together slightly, making the H–N–H angle less than that in a regular tetrahedron.

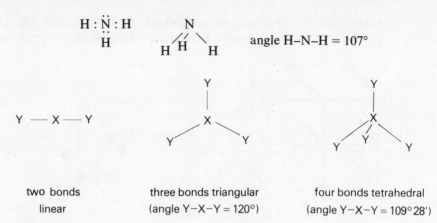

angle H–N–H = 107°

Y — X — Y

two bonds
linear

three bonds triangular
(angle Y–X–Y = 120°)

four bonds tetrahedral
(angle Y–X–Y = 109° 28′)

Figure 8.4

WATER, H_2O Once more there are four electron pairs arranged tetrahedrally around the central atom but this time only two of them are bonding pairs (see Figure 8.2.3). The molecule is thus *V-shaped*, the H–O–H angle being less than that in a regular tetrahedron because of the extra repulsion exerted by the lone pairs.

angle H–O–H = 105°

CARBON DIOXIDE, CO_2 Each oxygen atom in a carbon dioxide molecule is linked to the carbon atom by a double covalent bond. The two double bonds repel one another until they are as far apart as possible; hence the molecule is *linear*.

O=C=O

ETHENE, C_2H_4 Here we have three bonds, one double and two single, arranged around each carbon atom. They repel one another and take up the triangular arrangement shown, the whole molecule being planar, i.e. flat.

8.5 Crystal structure

Giant ionic lattices

Sodium chloride is crystalline, which means that the Na^+ and Cl^- ions must be arranged in a repeating pattern extending in three dimensions. Such three-dimensional arrangements are called *lattices*, the sodium chloride lattice being as shown in Figure 8.5.1. The diagram on the left makes the structure easy to see but as the ions are actually packed together as closely as possible, the right-hand one gives a truer picture.

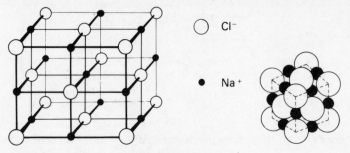

Figure 8.5.1 The sodium chloride lattice

The arrangement shown in Figure 8.5.1 extends throughout the crystal, involving millions of ions, but it does not exist separately on its own. Each cube is joined to its neighbours and the whole crystal is thus a continuous *giant structure of ions*, or *giant ionic lattice*.

This idea accounts for the high melting point of sodium chloride, since much energy has to be supplied to overcome the strong forces of electrostatic attraction which act between the oppositely charged ions throughout the whole lattice. Of course, ionic compounds do not all have the same crystal structure but they do all consist of giant ionic lattices of some kind.

Molecular lattices

The forces of attraction between the particles in an iodine crystal must be fairly weak since it readily vaporises (Experiment 8.5.a). The particles in this case are *molecules* consisting of pairs of iodine atoms joined together by strong covalent bonds. However, the individual molecules are only linked by very weak attractive forces, and so little heat energy is required to disrupt the lattice. These weak intermolecular forces are known as *van der Waals' forces*.

A similar situation occurs with solid carbon dioxide. Each molecule is made up of a carbon atom joined by strong covalent bonds to two oxygen atoms, but there are only weak van der Waals' forces acting between the separate molecules. The molecules pack together as closely as possible, just as ions do, but in this case the result is a *molecular lattice* (Figure 8.5.2).

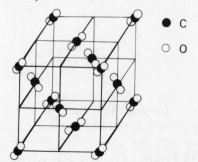

Figure 8.5.2 Solid carbon dioxide

The fact that the forces holding the molecules in a molecular lattice are so weak explains why most covalent substances are liquids, gases, or solids with comparatively low melting points.

Giant atomic lattices

There are a number of substances such as silicon(IV) oxide, diamond and graphite, which have very high melting points but which nevertheless do *not* consist of ions. How can we explain this? The particles in these cases are *atoms*, so they must be held together by covalent bonds but, clearly, there can be no separate molecules or the melting points would be much lower. The reason that these substances behave as they do is because they are examples of *giant structures of atoms*, or *giant atomic lattices*, with strong covalent bonds acting throughout the whole crystal. Like giant ionic lattices they have three-dimensional repeating patterns but no separate units. A great deal of heat energy has to be supplied to disrupt this type of lattice and melt the solid because millions of bonds have to be broken. Thus the melting points of such substances will be high.

In silicon(IV) oxide (silica) each silicon atom is at the centre of a regular tetrahedron and is joined by strong covalent bonds to four oxygen atoms positioned at the corners. These tetrahedra are linked together in a continuous three-dimensional network, giving a very rigid crystal (Figure 8.5.3).

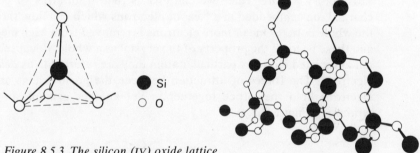

Figure 8.5.3 *The silicon (IV) oxide lattice*

The structure of graphite is compared with that of diamond in Chapter 16. Both substances have high melting points because they consist of giant structures of carbon atoms. However, their other physical properties are very different and any structures suggested for them must account for these differences.

Giant metallic lattices

Metals are crystalline. You may find this difficult to believe, since they do not appear to be so, but if you look at a newly galvanised piece of iron, such as a dustbin, you should be able to see crystals of zinc on its surface. Usually a microscope is needed to see metal crystals but larger ones can be specially grown, as the photograph shows.

The high melting point of most metals indicates that they have giant structures of some kind but they have unique properties, such as

Cubic crystals of the metal bismuth

flexibility and high electrical conductivity, which are not accounted for by any of the lattices described so far.

How do metal particles bind together in the crystals? The most satisfactory picture which we can give is that of an array of positively charged ions embedded in a 'sea' of electrons which can flow throughout the whole lattice. One or more electrons is removed from each metal atom and these become the property of the crystal as a whole. Because they are not associated with any particular atom they are referred to as *delocalised* electrons. The forces of attraction between the positive ions and these electrons hold the lattice together in yet another example of a *giant structure* (Figure 8.5.4).

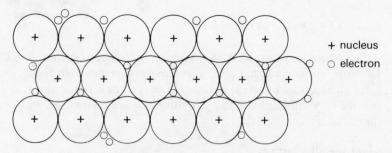

+ nucleus
○ electron

Figure 8.5.4 A metal lattice

In a metal lattice the ions are all spheres of the same size and so they can pack together very closely indeed. The two ways in which spheres can arrange themselves to take up the minimum space possible are known as *hexagonal close packing* and *cubic close packing*. Most metallic lattices fall into one of these types, or the slightly less close packed *body-centred cubic* structure. Units of each of these three lattices are illustrated in

Figure 8.5.5. Notice that there is no way in which the spheres in the first two arrangements could be any closer together.

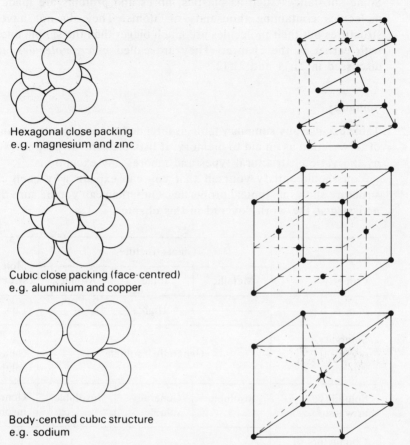

Hexagonal close packing
e.g. magnesium and zinc

Cubic close packing (face-centred)
e.g. aluminium and copper

Body-centred cubic structure
e.g. sodium

Figure 8.5.5

Now let us see how these ideas about the structure of metals can be used to account for their typical physical properties. The high densities of most metals are explained by the close packing of the particles, and their high melting points by the giant lattices. The less dense ones with lower melting points, such as sodium, are known to crystallise with the more open body-centred cubic structure. Metals are flexible because the layers of ions can slide past one another without greatly altering the forces of attraction between them and the mobile 'sea' of electrons. The delocalised electrons normally move in the lattice with random motion but if a potential difference is applied they will move through the metal in one direction, this flow of electrons being an electric current. Heat is also conducted by these electrons since those given extra kinetic energy at the hot end of the metal can move through the lattice, colliding with others and passing on the extra energy.

Thus you can see that the picture of a metallic lattice which has been developed in this section accounts for the common physical properties of metals very well.

Macromolecules

Some substances such as plastics, fibres and proteins are made up of molecules containing thousands of atoms. They do not have giant structures but their molecules are much bigger than the simple ones dealt with earlier in the chapter. They are called *macromolecules* and are discussed in 21.11 and 21.12.

Summary

Table 8.5, like any summary table, is intended to condense the main points of this section as an aid to memory. It lists the *general* properties shown by the various structural types and ignores any exceptions.

You should satisfy yourself that you can explain *why* each class of structure shows its typical properties. Only then can you be sure that you *understand* the work covered in this chapter.

Property	Giant structures			Molecular structures
	Metallic	Ionic	Atomic	
m.p. & b.p.	←——————— High ———————→			Low
Solubility in organic solvents	←———— Generally insoluble ————→			Generally soluble
Solubility in water	Insoluble	Generally soluble	Insoluble	Generally insoluble
Conductors of electricity (molten)	←——— Yes ———→		←——— No ———→	
Conductors of electricity (solid)	Yes	←——————— No ———————→		
Conduction of heat	Good	←——————— Bad ———————→		

Table 8.5 Structure and physical properties

Determination of structure

The ways in which the particles are arranged in a solid can be determined by the technique of X-ray diffraction, developed by Sir William Bragg and his son, Sir Lawrence Bragg. X-rays are reflected from a crystal face and allowed to fall on a photographic plate, where they produce a regular pattern of spots. Skilled scientists can relate this pattern to the lattice structure, but it is not an easy task.

X-ray diffraction photograph

8.6 Oxidation and reduction

In Chapters 3 and 4 you learned that oxidation was the addition of oxygen to, or the removal of hydrogen from a substance and that reduction was the opposite change, i.e. the removal of oxygen from, or the addition of hydrogen to a substance. We shall now take a fresh look at these topics in the light of our knowledge of chemical bonding.

Oxidation and reduction in terms of electron transfer

When magnesium burns in oxygen it forms magnesium oxide.

$$2Mg(s) + O_2(g) \rightarrow 2MgO(s)$$

According to the definition given earlier, the magnesium has been oxidised. However, we can also interpret the reaction in a different way and say that each oxygen atom has taken two electrons from the outermost shell of a magnesium atom, leaving both with a stable octet.

i.e. $$2Mg(s) + O_2(g) \rightarrow 2(Mg^{2+} + O^{2-})(s)$$

Similar changes occur when *any* metal combines with oxygen and so we arrive at a new definition of oxidation.

Oxidation is any change in which there is a loss of one or more electrons.

Thus an **oxidising agent** is a substance which takes electrons.

Since reduction is the opposite of oxidation it too may be redefined.

Reduction is any change in which there is a gain of one or more electrons.

It follows that a **reducing agent** is a substance which gives up electrons.

Electrons cannot simply disappear, and hence if one particle loses electrons (i.e. is oxidised), another must gain them (i.e. be reduced). In other words, oxidation cannot occur without reduction. For this reason, reactions involving electron transfer are often known as *redox* reactions (from *red*uction–*ox*idation).

These new definitions of oxidation and reduction give rise to many examples of redox reaction which do not involve oxygen or hydrogen at all. For example, when sodium burns in chlorine to form sodium chloride, sodium atoms lose electrons and therefore are oxidised, while chlorine atoms gain electrons and hence are reduced.

$$\overset{\displaystyle \ulcorner\text{reduction}\longrightarrow\urcorner}{2Na(s) + Cl_2(g) \rightarrow 2(Na^+ + Cl^-)(s)}$$
oxidation

Since the sodium atoms give up electrons, the sodium acts as a reducing agent; chlorine atoms take electrons and so the chlorine is an oxidising agent.

Consider the reaction between zinc and dilute sulphuric acid.

$$Zn(s) + H_2SO_4(aq) \rightarrow ZnSO_4(aq) + H_2(g)$$

or $$Zn(s) + 2H^+(aq) \rightarrow Zn^{2+}(aq) + H_2(g)$$

The ionic equation shows that this too is a redox reaction. Zinc is the reducing agent since its atoms give up electrons and are oxidised to zinc ions. The acid is the oxidising agent because its $H^+(aq)$ ions accept these electrons and are themselves reduced to hydrogen atoms.

When chlorine is bubbled into a solution of iron(II) chloride, the iron(II) chloride is changed to iron(III) chloride.

$$2FeCl_2(aq) + Cl_2(g) \rightarrow 2FeCl_3(aq)$$

Is this a redox reaction? The ionic equation leaves us in no doubt that it *is* because it shows clearly that electrons have been transferred.

$$2Fe^{2+}(aq) + Cl_2(g) \rightarrow 2Fe^{3+}(aq) + 2Cl^-(aq)$$

The iron(II) ions give up electrons and are oxidised; the chlorine atoms accept these electrons and are reduced. Which of the reacting substances is the oxidising agent and which the reducing agent?

Limitations of the electron transfer approach
Although the electron transfer approach to oxidation and reduction broadens the meaning of these terms, it is not without its drawbacks. For

example, consider what happens when sulphur burns in oxygen to form sulphur dioxide.

$$S(s) + O_2(g) \rightarrow SO_2(g)$$

There is no doubt that the sulphur has been oxidised, but since neither reactants nor products are ionic, no transfer of electrons has occurred. There are a large number of reactions of this type which fit the old definitions of oxidation and reduction but are not covered by the new ones. Thus we must continue to use *both* definitions or else look for a better one which will encompass *all* the examples.

Oxidation numbers

Oxidation numbers enable us to avoid the problems associated with using two separate definitions for oxidation. They may be used both for reactions which involve electron transfer and for those which do not.

OXIDATION NUMBERS AND IONS   The oxidation number of an ion is equal to the charge on it.

e.g.

$$Na^+ = +1; \quad Ca^{2+} = +2; \quad Cl^- = -1;$$
$$O^{2-} = -2; \quad NH_4^+ = +1; \quad SO_4^{2-} = -2$$

OXIDATION NUMBERS AND COVALENT MOLECULES   The oxidation number of an atom in a covalent molecule is the charge which would be left on that atom if the compound were ionic, e.g. in a *covalent* oxide the oxidation number of oxygen is -2 because the oxide *ion* is O^{2-}.

The following simple rules must be applied when using oxidation numbers with non-ionic substances.

1 The oxidation number of an atom in a molecule of a free element is taken to be zero.

e.g. in the molecules, O_2, N_2, Cl_2, the oxidation number of all the atoms is zero.

2 In neutral molecules the algebraic sum of the oxidation numbers of the atoms is zero.

e.g. in carbon dioxide, CO_2, (oxidation number of carbon) + $(2 \times$ oxidation number of oxygen) = 0.

3 Hydrogen has oxidation number $+1$ in covalent compounds.

4 Oxygen has oxidation number -1 in peroxides, instead of the usual -2.

Examples

(a) What is the oxidation number of sulphur in sulphur dioxide, SO_2?

$$\text{(oxidation number of S)} + (2 \times \text{oxidation number of O}) = 0 \text{ (see rule 2)}$$

∴   ? $+ (2 \times -2)$ $= 0$
∴   ? $+ (-4)$ $= 0$
∴   ? $= +4$

i.e. the oxidation number of sulphur in sulphur dioxide is $+4$.

(b) What is the oxidation number of sulphur in sulphuric acid, H_2SO_4?

The sum of the oxidation numbers in $H_2SO_4 = 0$ (see rule 2)

$$\therefore \quad (2\times +1) + ? + (4\times -2) = 0$$
$$\therefore \qquad\qquad ? \qquad\qquad = +6$$

i.e. the oxidation number of sulphur in sulphuric acid is $+6$.

(c) What is the oxidation number of manganese in the manganate(VII) ion, MnO_4^-?

The sum of the oxidation numbers in $MnO_4^- = -1$ (since the charge on
the ion is -1)

$$\therefore \quad ? + (4\times -2) = -1$$
$$\therefore \quad ? \qquad\qquad = +7$$

i.e. the oxidation number of manganese in this ion is $+7$.

(d) By how much do the oxidation numbers of sulphur and oxygen change when the elements combine to form sulphur dioxide?

$$S(s) + O_2(g) \rightarrow SO_2(g)$$
oxidation numbers: $\qquad 0 \;+\; 0 \;=\; ? + (2\times -2)$
$$\therefore \qquad\qquad\qquad\qquad ? = +4$$

Thus the oxidation number of sulphur increases from 0 to $+4$ while that of oxygen decreases from 0 to -2.

(e) By how much do the oxidation numbers of sodium and chlorine change when the elements combine to form sodium chloride?

$$2Na(s) + Cl_2(g) \rightarrow 2(Na^+ + Cl^-)(s)$$
oxidation numbers: $\quad 0 \;+\; 0 \;=\; (+1 \;+ -1)$

The oxidation number of the sodium increases from 0 to $+1$ while that of the chlorine decreases from 0 to -1.

(f) By how much do the oxidation numbers of copper, oxygen and hydrogen change when copper(II) oxide reacts with hydrogen?

$$CuO(s) + H_2(g) \rightarrow Cu(s) + H_2O(g)$$
$$(+2) + (-2) \;\; +0 \;\;= 0 \qquad + (2\times +1) + (-2)$$

The oxidation number of the copper decreases from $+2$ to 0 while that of the hydrogen increases from 0 to $+1$. The oxidation number of the oxygen is unchanged.

Look closely at the equations in the last three examples and pick out the substances which are oxidised. Does the oxidation number increase or decrease during oxidation? Now pick out the substances which are reduced in each reaction. What happens to oxidation numbers during reduction?

If you have worked carefully through this section it should be clear to you that the following definition, based on oxidation numbers, will cover *all* examples of oxidation and reduction.

An atom or ion undergoes **oxidation** when its oxidation number *increases* and undergoes **reduction** when its oxidation number *decreases*.

Questions on Chapter 8

1 Atoms of lithium, carbon and chlorine contain 3, 6 and 17 orbital electrons respectively.
(a) Describe and explain the electronic structure of the atoms of these three elements.
(b) What type of bonding would you expect when (i) lithium, (ii) carbon combine with chlorine? Describe the electronic structure of each compound and say what will happen when the compounds are added separately to water.
(c) Elements may be classified as metals or non-metals. How would you classify lithium and chlorine? In each case give *two* reasons for your answer.
(d) Name *two* other elements in each case which are in the same family as (i) lithium, (ii) chlorine. (O & C)
2 The number of protons, neutrons and orbital electrons in particles A to F are given in the following table.

Particle	Protons	Neutrons	Electrons
A	3	4	2
B	9	10	10
C	12	12	12
D	17	18	17
E	17	20	17
F	18	22	18

(a) Choose from the table the letters that represent (i) a neutral atom of a metal, (ii) a neutral atom of a non-metal, (iii) an atom of a noble gas (inert gas), (iv) a pair of isotopes, (v) a cation (positive ion), (vi) an anion (negative ion).
(b) Give the formulae of the compounds you would expect to be formed between (i) C and D, (ii) E and hydrogen, (iii) C and oxygen. (You may use the letters C, D and E as symbols for the elements.)
   In *each* case, say whether the compound will be an acid, a base or a salt.
(c) What will be the formula of a compound containing only particles of A and B? Give *two* physical properties you would expect this compound to have. (C)
3 (a) What is thought to happen when atoms are bonded (i) by an ionic (electrovalent) bond, (ii) by a covalent bond?
(b) From the following list name one compound which is ionic and one which is covalent: ethanol, potassium chloride, calcium nitrate, starch, copper(II) sulphate. Describe one experiment applied to each of the chosen compounds which confirms your choice.
(c) What is meant by the term *molecule*? Explain why the term has no meaning for an electrovalent compound. (SCE)
4 Show by means of diagrams the three-dimensional structures or shapes of the following substances: a tetrachloromethane molecule, a water molecule, and part of a sodium chloride crystal. (SCE)

5   Write out and complete the following table.
(a) Which particles, apart from a neon atom, have the same total number of electrons as a noble gas?
(b) The electron configuration of the carbon atom can be written as C: 2.4. Show in a similar way the configurations of (i) the Li atom, (ii) the S^{2-} ion.
(c) Write the formula of (i) a compound containing the S^{2-} ion, (ii) a covalent compound of sulphur, (iii) a compound composed only of Li^+ and N^{3-} ions.

Particle	Mass number	Number of protons	Number of neutrons	Number of electrons
Li atom		3	4	
Li^+ ion				
^{12}C isotope	12			6
^{13}C isotope	13	6		
N^{3-} ion	14	7		
Ne atom			10	10
S^{2-} ion	32			18

(C)

6   What explanation can you offer for each of the following statements?
(a) Crystalline solids, when broken, tend to cleave along well-defined planes.
(b) Tetrachloromethane is a liquid of low boiling point which is a non-electrolyte, and has no reaction with aqueous silver nitrate.
(c) Metals in the solid state are electrical conductors.   (W)
7   (a) How do you account for the fact that the formula of calcium fluoride is CaF_2 and that the compound is a solid of high melting point?
(b) Show diagrammatically a likely structure for a compound of nitrogen and fluorine.
(c) Why does neon exist naturally as single atoms whereas fluorine exists as diatomic molecules, F_2?   (W)
8   The following table gives some information about the physical properties of four substances which are represented here by the letters W, X, Y and Z.

Substance	Melting point	Boiling point	Heat of vaporisation	Electrical conductivity of substance when solid	Electrical conductivity of substance when liquid (i.e. molten)
W	High	High	High	Poor	Poor
X	High	High	High	Good	Good
Y	High	High	High	Poor	Good
Z	Low	Low	Low	Poor	Poor

(a) Give the letter of the substance in the table which in each case fits the description below.

(i) A compound with ionic structure.

(ii) A substance composed of simple molecules (a small number of atoms held together by covalent bonding).

(iii) A substance with a giant atomic structure.

(iv) An element with a metallic structure.

(b) Give the chemical name and formula of any compound which you know has an ionic structure. Show by a suitable diagram how the ions in the compound are formed from neutral atoms. Explain briefly how an ionic structure accounts for the typical physical properties given in the table.

(c) Give the chemical name and formula of any substance which you know is composed of simple molecules. Show by a suitable diagram how the atoms are held together in a molecule of the substance. Explain briefly how a simple molecular structure accounts for the typical physical properties given in the table.

(d) Give the name of any substance which you know has a giant atomic structure. Show by a simple diagram the arrangement of the atoms in this structure and explain briefly how the structure accounts for the typical physical properties given in the table.

(e) Give the name of any substance with a metallic structure and explain briefly how this structure accounts for the typical physical properties given in the table.

(L)

9   Describe and explain, in terms of reduction/oxidation, the following reactions: (a) hydrogen reacting with chlorine, (b) a *named* metal reacting with dilute sulphuric acid, (c) a *named* metallic oxide converted into a metal, (d) chlorine reacting with potassium iodide solution, (e) a *named* metal reacting with copper(II) sulphate solution to give a precipitate of copper.

State clearly in each reaction which substance is oxidised and which is reduced. Give an ionic explanation where appropriate. (O & C)

10   From the following list select two examples of oxidation and two of reduction.

(a) $Mg(s) \rightarrow Mg^{2+}(aq) + 2e$

(b) $Cu^{2+}(aq) + CO_3^{2-}(aq) \rightarrow Cu^{2+} CO_3^{2-}(s)$

(c) $Fe^{2+}(aq) \rightarrow Fe^{3+}(aq) + e$

(d) $Cl_2(g) + 2e \rightarrow 2Cl^-(aq)$

(e) $H^+(aq) + OH^-(aq) \rightarrow H_2O(l)$

(f) $Ag^+(aq) + e \rightarrow Ag(s)$

Give one word in each case which describes the type of reaction taking place in the two examples you have not used. (SCE)

9 Moles, formulae and equations

Chemists not only need to know *which* substances react with one another but also *how much* of each is required. For instance, it would be rather foolish to build a huge chemical plant to make a particular product and then simply mix tons of materials and hope for the best. Reactions take place between atoms and molecules in ratios shown by *chemical equations*. If we can find a link between the numbers of particles taking part in a reaction and the masses of reactants and products we can use the chemical equation to predict how much material is required. Alternatively, if the equation for a reaction is not known we should be able to work it out by measuring the masses of the various substances involved.

Experiments

9.1.a Relative masses of atoms ★★

For this experiment you will need a supply of ball bearings or similar uniform spheres in three sizes. Size A should each have a mass less than 0.5 g, size B about 10 g and size C about 15 g.

PROCEDURE (a) Place a size B sphere in one pan of a simple balance and add size A ones to the other pan until the two masses are equal. Note the number of size A spheres used.
(b) Repeat the experiment and find how many size A spheres have the same mass as one of size C.
(c) Suppose that the numbers of size A spheres used to balance one of B and one of C are x and y respectively, then the *relative* masses of A, B and C must be 1:x:y. Weigh out 1 g of A, x g of B and y g of C and count the number of spheres used in each case.

Questions

1 How many of the spheres of size A, B and C were used in (c)?
2 Suppose that you had some spheres which were 100 times heavier than those of size A. How many of these would be required to make up 100 g?
3 If you had atoms with masses in the same *ratio* as those of spheres A, B and C (i.e. 1:x:y) and you weighed out 1 g, x g and y g of each respectively, would the numbers of atoms be equal? Would the number be (a) much larger than, (b) the same as, (c) much smaller than that obtained in the experiment with the spheres?

9.1.b To measure the volumes of one mole of atoms of various elements ★

PROCEDURE (a) Pour propanone into a 100 cm³ measuring cylinder until the bottom of the meniscus is level with the 50 cm³ mark. Weigh out about 12 g of magnesium turnings and add them to the propanone. Stir with a glass rod (why?) and note the increase in volume.

(b) Repeat the experiment using about 30 g of copper turnings, 30 g of granulated zinc and 16 g of roll sulphur chips. (Water may be used instead of propanone with these substances.)

Questions

1 What fraction of a mole of atoms has been used in each case?
2 Calculate the volume of 1 mole of atoms of each element.
3 How many atoms are present in 1 mole of any element?
4 Do these atoms occupy the same volume?

9.2.a To find the empirical formula of magnesium oxide

PROCEDURE Find the mass of a clean, dry crucible and lid. Cut off a 15 cm length of magnesium ribbon, remove any surface coating of oxide with sandpaper and coil up the ribbon so that it fits in the crucible. Place the lid in position and find the mass again. Heat the whole apparatus on a pipe-clay triangle until the bottom of the crucible is red-hot and then lift the lid slightly for a second or so to admit air. Replace the lid so that no smoke escapes and repeat this procedure until no further combustion occurs. Finally heat for about a minute with the lid off. Allow the crucible to cool with the lid in position and then find the mass of the apparatus and contents. Repeat the final heating, cooling and weighing until the mass is constant.

Questions

1 Why was the final heating, cooling and weighing repeated until the mass was constant?

2 Calculate the masses of magnesium oxide, magnesium and oxygen as shown below.

Mass of crucible + lid	=	a g
Mass of crucible + lid + magnesium	=	b g
Mass of crucible + lid + magnesium oxide	=	c g
∴ Mass of magnesium oxide	=	(c − a) g
∴ Mass of magnesium	=	(b − a) g
∴ Mass of oxygen	=	(c − a) − (b − a) g

3 Calculate the empirical formula of magnesium oxide by the method used in the example on page 162 and compare your result with those obtained by the rest of the class.

9.2.b To find the empirical formula of copper(II) oxide

PROCEDURE The procedure for this experiment is exactly the same as for Experiment 6.4.b on page 92.

Questions

1 Why was it essential to keep passing the gas until the copper had cooled down?
2 Why was the last part of the experiment repeated until the mass was constant?

3 Set out your results in a similar way to that shown in Experiment 9.2.a and find the masses of copper and oxygen in the sample.

4 Calculate the empirical formula of copper(II) oxide and compare it with the results obtained by the rest of the class.

9.2.c To find the empirical formula of zinc iodide

PROCEDURE Find the mass of a test tube, add about 0.5 g of zinc powder and find the mass again. Place about 1.0 g of iodine (caution: corrosive, harmful vapour) on top of the zinc and once more find the mass. Add 2.0 cm³ of ethanol *dropwise* (the reaction is violent at first) and when all has been added, shake the tube gently until the liquid becomes colourless. At this point all the iodine has been used up and only excess zinc remains. Allow the contents of the tube to settle, and then suck off as much of the clear liquid as possible with a dropping pipette. Make sure that no zinc is removed during this operation. Add 1 cm³ of ethanol to the metal and shake to ensure that the particles are cleaned. Allow to settle, suck off the liquid and dry the tube and contents in a warm oven. Find the mass when completely dry.

Questions

1 Set out your results in the usual way and find the masses of zinc and iodine which combined together.

2 Calculate the empirical formula of zinc iodide and compare it with the results obtained by the rest of the class.

9.3.a To find the empirical formula of hydrated magnesium sulphate

This experiment may also be carried out using hydrated barium chloride or hydrated zinc sulphate.

PROCEDURE Find the mass of an evaporating basin, add about 5 g of hydrated magnesium sulphate and find the mass again. Heat gently and stir the solid with a glass rod to avoid loss by spitting. Some crystals may stick to the rod but these may be carefully scraped back into the basin by means of a spatula. When the dehydration seems complete, allow the basin and anhydrous salt to cool and find the mass. Repeat the heating and cooling until the mass is constant (why?). There is no need to stir during this final heating.

Questions

1 Set out your results in the usual way and find the masses of anhydrous magnesium sulphate and water in the sample.

2 Given that the empirical formula of anhydrous magnesium sulphate is $MgSO_4$, calculate the empirical formula of the hydrated salt by the method used in the example on page 166.

9.4.a To find the equation for the reaction between iron and copper(II) sulphate solution

It may be shown that when iron reacts with a solution of copper(II) sulphate the products formed are iron(II) sulphate solution and copper. Using this information, we can determine the equation for the reaction.

PROCEDURE Find the mass of a clean, dry crucible, add about 0.5 g of iron filings and find the mass again. Heat about 10 cm³ of copper(II) sulphate solution (mass concentration 300 g dm⁻³) in a test tube, with shaking, until nearly boiling. Add this solution to the crucible and stir with a glass rod. Allow the resultant copper to settle and check that the liquid above it is blue. If it is not, add more copper(II) sulphate solution until it is. Carefully suck off as much of the liquid as possible with a dropping pipette, making sure that none of the copper is removed. Wash the solid twice with distilled water to ensure that it is clean, sucking off the liquid each time. Transfer the crucible and contents to a warm oven and find the mass when completely dry. (If the solid is given a final rinse with propanone it will dry more quickly.)

Questions

1 Why is it essential that the solution be blue at the end of the reaction?
2 Set out your results as usual and find the masses of iron and copper involved in the reaction.
3 Work out the equation for the reaction as shown, substituting your own values for x and y. (Relative atomic masses: $Fe = 56$, $Cu = 63.5$.)

x g of iron   displaces y g of copper   from copper(II) sulphate solution

$\therefore \quad \dfrac{x}{56}$ mol of Fe atoms   displaces $\dfrac{y}{63.5}$ mol of Cu atoms   from copper(II) sulphate solution

$\therefore \quad$ 1 mol of Fe atoms   displaces $\dfrac{y}{x} \dfrac{56}{63.5}$ mol of Cu atoms   from copper(II) sulphate solution

$\qquad\qquad\qquad\qquad\qquad\qquad = z$ mol of Cu atoms

$\therefore \quad$ 1 atom of Fe displaces z atoms of Cu from copper(II) sulphate solution. Since the other product of the reaction is iron(II) sulphate solution the equation must be of the form

$$Fe(s) + ?CuSO_4(aq) \rightarrow zCu(s) + ?FeSO_4(aq).$$

The number of atoms present before the reaction must be the same as the number present at the end (law of conservation of mass). Use this fact to complete the equation.

9.4.b To find the equation for the reaction between lead(II) nitrate and potassium iodide

When solutions of lead(II) nitrate and potassium iodide react together a yellow precipitate is formed. Chemical analysis shows that this is lead(II) iodide, the other product being potassium nitrate, which stays in solution.

PROCEDURE   Run 5.0 cm³ of potassium iodide solution (concentration 1 mol KI in 1 dm³) from a burette into a test tube. Add 1.0 cm³ of lead(II) nitrate solution (concentration 1 mol $Pb(NO_3)_2$ in 1 dm³) from a second burette and stir well with a glass rod. Add 5 drops of ethanol to the tube to give a better separation of the precipitate. Make up a similar mixture in another tube. (This will serve as a balancing tube in the centrifuge and will enable you to obtain a second set of results to check the accuracy of the experiment.) Number the tubes, centrifuge them for 15 seconds and then measure the height of the precipitate of lead(II) iodide in each one. Add 0.5 cm³ of lead(II) nitrate solution to both tubes, stir well, centrifuge for 15 seconds and measure the heights of the precipitates again. Repeat this procedure until a total of 4.0 cm³ of lead(II) nitrate solution has been added to each tube.

As more lead(II) nitrate solution is added, the level of the precipitate rises until all of the potassium iodide solution is used up. Further portions of lead(II) nitrate solution remain as excess and the level of the precipitate does not change. Use your results to find the volume of lead(II) nitrate solution which just reacts with 5 cm³ of potassium iodide solution.

Question

1   Work out the equation for the reaction as follows, substituting your own value for x.

x cm³ $Pb(NO_3)_2$ solution reacts with 5 cm³ KI solution

$\therefore$ 1 cm³ $Pb(NO_3)_2$ solution reacts with $\frac{5}{x}$ cm³ KI solution

$\therefore$ 1 dm³ $Pb(NO_3)_2$ solution reacts with $\frac{5}{x}$ dm³ KI solution

But 1 dm³ of each solution contains 1 mole of solute.
$\therefore$ 1 mol $Pb(NO_3)_2$ reacts with $\frac{5}{x}$ mol KI.

Since the products of the reaction are known to be $PbI_2(s)$ and $KNO_3(aq)$, the equation must be of the form

$$Pb(NO_3)_2(aq) + \tfrac{5}{x} KI(aq) \rightarrow ?PbI_2(s) + ?KNO_3(aq).$$

Use the fact that atoms can be neither created nor destroyed during a chemical reaction to balance the equation.

9.4.c   To find the equation for the reaction between sodium carbonate and dilute hydrochloric acid

You probably already know that when an acid is added to a carbonate, carbon dioxide is evolved. This experiment is divided into two parts, the first giving the relative numbers of moles of reactants and the second the number of moles of carbon dioxide produced.

PROCEDURE   (a) Using a measuring cylinder or pipette, place 10 cm³ of sodium carbonate solution (concentration 1 mole Na_2CO_3 in 1 dm³) in a conical flask. Add 2–3 drops of screened methyl orange indicator and then run in 1 cm³ portions of dilute hydrochloric acid (concentration 1 mole HCl in 1 dm³) from a burette. Shake the flask after each addition and note the volume of acid required to neutralise the alkali. (The indicator is green in alkalis, buff in neutral solutions and red in acids.)

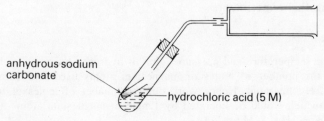

Figure 9.4.c

(b) Place about 15 cm³ of hydrochloric acid (concentration 5 mol dm⁻³) in a 25×150 mm test tube and add two spatula measures of anhydrous sodium carbonate to saturate it with carbon dioxide. (This prevents any gas from dissolving in it later in the experiment.) Make sure that none of the carbonate remains on the side of the tube (why?). Find the mass of a small test tube $(12 \times 75$ mm), add 0.21–0.22 g of anhydrous sodium carbonate and find the mass again. Carefully place the small test tube inside the large one, ensuring that the reactants do not come into contact with one another, and set up the apparatus as shown in Figure 9.4.c. Twist the syringe plunger (why?), note the reading and tilt the tube so that the acid and carbonate mix. When the reaction has ceased, allow the tube and contents to cool to room temperature, twist the syringe plunger again and once more note the reading.

Questions

1 Using similar reasoning to that in the previous experiment, find the number of moles of HCl(aq) which react with 1 mole of Na_2CO_3 in (a).
2 Calculate the number of moles of sodium carbonate used in (b) and also the number of moles of carbon dioxide produced. (Relative atomic masses: Na = 23, C = 12, O = 16. At room temperature and pressure 1 mole of carbon dioxide molecules has a volume of approximately 24 dm³.) Hence work out the number of moles of carbon dioxide which would be produced from 1 mole of sodium carbonate.
3 The other products of the reaction may be shown to be sodium chloride and water. If the answers to questions 1 and 2 are x mol and y mol respectively, then the equation must be of the form

$$Na_2CO_3(s) + xHCl(aq) \rightarrow ?NaCl(aq) + ?H_2O(l) + yCO_2(g).$$

Balance this equation.

9.4.d To find the equation for the reaction between magnesium and dilute hydrochloric acid

The equation for this reaction is found by measuring the volume of hydrogen displaced from the acid by a known mass of magnesium.

PROCEDURE   The procedure is similar to that in part (b) of the previous experiment. Use about 20 cm³ of dilute hydrochloric acid (concentration 2 mol dm⁻³) in the large tube and a piece of clean magnesium ribbon of mass 0.070–0.080 g in the small tube. Do not, of course, add sodium carbonate to the acid in this experiment.

Questions

1   At room temperature and pressure 1 dm³ of hydrogen has a mass of 0.083 g. Calculate the number of moles of magnesium atoms used and of hydrogen molecules, H_2, liberated. Hence work out the number of moles of hydrogen molecules which would be liberated by 1 mole of magnesium atoms. (Relative atomic masses: H = 1, Mg = 24.)

2   The other product of the reaction may be shown to be magnesium chloride. If 1 mole of magnesium atoms liberates x moles of hydrogen molecules from the acid, the equation must be of the form

$$Mg(s) + ?HCl(aq) \rightarrow ?MgCl_2(aq) + xH_2(g).$$

Balance this equation.

9.1   Masses of atoms and the mole

Relative atomic masses

As you will find out later in the chapter, we need to be able to compare the masses of different atoms with one another. If we take oxygen as an example, the mass of one atom is only 2.7×10^{-23} g (that is, 0.000 000 000 000 000 000 000 027 g), so you can see that something very small indeed must be chosen as a basis for comparison. Originally the masses of all atoms were found relative to that of a hydrogen atom, which is the smallest atom of all. (This corresponds to finding the masses of large spheres relative to that of small ones, as in Experiment 9.1.a.) Of course, the individual atoms could not be weighed or counted out but the methods which were actually used are only of historical interest and will not be discussed here.

Nowadays the masses of atoms are found relative to 1/12 of the mass of a carbon atom. This is because the measuring instrument, called a mass spectrometer, is calibrated using carbon compounds. A simple definition of *relative atomic mass* is given below and will be adequate for the work covered in this chapter. However, if you have studied atomic structure (6.5) you will realise that this definition is not completely correct (How should it be modified?).

The **relative atomic mass** of an element is the number of times that the mass of one of its atoms is greater than 1/12 of the mass of an atom of carbon.

Thus when we say that the relative atomic mass of oxygen is 16, we mean that one atom of oxygen has a mass 16/12 times that of a carbon atom. Similarly the relative atomic mass of sodium is 23 because the mass of a sodium atom is 23/12 times that of an atom of carbon.

Notice that relative atomic mass (formerly called 'atomic weight') is a *ratio* and therefore has no units.

The mole, Avogadro constant and molar mass

The relative atomic masses of hydrogen, carbon and sulphur are 1, 12 and 32 respectively. Thus, we can write the following.

	hydrogen	:	carbon	:	sulphur
For 1 atom of each, ratio of masses =	1	:	12	:	32
For 2 atoms of each, ratio of masses =	2	:	24	:	64
i.e.	1	:	12	:	32
For 3 atoms of each, ratio of masses =	3	:	36	:	96
i.e.	1	:	12	:	32

We could continue in this way for ever, but you can see that the ratio of masses of the three elements will always be the same as the ratio of their relative atomic masses (i.e. 1:12:32), *provided that the numbers of atoms of each element are equal.*

If we went on adding atoms, one at a time, we would eventually reach the point where we had 1 g of hydrogen. What masses of carbon and sulphur would we have? The answer, of course, is 12 g and 32 g respectively, to keep the ratio at 1:12:32. Thus a certain number of atoms of each of these elements will have a mass in grams which is numerically equal to its relative atomic mass. (This same conclusion is drawn, using a rather different approach, in Experiment 9.1.a). As you can imagine, the actual number of atoms required to give us these masses is huge, its value being 6.02×10^{23}, i.e. 602 000 000 000 000 000 000 000!

Since we chose hydrogen, carbon and sulphur at random in our example, it follows that for *any* element, 6.02×10^{23} atoms must have a mass in grams which is numerically equal to its relative atomic mass. This amount of the element is referred to as a *mole* of atoms. (The term 'mole' can also be applied to molecules, electrons and other particles, as you will discover later.) Like relative atomic mass, the mole is defined in terms of carbon atoms.

A **mole** of any substance is the amount of it which contains as many particles (atoms, molecules, etc.) as there are carbon atoms in 12 g of carbon.* The abbreviation for mole is 'mol'.

The number of particles in one mole of any substance is called the **Avogadro constant**, L. Its value is 6.02×10^{23} mol^{-1}.

The **molar mass** of a substance is the mass of one mole of it.

The units of molar mass used in this book are g mol^{-1}; this means that the relative atomic mass of any element and the molar mass of its atoms are *numerically* equal. Thus the molar mass of oxygen atoms, relative

*Strictly this should be 12 g of carbon-12 (see 6.5).

atomic mass 16, is $16\,g\,mol^{-1}$, while that of phosphorus atoms, relative atomic mass 31, is $31\,g\,mol^{-1}$. What is the molar mass of aluminium atoms, relative atomic mass 27?

The following simple examples have been chosen to illustrate the relationship between the mole, Avogadro constant and molar mass.

Examples

1 What is (a) the mass in grams, (b) the number of atoms, in 2.5 moles of magnesium atoms? (Relative atomic mass of Mg = 24.)

(a)   The molar mass of magnesium atoms is $24\,g\,mol^{-1}$
∴   1 mol of Mg atoms has a mass of 24 g
∴   2.5 mol of Mg atoms has a mass of $2.5 \times 24\,g$
$$= \underline{60\,g}$$

(b)   1 mol of Mg atoms contains 6.02×10^{23} atoms
∴   2.5 mol of Mg atoms contains $2.5 \times 6.02 \times 10^{23}$ atoms
$$= \underline{1.505 \times 10^{24} \text{ atoms}}$$

2 (a) What fraction of a mole is 3 g of calcium? (b) How many atoms does it contain? (Relative atomic mass of Ca = 40.)

(a)   The molar mass of calcium atoms is $40\,g\,mol^{-1}$
∴   40 g of Ca is 1 mol of Ca atoms
∴   1 g of Ca is $\frac{1}{40}$ mol of Ca atoms
    3 g of Ca is $\frac{3}{40}$ mol of Ca atoms
$$= \underline{0.075 \text{ mol of Ca atoms}}$$

(b)   1 mol of Ca atoms contains 6.02×10^{23} atoms
∴ $\frac{3}{40}$ mol of Ca atoms contains $\frac{3}{40} \times 6.02 \times 10^{23}$ atoms
$$= \underline{4.52 \times 10^{22} \text{ atoms}}$$

3 (a) How many moles of sulphur atoms do 3.01×10^{23} atoms represent? (b) What is the mass of these atoms in grams? (Relative atomic mass of S = 32.)

(a) 6.02×10^{23} atoms represent   1   mol of S atoms
∴   1   atom represents $\dfrac{1}{6.02 \times 10^{23}}$ mol of S atoms
∴ 3.01×10^{23} atoms represent $\dfrac{3.01 \times 10^{23}}{6.02 \times 10^{23}}$ mol of S atoms
$$= \underline{\tfrac{1}{2} \text{ mol of S atoms}}$$

(b)   The molar mass of sulphur atoms is $32\,g\,mol^{-1}$
∴   1 mol of S atoms has a mass of 32 g
    $\frac{1}{2}$ mol of S atoms has a mass of $\frac{1}{2} \times 32\,g$
$$= \underline{16\,g}$$

Conversion rules

The following conversion rules may be deduced from the examples given above.

1 *Moles of atoms → mass*

Multiply the number of moles by the molar mass – see Example 1(a). Compare this with changing money from pounds to pence by multiplying the number of pounds by the number of pence per pound.

2 *Mass → moles of atoms*

Divide the mass in grams by the molar mass – see Example 2(a). This may be compared with changing pence to pounds by dividing the number of pence by the number of pence per pound.

3 *Moles of atoms → number of atoms*

Multiply the number of moles by the Avogadro constant – see Example 1(b).

4 *Numbers of atoms → moles of atoms*

Divide the number of atoms by the Avogadro constant – see Example 3(a).

5 *Mass → number of atoms*

Convert the mass in grams to a number of moles of atoms (rule 2) and then to a number of atoms (rule 3) – see Example 2.

6 *Numbers of atoms → mass*

Convert the number of atoms to a number of moles of atoms (rule 4) and then to a mass in grams (rule 1) – see Example 3.

Summary

In all conversions remember that one mole of atoms of any element
(a) contains 6.02×10^{23} atoms,
(b) has a mass in grams which is numerically equal to the relative atomic mass of the element.

Molar volume

> The **molar volume** of a substance is the volume of one mole of it.

The units of molar volume used in this book are $dm^3 mol^{-1}$.

If the volumes of one mole of atoms of various solid elements are compared, as in Experiment 9.1.b, it is found that even though the number of atoms involved is the same in each case (how many?), the volumes differ. This is due to the variation both in the sizes of the atoms and in the ways in which they pack together.

9.2   Empirical formulae

> The **empirical formula** of a compound shows the simplest ratio of the numbers of atoms of the different elements in it.

Suppose element A combines with element B to give a compound with empirical formula AB_2. This means that

for every   1   atom of A there are   2   atoms of B . . . (i)
or for every   2   atoms of A there are   4   atoms of B
or for every   3   atoms of A there are   6   atoms of B
or for every 6.02×10^{23} atoms of A there are 12.04×10^{23} atoms of B
i.e. for every 1 mol of atoms of A there are 2 mol of atoms of B . . . (ii)

Thus you can see from lines (i) and (ii) that the ratio of *numbers of atoms* of the different elements in the compound is the same as the ratio of *moles of atoms*.

If we wish to find the empirical formula of a compound, we can carry out an analysis to find the ratio of grams of each element in it, and then convert this to a ratio of moles of atoms by using rule 2 in the previous section. Since we have just shown that the ratio of moles of atoms is the same as the ratio of numbers of atoms, we will have found the empirical formula of the compound.

Example

10.00 g of a compound was found to consist of 4.34 g of sodium, 1.13 g of carbon and 4.53 g of oxygen. Find its empirical formula. (Relative atomic masses: $Na = 23$, $C = 12$, $O = 16$.)

	Na	:	C	:	O
Ratio of grams	4.34	:	1.13	:	4.53
Ratio of mol of atoms (= ratio of *numbers* of atoms)	$\dfrac{4.34}{23}$	:	$\dfrac{1.13}{12}$	:	$\dfrac{4.53}{16}$
	= 0.189	:	0.094	:	0.283
(Divide by the smallest to convert to whole numbers)	$\approx$ 2	:	1	:	3

∴ The empirical formula of the compound is Na_2CO_3.

Note that the numbers of atoms of each element are written as subscripts after the symbols, *not* in front of them, and if there is only one atom, no number is needed.

Radicals

You may have noticed that certain groups of atoms appear quite often in chemical formulae. For example, all nitrates contain NO_3 groups, all carbonates contain CO_3 groups and all ammonium compounds contain NH_4 groups. Such groups are examples of *radicals* and they only exist combined with other atoms. Thus, though sodium sulphate, magnesium sulphate and zinc sulphate all contain the sulphate radical, SO_4, there is no substance which is just called 'sulphate'.

A **radical** is a group of atoms which occurs in compounds but which cannot exist on its own.

Working out empirical formulae from oxidation numbers (or from valencies)

The only sure way of finding an empirical formula is to calculate it from the results of chemical analysis. However, you cannot possibly do this for every compound you encounter. You could, of course, learn long lists of formulae but this would not be very interesting and it would take a considerable time to accomplish. The alternative is to learn a list of *oxidation numbers* (or a list of *valencies*). The meaning of these terms is explained in Chapter 8 but you can still make use of them even if at this stage you do not understand their full significance.

The oxidation number of an atom or radical shows its combining power, and combination between atoms and radicals takes place in such proportions that the algebraic sum of the oxidation numbers is *zero*. Thus in magnesium oxide, the ratio of magnesium atoms (oxidation number $+2$) to oxygen atoms (oxidation number -2) is $1:1$, making the empirical formula MgO.

$$Mg \quad O$$
oxidation numbers: $\quad (+2)+(-2)=0$

Similarly in aluminium chloride, the ratio of aluminium atoms (oxidation number $+3$) to chlorine atoms (oxidation number -1) is $1:3$ and so the empirical formula is $AlCl_3$.

$$Al \quad Cl_3$$
oxidation numbers: $\quad (+3)+3(-1)=0$

This simple method may be used to work out the empirical formulae of most of the compounds which you are likely to encounter. Table 9.2 shows the oxidation numbers of some common atoms and radicals and you should try to commit it to memory (see overleaf).

(If you use valencies, they are numerically equal to the oxidation numbers but have no sign. The sum of the valencies of the metal atoms in a formula must be *equal* to the sum of the valencies of the non-metal atoms or radicals.)

Can you see why the empirical formulae of calcium chloride, silver sulphate and iron(III) oxide are $CaCl_2$, Ag_2SO_4 and Fe_2O_3 respectively? Use the table to work out empirical formulae for potassium chloride, zinc carbonate, magnesium nitrate and calcium phosphate. Remember that if a radical appears more than once in a formula it must be enclosed in a bracket; for example, zinc nitrate is represented by $Zn(NO_3)_2$ rather than ZnN_2O_6.

A word of warning

Pupils who work out formulae by balancing valencies sometimes write nonsense such as Cl_2SO_4. Anyone using oxidation numbers would know that this is incorrect because the signs for both Cl and SO_4 are negative and so the algebraic sum cannot possibly be zero. However, there are still problems, even with oxidation numbers. For example, the oxidation number of Ag is $+1$ and that of OH is -1, therefore AgOH is the empirical formula of silver hydroxide. Unfortunately, no such substance exists. Only experience will tell you whether the compound whose formula you have worked out can actually be prepared.

Metal		Oxidation number		Non-metal or radical	
Ammonium*	NH₄			Br	Bromide
Hydrogen†	H			Cl	Chloride
Potassium	K	+1	−1	OH	Hydroxide
Silver	Ag			I	Iodide
Sodium	Na			NO₃	Nitrate
Barium	Ba				
Calcium	Ca				
Copper	Cu				
(in copper(II) compounds)					
Iron	Fe			CO₃	Carbonate
(in iron(II) compounds)		+2	−2	O	Oxide
Lead	Pb			SO₄	Sulphate
(in lead(II) compounds)					
Magnesium	Mg				
Zinc	Zn				
Aluminium	Al				
Iron	Fe	+3	−3	PO₄	Phosphate(v)
(in iron(III) compounds)					

*The ammonium radical behaves like a metal in many ways.
†Hydrogen is not a metal but is included in the table for completeness.

Table 9.2 The oxidation numbers of some common atoms and radicals

9.3 Masses of molecules, molecular formulae and percentage composition

Relative molecular mass

The **relative molecular mass** of an element or compound is the number of times that the mass of one of its molecules is greater than 1/12 of the mass of a carbon atom.*

The relative molecular mass of a substance may be calculated by adding together the relative atomic masses of the individual atoms in one

*A more accurate definition is given in 6.5.

molecule of it. For example, the relative molecular mass of carbon dioxide, CO_2, is $(12 + 16 + 16) = 44$.

Many compounds consist not of molecules but of giant structures of atoms or ions (8.5). In cases of this sort, the term 'relative molecular mass' applies to a *formula unit* of the substance. Thus for potassium chloride, empirical formula KCl, the relative molecular mass is $(39 + 35.5) = 74.5$, while for calcium carbonate, empirical formula $CaCO_3$, it is $(40 + 12 + 48) = 100$.

Molecular formulae

> The **molecular formula** of a compound shows the actual numbers of atoms of the different elements in one molecule of it.

The molecular and empirical formulae of a particular compound may be the same or may differ, as the following examples show.

Substance	Molecular formula	Empirical formula
Hydrogen chloride	HCl	HCl
Ammonia	NH_3	NH_3
Hydrogen peroxide	H_2O_2	HO
Glucose	$C_6H_{12}O_6$	CH_2O

Note: unlike relative molecular mass, the term 'molecular formula' must *only* be applied to substances which consist of separate molecules.

The connection between empirical formulae and molecular formulae

From the examples given above you can see that the molecular formula is the empirical formula multiplied by some whole number. In order to find the molecular formula of a compound it is necessary to determine its empirical formula and its relative molecular mass. Once these have been found, the calculation of the molecular formula is an easy task.

Example

A compound consists of 80% carbon and 20% hydrogen by mass and its relative molecular mass is 30. Find its empirical formula and hence its molecular formula. (Relative atomic masses: $C = 12$, $H = 1$.)

	C	:	H
Ratio of grams	80	:	20
Ratio of mol of atoms	80/12	:	20/1
(= ratio of *numbers* of atoms)			
	= 6.67	:	20
	= 1	:	3

$\therefore$ Empirical formula is CH_3
The 'relative empirical formula mass' $= (12 + 3) = 15$
The relative molecular mass $= 30$ (given)
$\therefore$ The molecular formula is $2 \times$ the empirical formula
$\therefore$ The molecular formula is C_2H_6.

Moles of molecules

The definitions of the mole and Avogadro constant given on page 159 apply to molecules and formula units as well as to atoms. Thus, just as a mole of atoms of an element contains 6.02×10^{23} atoms and has a mass in grams which is numerically equal to its relative atomic mass, so a mole of a *compound* will contain 6.02×10^{23} *molecules* (or formula units) and its mass in grams will be numerically equal to its relative *molecular* mass.

Examples

1 What is the mass of 1.5 moles of water molecules? (Relative molecular mass of $H_2O = 18$.)
The molar mass of H_2O molecules $= 18 \, g \, mol^{-1}$
∴   1 mol of H_2O molecules has a mass of 18 g
∴ 1.5 mol of H_2O molecules has a mass of $1.5 \times 18 \, g$
$$= \underline{27 \, g}$$

2 What fraction of a mole is 8 g of sulphur dioxide? (Relative molecular mass of $SO_2 = 64$.)
The molar mass of SO_2 molecules is $64 \, g \, mol^{-1}$
∴ 64 g of SO_2 is 1 mol of SO_2 molecules
∴ 1 g of SO_2 is $\frac{1}{64}$ mol of SO_2 molecules
∴ 8 g of SO_2 is $\frac{8}{64}$ mol of SO_2 molecules
$$= \underline{0.125 \, mol \, of \, SO_2 \, molecules}$$

The empirical formulae of hydrated salts

By chemical analysis, or by using the oxidation numbers given in Table 9.2, we can work out that the empirical formula of anhydrous copper(II) sulphate is $CuSO_4$. This means that the empirical formula of the hydrated salt must be $CuSO_4 \cdot xH_2O$. The value of x may be found by a simple experiment, as the following example demonstrates.

Example

When 8.0 g of hydrated copper(II) sulphate, $CuSO_4 \cdot xH_2O$, was heated to drive off the water of crystallisation, a residue of 5.1 g of anhydrous salt remained. Calculate the empirical formula of the hydrated salt. (Relative atomic masses: $Cu = 63.5$, $S = 32$, $O = 16$, $H = 1$.)

The mass of water of crystallisation evolved $= (8.0 - 5.1) = 2.9 \, g$
The molar mass of anhydrous $CuSO_4 = (63.5 + 32 + 64) = 159.5 \, g \, mol^{-1}$
The molar mass of water $= (2 + 16) = 18 \, g \, mol^{-1}$

	$CuSO_4$	:	H_2O
Ratio of grams	5.1	:	2.9
Ratio of moles	5.1/159.5	:	2.9/18
=	0.032	:	0.16
=	1	:	5

∴ Empirical formula is $\underline{CuSO_4 \cdot 5H_2O}$.

Percentage composition of compounds

So far we have used the percentage composition by mass to calculate the empirical formula of a compound. However, it is sometimes useful to find the percentage of a particular element in a compound of *known* formula. The method of working may be divided into three stages.

1 Work out the relative molecular mass of the compound.
2 Write down the fraction by mass of each element (or water of crystallisation) and convert this to a percentage.
3 Check that the percentages add up to 100.

Example

Calculate the percentage composition by mass of hydrated sodium sulphate, $Na_2SO_4 \cdot 10H_2O$. (Relative atomic masses: $Na = 23$, $S = 32$, $O = 16$, $H = 1$.)

Relative molecular mass of $Na_2SO_4 \cdot 10H_2O$ $= (2 \times 23) + 32 + (4 \times 16) + (10 \times 18) = 322$

Percentage of Na	$= \frac{46}{322} \times 100 =$	14.3
Percentage of S	$= \frac{32}{322} \times 100 =$	9.9
Percentage of O	$= \frac{64}{322} \times 100 =$	19.9
Percentage of H_2O	$= \frac{180}{322} \times 100 =$	55.9
	Total $=$	100.0

Notice that the water is treated as a separate unit and its oxygen is not added to that of the Na_2SO_4.

9.4 Chemical equations

A chemical equation shows the relative numbers of atoms and molecules taking part in a chemical reaction. It also shows whether the various substances involved are in the solid, liquid or gaseous state, or are dissolved in water.

The only sure way of determining an equation is to carry out an experiment to find the relative numbers of moles of reactants and products (Experiments 9.4.a to 9.4.d). However, since you cannot possibly do this for every reaction you encounter you should learn how to work out what the equation *should* be, given the names of the various substances involved.

Steps in writing equations

1 You must *know* what the reactants and products are. It is useless to try to write an equation by guesswork since, even if you manage to balance it, the reaction which it represents may not take place.
2 In the early stages you will find it helps to write a word equation for the reaction. Include the state symbols in this equation, i.e. (s) for solid, (l) for liquid, (g) for gas and (aq) for aqueous solution.

3   Write down the formulae and states of the reactants and products underneath the word equation, leaving a space in front of each formula for balancing.

4   Work systematically through the equation, from left to right, and check that the same numbers of atoms of the various elements appear on both sides.

Examples

(a) Write an equation for the reaction between magnesium and oxygen.

1   When magnesium burns in oxygen the only product is magnesium oxide.

2   magnesium(s) + oxygen(g) → magnesium oxide(s)

3      $Mg(s)$ + $O_2(g)$ → $MgO(s)$

4   (i) There is one atom of magnesium on each side of the equation.

(ii) There are two atoms of oxygen on the left but only one on the right; therefore try a 2 in front of the MgO(s).

$$Mg(s) + O_2(g) \rightarrow 2MgO(s)$$

(iii) Since the 2 in front of the MgO(s) has doubled the number of magnesium atoms as well as the number of oxygen atoms, a 2 must be placed in front of the Mg(s).

$$2Mg(s) + O_2(g) \rightarrow 2MgO(s)$$

A final check from left to right shows that the equation is now balanced.

(b) Write an equation for the reaction between solutions of zinc chloride and silver nitrate.

1   When solutions of zinc chloride and silver nitrate are mixed a precipitate of silver chloride is formed, together with a solution of zinc nitrate.

2   zinc chloride(aq) + silver nitrate(aq) → silver chloride(s) + zinc nitrate(aq)

3   $ZnCl_2(aq)$ + $AgNO_3(aq)$ → $AgCl(s)$ + $Zn(NO_3)_2(aq)$

4   (i) There is one atom of zinc on each side of the equation.

(ii) There are two chlorine atoms on the left but only one on the right. Try a 2 in front of the AgCl(s).

$$ZnCl_2(aq) + AgNO_3(aq) \rightarrow 2AgCl(s) + Zn(NO_3)_2(aq)$$

(iii) There is one Ag atom on the left but two on the right. Try a 2 in front of the AgNO_3(aq).

$$ZnCl_2(aq) + 2AgNO_3(aq) \rightarrow 2AgCl(s) + Zn(NO_3)_2(aq)$$

(iv) There are two NO_3 radicals on each side. A final check of all the atoms shows that the equation is now balanced.

9.5   Ionic equations

Consider the reaction between aqueous solutions of magnesium sulphate and barium chloride. This produces a precipitate of barium sulphate and a solution of magnesium chloride, the chemical equation being

$$MgSO_4(aq) + BaCl_2(aq) \rightarrow BaSO_4(s) + MgCl_2(aq).$$

All of the compounds involved are ionic. In solution, their ions are separated from one another and spread evenly throughout the liquid. Thus, although the barium and sulphate ions have come together to form a solid, the magnesium and chloride ions are as free at the end of the reaction as they were at the beginning. As they have not taken part in the reaction they are referred to as *spectator ions*.

An ionic equation includes only those ions which *change* in some way during the reaction. Spectator ions are omitted.

Steps in writing ionic equations

1   Write a normal chemical equation for the reaction.
2   Write the equation in terms of ions. As a general guide you should remember that *metallic compounds*, *ammonium compounds* and *dilute acids* are *ionic*, all other compounds being covalent. (This is an oversimplification but it will suffice at this level.)
3   Rewrite the equation, omitting the spectator ions.

Example

Write an ionic equation for the reaction between aqueous solutions of magnesium sulphate and barium chloride.
1   $MgSO_4(aq) + BaCl_2(aq) \rightarrow BaSO_4(s) + MgCl_2(aq)$
2   $Mg^{2+}(aq) + SO_4^{2-}(aq) + Ba^{2+}(aq) + 2Cl^-(aq) \rightarrow$
$$BaSO_4(s) + Mg^{2+}(aq) + 2Cl^-(aq)$$
3   $SO_4^{2-}(aq) + Ba^{2+}(aq) \rightarrow BaSO_4(s)$
Note: the charges on the ions in an ionic solid are generally omitted for simplicity. However, if the ionic nature of the barium sulphate needs to be stressed for some reason, its formula may be written as $(Ba^{2+} + SO_4^{2-})(s)$.

9.6   Calculating reacting masses from chemical equations

In section 9.4 you learned that an equation is derived by measuring reacting masses and converting these to relative numbers of atoms, molecules or formula units. However, if the equation for a reaction is already known we can reverse this process and calculate the relative masses of reactants and products involved. The importance of calculations of this type was mentioned at the beginning of the chapter.

Steps in calculating reacting masses

1   Write a balanced equation for the reaction.
2   Pick out the substances mentioned in the question and write down the ratio of moles of each as they appear in the equation. Always write the ratio so that the substance whose mass is known is on the left and the one whose mass is required is on the right. Convert the ratio, if necessary, so

that only *one* mole of the first substance is involved (see below).
3 Set out the calculation as shown in the example.

Example

What mass of zinc oxide will just neutralise dilute nitric acid containing
12.6 g of pure HNO_3 dissolved in water? (Relative atomic masses:
$Zn = 65$, $O = 16$, $H = 1$, $N = 14$.)

1            $2HNO_3(aq) + ZnO(s) \rightarrow Zn(NO_3)_2(aq) + H_2O(l)$

2   mol $HNO_3(aq)$: mol $ZnO(s) = 2 : 1 = 1 : \frac{1}{2}$ ... A

3   g of $HNO_3 = 12.6$

$$\therefore \quad \text{mol of } HNO_3 = \frac{12.6}{63} \quad \begin{array}{l}\text{(molar mass of}\\ HNO_3 = 63 \text{ g mol}^{-1})\end{array}$$

$$\therefore \quad \text{mol of } ZnO = \frac{12.6}{63} \times \frac{1}{2} \qquad \qquad \text{... from A}$$

$$\therefore \quad \text{g of } ZnO = \frac{12.6}{63} \times \frac{1}{2} \times 81 \quad \begin{array}{l}\text{(molar mass of}\\ ZnO = 81 \text{ g mol}^{-1})\end{array}$$

$$= \underline{8.1}$$

Notice the symmetrical layout of the units, i.e. g, mol, mol, g.

Final note

Chemical calculations almost always involve the mole and molar mass.
You *must* fully understand the meanings of these terms and the relationship between them.

Questions on Chapter 9

1 What is the mass of (a) 0.1 mol of magnesium atoms, (b) 0.25 mol of oxygen
molecules, (c) 0.5 mol of aluminium oxide, (d) 0.6 mol of calcium carbonate?
2 What fraction of a mole is (a) 5 g of calcium atoms, (b) 35.5 g of chlorine
molecules, (c) 6.8 g of anhydrous zinc chloride, (d) 2.12 g of anhydrous sodium
carbonate?
3 How many atoms are present in (a) 2.8 g of iron, (b) 7.8 g of potassium, (c) 7 g
of nitrogen molecules, (d) 4 g of bromine molecules? (Avogadro constant $= 6.02 \times 10^{23}$ mol^{-1}).
4 Calculate the empirical formulae of the compounds having the following
composition by mass: (a) $Na = 32.4\%$, $S = 22.6\%$, $O = 45.0\%$; (b) $K = 23.5\%$,
$I = 76.5\%$; (c) $Fe = 36.9\%$, $S = 21.0\%$, $O = 42.1\%$; (d) $Na = 18.5\%$, $S = 25.8\%$,
$O = 19.4\%$, $H_2O = 36.3\%$.
5 Use the following data to determine the molecular formulae of the compounds
concerned: (a) $C = 92.3\%$, $H = 7.7\%$, relative molecular mass $= 78$; (b) $N = 87.5\%$,
$H = 12.5\%$, relative molecular mass $= 32$; (c) $C = 40.0\%$, $H = 6.7\%$, $O = 53.3\%$,
relative molecular mass $= 180$; (d) $P = 43.7\%$, $O = 56.3\%$, relative molecular
mass $= 284$. (All percentages are by mass.)
6 Work out the percentage composition by mass of the following compounds: (a)
ammonium nitrate, NH_4NO_3; (b) anhydrous iron(III) sulphate, $Fe_2(SO_4)_3$; (c)
calcium bromide, $CaBr_2$; (d) hydrated copper(II) sulphate, $CuSO_4 \cdot 5H_2O$.
7 What mass of silver chloride will be precipitated when a solution containing

3.4 g of zinc chloride is mixed with excess silver nitrate solution?

$$ZnCl_2(aq) + 2AgNO_3(aq) \rightarrow 2AgCl(s) + Zn(NO_3)_2(aq)$$

8   Find the mass of potassium hydroxide which must be dissolved in water to give a solution which will neutralise a sample of dilute sulphuric acid containing 14 g of pure H_2SO_4.

$$2KOH(aq) + H_2SO_4(aq) \rightarrow K_2SO_4(aq) + 2H_2O(l)$$

9   What mass of sodium nitrate must be heated in order to obtain 8 g of oxygen?

$$2NaNO_3(s) \rightarrow 2NaNO_2(s) + O_2(g)$$

10   What is the maximum mass of hydrated magnesium sulphate, $MgSO_4 \cdot 7H_2O$, which could be obtained by reacting 3 g of magnesium with dilute sulphuric acid?

$$Mg(s) + H_2SO_4(aq) \rightarrow MgSO_4(aq) + H_2(g)$$

11   Write ionic equations for the following reactions.
(a) $Fe(s) + H_2SO_4(aq) \rightarrow FeSO_4(aq) + H_2(g)$
(b) $2NaCl(aq) + Ag_2SO_4(aq) \rightarrow 2AgCl(s) + Na_2SO_4(aq)$
(c) $Ca(OH)_2(aq) + 2HCl(aq) \rightarrow CaCl_2(aq) + 2H_2O(l)$
(d) $Na_2CO_3(aq) + 2HNO_3(aq) \rightarrow 2NaNO_3(aq) + H_2O(l) + CO_2(g)$

12   Metal X reacted with chlorine to form a chloride which had a relative molecular mass of 42.5. State, giving your reasons, the name of X. Write the formula of the chloride.   (SCE)

13   A hydrate of sodium carbonate, $Na_2CO_3 \cdot xH_2O$, weighing 2.48 g loses 0.36 g of water on heating to constant mass. What is the value of x?   (SUJB)

14   Ammonium nitrate and ammonium sulphate are both used as fertilisers because of their high nitrogen content. The cost of ammonium nitrate is twice that of ammonium sulphate. Considering nitrogen content and price only, which fertiliser would you use for economy? Justify your answer.   (SUJB)

15   (a) In the following account there are certain errors and omissions. Make a list of these and give the corrections required.

'From the formula PbO it can be calculated that 1 g oxygen combines with 10.25 g lead. To check this experimentally the following procedure should be carried out.

From the bottle labelled lead oxide (technical) carefully weigh a sample in a container. Place the container in a glass tube and pass hydrogen over it. As soon as some lead is seen in the container, reweigh.'

(b) From a properly conducted experiment on an oxide of lead the following results were obtained.

Mass of porcelain boat   = 5.20 g
Mass of porcelain boat + the oxide of lead = 7.60 g
Mass of porcelain boat + lead   = 7.28 g

What mass of lead combines with 1 g of oxygen?
(c) Given that one atom of lead is thirteen times as heavy as one atom of oxygen what is the simplest formula for this oxide?
(d) White lead, $Pb_3(OH)_2(CO_3)_2$, and chrome yellow, $PbCrO_4$, are used as pigments in paint. Calculate which has the higher percentage of lead. Show your working clearly.   (SCE)

16   Lead iodide (lead(II) iodide) and barium sulphate are formed as precipitates on mixing solutions of the substances shown on the left in the following equations:

$$Pb(NO_3)_2 + 2KI \rightarrow PbI_2 + 2KNO_3$$
$$Ba(OH)_2 + H_2SO_4 \rightarrow BaSO_4 + 2H_2O$$

(A '0.1 M solution' contains one tenth of a mole of solute per dm^3. Thus, 0.1 M HCl contains 3.65 g of HCl – relative molecular mass 36.5 – per dm^3.)

(a) Calculate the relative molecular masses of PbI_2 and $BaSO_4$.

(b) What fraction of a mole of solute is present in 100 cm^3 of a 0.1 M solution?

(c) Calculate the masses of precipitate formed when the following solutions are mixed.

Experiment A   100 cm^3 of 0.1 M $Pb(NO_3)_2$ + 100 cm^3 of 0.1 M KI

Experiment B   100 cm^3 of 0.1 M $Pb(NO_3)_2$ + 200 cm^3 of 0.1 M KI

Experiment C   100 cm^3 of 0.1 M $Ba(OH)_2$ + 100 cm^3 of 0.1 M H_2SO_4

Experiment D   100 cm^3 of 0.1 M $Ba(OH)_2$ + 200 cm^3 of 0.1 M H_2SO_4

(d) It is found that the liquids formed in experiments B and D are good conductors of electricity while the liquid formed in experiment C will not conduct. Explain why this is so.

(e) For *each* of the *four* experiments, say whether the liquid formed will be acidic, alkaline or neutral.

(f) If 100 cm^3 of 0.1 M $Ba(OH)_2$ is mixed with 100 cm^3 of 0.1 M HCl, will the liquid formed be acidic, alkaline or neutral? Give reasons for your answer.   (C)

17   0.1 g samples of magnesium carbonate, calcium carbonate and sodium carbonate were dissolved in excess dilute hydrochloric acid. Which of these samples would evolve the most carbon dioxide? What is the minimum volume of hydrochloric acid (containing 1.825 g dm^{-3}) necessary to dissolve each of the samples? Indicate, with the aid of a sketch, how you would carry out the experiment in the laboratory.   (SUJB)

18   What mass of silver is obtained if 1 g of magnesium is added to excess silver nitrate solution?

$$Mg(s) + 2Ag^+(aq) \rightarrow Mg^{2+}(aq) + 2Ag(s)$$

19   In an experiment 20 dm^3 of hydrogen and 10 dm^3 of oxygen are obtained from water.

(a) The density of hydrogen is 0.1 g dm^{-3}. What mass of hydrogen is obtained in the experiment?

(b) The density of oxygen is 1.6 g dm^{-3}. What mass of oxygen is obtained in the experiment?

(c) How many moles of hydrogen are obtained in the experiment?

(d) How many moles of oxygen are obtained in the experiment?

(e) Use your answers to justify that the formula for water is H_2O.   (SCE)

10 The molecular theory of gases

The word 'gas' was invented by the Dutchman Johann van Helmont, early in the seventeenth century. He derived it from the Greek word 'chaos', which aptly describes our present-day view of the motion of gaseous molecules. The fact that the molecules of *all* gases share the same chaotic motion and wide separation from one another results in some striking similarities between them, as the following experiments will show.

Experiments

10.1.a   To find the relative volumes of sulphur dioxide and oxygen which combine together ★

This experiment must be carried out in a fume cupboard.

In this experiment sulphur dioxide and oxygen react together, in the presence of a heated catalyst of platinised asbestos, to form solid sulphur trioxide. The sulphur trioxide is absorbed by traces of moisture in the apparatus.

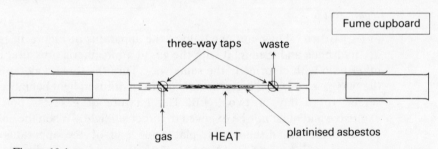

Figure 10.1.a

PROCEDURE   Set up the apparatus as shown in Figure 10.1.a. Flush out one syringe with dry sulphur dioxide and collect 60 cm³ of the gas in it. Collect an equal volume of dry oxygen in the other syringe in a similar manner. Heat the platinised asbestos to red heat and pass the gases from one syringe to the other until no further change in volume is detected. Push all the residual gas into one syringe and allow the apparatus to cool. Twist the plunger of the syringe (why?) and note the reading. Finally, expel the residual gas on to a glowing splint.

Questions

1   What was the residual gas? How could you tell?
2   What volume of (a) sulphur dioxide, (b) oxygen was used up?

3   Allowing for experimental error, is there any relationship between the two reacting volumes?

Note: this experiment may be modified to investigate a number of reactions involving gases, for example:

(a) a known volume of pure nitrogen oxide (or dinitrogen oxide) could be reduced to nitrogen by passing it over heated copper,

(b) a known volume of oxygen could be passed over heated graphite to produce carbon dioxide (charcoal sometimes causes explosions),

(c) a known volume of oxygen could be passed over heated sulphur to produce sulphur dioxide,

(d) a known volume of ammonia could be reacted with a known volume of hydrogen chloride to form solid ammonium chloride (no silica tube or heating required),

(e) a known volume of nitrogen oxide could be reacted with a known volume of oxygen to form nitrogen dioxide (no silica tube or heating required).

10.1.b   To find the relative volumes of hydrogen and chlorine which combine together to form hydrogen chloride ★ ★

This experiment must not be performed by pupils.

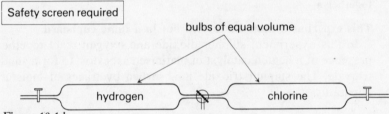

Safety screen required

bulbs of equal volume

hydrogen   chlorine

Figure 10.1.b

PROCEDURE   Flush out one bulb of the apparatus of Figure 10.1.b with dry hydrogen and then fill it with the gas at atmospheric pressure. Fill the other bulb with chlorine in the same way. Turn the three-way tap so that the two gases mix and leave the apparatus in diffused light behind a safety screen for a day or two. (**N.B.** The mixture of gases is potentially explosive and must *not* be exposed to direct sunlight.) When the colour of the chlorine has disappeared, place one end of the apparatus under mercury and open the tap. Now close the tap and repeat the procedure using water in place of the mercury.

Questions

1   Did any mercury enter the apparatus? How did the volume of product compare with that of the reactants?

2   How much gas remained after opening the tap under water? Do you think that any hydrogen or chlorine remained uncombined?

3   Is there a simple relationship between the volumes of the various gases?

10.2.a   To find the volume of one mole of hydrogen molecules at room temperature and pressure

PROCEDURE   The procedure for this experiment is exactly the same as that for Experiment 9.4.d on page 157.

Questions

1   Write an equation for the reaction and state how many moles of hydrogen molecules are liberated from the acid by 1 mole of magnesium atoms.
2   What were (a) the number of grams, (b) the number of moles of magnesium atoms used? What volume of hydrogen was obtained?
3   What volume of hydrogen would be obtained by using 1 mole of magnesium atoms?
4   Use your answers to questions 1 and 3 to find the volume of 1 mole of hydrogen molecules at room temperature and pressure.

10.2.b   To find the molar volumes of several gases under the same conditions of temperature and pressure

PROCEDURE   Calculate the volume occupied by 1 mole of each of the gases in Table 10.2.b.

Gas	Mass of 1 dm³ of gas/g (at 298 K and 101325 Pa)	Molar mass/g mol⁻¹
O_2	1.33	32
N_2	1.17	28
CO_2	1.81	44
HCl	1.50	36.5
SO_2	2.68	64

Table 10.2.b

Questions

1   What do you notice about the figures obtained?
2   What can you say about the number of molecules found in 1 mole of any gas?
3   What can you say about the volume occupied by equal numbers of gas molecules at the same temperature and pressure?

10.2.c   To find the relative molecular mass of a gas by direct weighing★

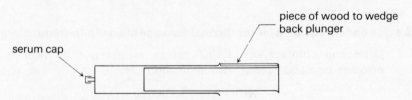

Figure 10.2.c

PROCEDURE   Close the end of a 50 cm³ plastic syringe with a serum cap or a piece of rubber tubing and a clip. Pull the plunger back to the 50 cm³ mark and wedge it in position with a piece of wood as in Figure 10.2.c. Find the mass of the empty syringe as quickly as possible. (Alternatively, find the mass of the syringe when full of air and then subtract the mass of the air. You can calculate the latter from the fact that $1 cm^3$ of air at room temperature and pressure has a mass of $0.0012 g$.) Now flush out the syringe several times with dry sulphur dioxide, fill it, and find the mass again. Repeat the experiment with other gases of your choice.

Questions

1 Why was it necessary to find the mass of the empty syringe in the *open* position?
2 From your results, calculate the relative molecular mass of the gas. (1 mole of any gas occupies approximately $24 dm^3$ at room temperature and pressure, or $22.4 dm^3$ at s.t.p.)

10.2.d To find the molecular formula of a gas ★

The general method outlined here may be used to determine the molecular formulae of several other gases.

PROCEDURE (a) Modify Experiment 10.1.a to find the volume of sulphur dioxide formed by heating sulphur with $80 cm^3$ of oxygen.
(b) Find the relative molecular mass of sulphur dioxide as in Experiment 10.2.c.

Questions

1 What volume of sulphur dioxide was formed from $80 cm^3$ of oxygen in (a)?
2 Use Avogadro's principle to calculate the number of oxygen molecules which are used in making one molecule of sulphur dioxide. Since oxygen molecules are diatomic, how many oxygen *atoms* are present in one molecule of sulphur dioxide?
3 The relative atomic masses of sulphur and oxygen are 32 and 16 respectively. Thus, if the molecular formula of sulphur dioxide is written as S_xO_y, its relative molecular mass must be $(32x + 16y)$. You have already found the value of y in question 2. Use the relative molecular mass determined in part (b) of the experiment to calculate the value of x.
4 What is the molecular formula of sulphur dioxide?

10.2.e To find the equation for the thermal decomposition of potassium chlorate(vii)

Potassium chlorate(vii), $KClO_4$, gives off oxygen on heating. Four possible equations for its decomposition are:

$$2KClO_4(s) \rightarrow 2KClO_3(s) + O_2(g)$$
$$KClO_4(s) \rightarrow KClO_2(s) + O_2(g)$$
$$2KClO_4(s) \rightarrow 2KClO(s) + 3O_2(g)$$
$$KClO_4(s) \rightarrow KCl(s) + 2O_2(g)$$

If the number of moles of oxygen molecules obtained from 1 mole of potassium chlorate(vii) is found then the correct equation may be selected.

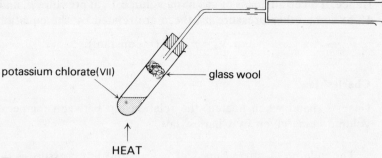

potassium chlorate(VII)   glass wool

HEAT

Figure 10.2.e

PROCEDURE Find the mass of a 19×125 mm Pyrex test tube, add 0.25–0.28 g of potassium chlorate(VII) and find the mass again. Set up the apparatus as shown in Figure 10.2.e, twist the syringe plunger and note the reading. Heat the test tube gently, and then more strongly, until no more gas is evolved. Allow the apparatus to cool, twist the syringe plunger and note the final reading. Expel the gas on to a glowing splint to check that it is oxygen.

Questions

1 What was (a) the number of grams, (b) the number of moles of potassium chlorate(VII) used? What volume of oxygen was evolved?
2 What volume of oxygen would be obtained from 1 mole of potassium chlorate(VII)?
3 Using the fact that the molar volume of any gas at room temperature and pressure is about 24 dm³ mol⁻¹, calculate the number of moles of oxygen produced from one mole of potassium chlorate(VII).
4 Select the equation which fits these results.
5 What further tests should be carried out to *prove* that this equation is correct?

10.1 The gas laws

Boyle's law

Anyone who has used a bicycle pump or a gas syringe is familiar with the fact that if the pressure exerted on a gas is increased, its volume decreases. Closer investigation shows that provided the mass and temperature remain constant, doubling the pressure will halve the volume, trebling the pressure will reduce the volume to one-third, and so on. This relationship was discovered by Robert Boyle in 1662 and he expressed his results in the law named after him.

The volume of a fixed mass of gas at constant temperature is inversely proportional to its pressure.

i.e. $p \propto \dfrac{1}{V}$ or $pV = \text{constant}$

Hence, if a certain mass of gas has a volume V_1 at pressure p_1 and volume V_2 at some other pressure p_2, these are related by the equation,

$$p_1 V_1 = p_2 V_2 \quad (= \text{constant}).$$

Charles' law

Gases expand when heated, the relationship between temperature and volume being given by Charles' law.

The volume of a fixed mass of gas at constant pressure is directly proportional to its temperature in kelvins.

i.e. $\qquad\qquad V \propto T \qquad$ or $\qquad V/T = \text{constant}$

This means that if a certain mass of gas with volume V_1 at temperature T_1 has its temperature changed to T_2, then its volume will change to V_2 such that

$$V_1/T_1 = V_2/T_2 \quad (= \text{constant}).$$

Note: a temperature in degrees Celsius is converted to one in kelvins by adding 273. Thus,

$$25\,°\text{C} = (25 + 273)\,\text{K} = 298\,\text{K}.$$

The ideal gas equation

If the equations which summarise the laws of Boyle and Charles are combined then the *ideal gas equation* is obtained:

i.e. $\qquad\qquad\qquad \dfrac{pV}{T} = \text{constant}.$

Thus if the pressure, volume and temperature of a fixed mass of gas change from p_1, V_1 and T_1 to p_2, V_2 and T_2, it follows that

$$\frac{p_1 V_1}{T_1} = \frac{p_2 V_2}{T_2} \quad (= \text{constant}).$$

Example

A certain mass of gas occupies $500\,\text{cm}^3$ at $27\,°\text{C}$ and $100\,000\,\text{Pa}$. What will be its volume at $0\,°\text{C}$ and $101\,325\,\text{Pa}$?

$$\frac{p_1 V_1}{T_1} = \frac{p_2 V_2}{T_2}$$

$$\therefore \frac{100\,000 \times 500}{(273 + 27)} = \frac{101\,325 \times V_2}{(273 + 0)}$$

$$\therefore V_2 = \frac{100\,000 \times 500 \times 273}{101\,325 \times 300}$$

$$= \underline{449\,\text{cm}^3}$$

Standard temperature and pressure

Since the volume of a gas, and related properties such as density, vary enormously with changes in temperature and pressure, it is necessary to specify certain standard conditions under which such properties are compared. These conditions have been chosen arbitrarily as 273 K (0 °C) and 101 325 Pa (760 mm of mercury). They are referred to as *standard temperature and pressure*, or s.t.p.

Gay-Lussac's law

When sulphur dioxide and oxygen are passed over a heated platinised asbestos catalyst they combine to form solid sulphur trioxide, the volume of which is negligible (Experiment 10.1.a). If the volumes of the two gases are measured at the same temperature and pressure a typical result is:

60 cm^3 sulphur dioxide(g) + 30 cm^3 oxygen(g) → sulphur trioxide(s),

i.e.   2 volumes of   +   1 volume of   →   sulphur trioxide.
       sulphur dioxide   oxygen

Joseph Gay-Lussac, a professor in Paris, studied many reactions between gases and obtained results such as the following.

2 volumes of hydrogen + 1 volume of oxygen → 2 volumes of steam

1 volume of hydrogen + 1 volume of chlorine →
                                    2 volumes of hydrogen chloride

2 volumes of carbon monoxide + 1 volume of oxygen →
                                    2 volumes of carbon dioxide

(All volumes were measured at the same temperature and pressure.)

The simple relationships shown by these results led Gay-Lussac, in 1808, to put forward his *law of gaseous volumes*.

> When gases combine, they do so in volumes which bear a simple ratio to one another and to the volume of the product, if gaseous, all volumes being measured at the same temperature and pressure.

You must realise that the law applies only to gases. For example, if measurements in the experiment where hydrogen combines with oxygen (see above) are made below 100 °C then the final product is water, not steam. The volume of this water is negligible compared with that of the reacting gases.

Avogadro's principle

At about the same time as Gay-Lussac was developing his law of gaseous volumes, John Dalton introduced his atomic theory. One of the main points of this theory is that *atoms* combine together in a simple ratio. Now Gay-Lussac's law states that *volumes of gases* combine together in a simple ratio and the similarity between the two ideas led Jöns Berzelius, a Swedish chemist, to suggest that equal volumes of all gases at the same temperature and pressure contain the same number of atoms. Unfortunately, this suggestion is incompatible with Dalton's theory, as the following example shows.

Consider the combination of hydrogen and chlorine. Experiment 10.1.b shows that if all measurements are made at the same temperature and pressure,

1 volume of hydrogen + 1 volume of chlorine →

2 volumes of hydrogen chloride.

Thus, if one volume of any gas contains n atoms, it follows that

n atoms of hydrogen + n atoms of chlorine →

2n 'compound atoms' of hydrogen chloride.

∴ 1 atom of hydrogen + 1 atom of chlorine →

2 'compound atoms' of hydrogen chloride.

This means that one 'compound atom' of hydrogen chloride must contain half an atom of hydrogen and half an atom of chlorine. But atoms, according to Dalton, are indivisible. Clearly, something must be wrong!

This problem was solved in 1811 by the Italian scientist Amadeo Avogadro, who suggested that the smallest particle of a gas which could exist independently was not an atom, but a group of atoms combined together, i.e. *a molecule*. He was then able to link the work of Dalton and Gay-Lussac by the simple, yet extremely important statement which we now know as *Avogadro's principle*.

Equal volumes of all gases at the same temperature and pressure contain the same number of molecules.

Consider again the reaction between hydrogen and chlorine.

1 volume of hydrogen + 1 volume of chlorine →

2 volumes of hydrogen chloride.

. . . (i)

According to Avogadro's principle, if one volume of gas contains n molecules, we can write

n molecules of hydrogen + n molecules of chlorine →

2n molecules of hydrogen chloride.

∴ 1 molecule of hydrogen + 1 molecule of chlorine →

2 molecules of hydrogen chloride

. . . (ii)

This means that one molecule of hydrogen chloride contains half a molecule of hydrogen and half a molecule of chlorine. Provided that molecules of hydrogen and chlorine contain *even* numbers of atoms, Dalton's theory is not invalidated, since half a molecule will contain a whole number of atoms.

We now know that all hydrogen and chlorine molecules contain two atoms. Thus one molecule of hydrogen chloride contains one atom of each element and its molecular formula must be HCl.

Lines (i) and (ii) illustrate an extremely important point, namely that the ratio of *volumes* of gases involved in a reaction is the same as the ratio of *numbers of molecules*. Amazing though it may seem, Avogadro's principle was largely ignored for almost half a century and it was not until 1858 that another Italian, Stanislao Cannizzaro, showed just how useful an idea it is.

Example

If 20 cm³ of hydrogen sulphide are burned in 80 cm³ of oxygen, what will be the volume and composition of the residual gas? (All volumes are measured at room temperature and pressure.)

When hydrogen sulphide burns in a plentiful supply of oxygen the products are sulphur dioxide and steam. However, the volumes in this case are to be measured at *room* temperature and pressure, and under these conditions the steam will condense to form water of negligible volume.

Remember, Avogadro's principle applies only to *gases*.

$$2H_2S(g) + 3O_2(g) \rightarrow 2H_2O(l) + 2SO_2(g)$$

By Avogadro's principle	2 molecules	3 molecules	2 molecules	2 molecules
	2 volumes	3 volumes	negligible	2 volumes
i.e.	20 cm³	30 cm³	–	20 cm³
		(+ 50 cm³ excess)		

Thus the residual gas will be made up of 50 cm³ of oxygen and 20 cm³ of sulphur dioxide.

10.2 Atomicity and molar volume of a gas

Atomicity

The **atomicity** of an element is the number of atoms in one molecule of it.

Most of the elements which are gaseous at room temperature and pressure consist of molecules containing two atoms. They are said to be *diatomic*, e.g. H_2, O_2, N_2. The noble gases are *monatomic*, while ozone, O_3, is *triatomic*.

Knowledge of the atomicity of hydrogen and chlorine enables us to determine the molecular formula of hydrogen chloride, from measurements of reacting volumes and an application of Avogadro's principle (see page 180). This method may also be used for other gaseous compounds, as the following example shows.

Example

One volume of nitrogen combines with three volumes of hydrogen to form two volumes of a gaseous compound, X. What is the molecular formula of X? (Assume that all volumes are measured at the same temperature and pressure.)

1 volume of nitrogen + 3 volumes of hydrogen → 2 volumes of X

∴ By Avogadro's principle,

1 molecule of nitrogen + 3 molecules of hydrogen → 2 molecules of X

Nitrogen and hydrogen are both diatomic.

∴ $$N_2(g) + 3H_2(g) \rightarrow 2X(g)$$

∴ X must be NH_3

Molar volume of a gas

The **molar volume of a gas** is the volume of one mole of it.

The units of molar volume are $dm^3\,mol^{-1}$ and, because the volume of a gas varies greatly with changes in temperature and pressure, it is essential to state the conditions under which it is measured.

Example

Find the molar volume of hydrogen gas, $H_2(g)$, at s.t.p., given that $1\,dm^3$ of it under these conditions has a mass of $0.09\,g$ and that the relative atomic mass of hydrogen is 1.008.

The molar mass of $H_2(g)$ is $2.016\,g\,mol^{-1}$
$0.09\,g$ of $H_2(g)$ has a volume of $1\,dm^3$ at s.t.p.

∴ $2.016\,g$ of $H_2(g)$ has a volume of $\frac{1}{0.09} \times 2.016 = 22.4\,dm^3$ at s.t.p.

i.e. the molar volume of hydrogen at s.t.p. is $22.4\,dm^3\,mol^{-1}$.

Now, one mole of *any* gas contains 6.02×10^{23} molecules and according to Avogadro, equal numbers of gas molecules occupy equal volumes at the same temperature and pressure. Thus under given conditions the molar volume must be the same for *all* gases. As the example shows, its value at s.t.p. is $22.4\,dm^3\,mol^{-1}$; at room temperature and pressure it is approximately $24\,dm^3\,mol^{-1}$ (see Experiments 10.2.a and 10.2.b).

We can make use of the molar volume of a gas in a number of different ways. It may help us to determine the relative molecular mass of a gas, as in Experiment 10.2.c, or to determine the equation for a reaction (Experiments 9.4.c and 10.2.e). In addition, if the equation for a reaction is known, then the volume of gas which should be produced from given masses of reactants may be calculated. The following examples should make the principles of the various methods clear.

Examples

1   Calculate the relative molecular mass of methane, given that $100\,cm^3$ of it at s.t.p. has a mass of $0.072\,g$.

100 cm³ of methane at s.t.p. has a mass of 0.072 g

1 dm³ of methane at s.t.p. has a mass of 0.72 g

∴   22.4 dm³ of methane at s.t.p. has a mass of $0.72 \times 22.4 = 16.1$ g

∴   The molar mass of methane is 16.1 g mol⁻¹

∴   Relative molecular mass of methane is 16.1

2   When 0.306 g of potassium chlorate(v), $KClO_3$, is heated it gives off 84 cm³ of oxygen (measured at s.t.p.) and leaves a residue of solid potassium chloride. What is the equation for the reaction? (Molar mass of potassium chlorate(v) = 122.5 g mol⁻¹; molar volume of any gas at s.t.p. = 22.4 dm³ mol⁻¹.)

Note: you will remember from Chapter 9 that to convert a mass in grams to a number of moles you must divide by the molar *mass*. Similarly, to convert a volume in dm³ to a number of moles you must divide by the molar *volume*. These principles are used in the second line of the calculation.

	$KClO_3(s)$	:	$O_2(g)$
Relative quantities	0.306 g	:	0.084 dm³ (s.t.p.)
Ratio of moles	$\dfrac{0.306}{122.5}$	:	$\dfrac{0.084}{22.4}$
=	0.002 5	:	0.003 75
=	1	:	1.5
=	2	:	3

∴   The equation is of the form

$$2KClO_3(s) \rightarrow ?KCl(s) + 3O_2(g).$$

Clearly, the balanced equation must be

$$2KClO_3(s) \rightarrow 2KCl(s) + 3O_2(g).$$

3   What volume of hydrogen, measured at s.t.p., will be given off when 5.4 g of aluminium dissolves in excess dilute hydrochloric acid? (Al = 27).

As you know, to convert a number of moles to a mass in grams you must multiply by the molar *mass*. To convert a number of moles to a volume in dm³ you must multiply by the molar *volume*.

$$2Al(s) + 6HCl(aq) \rightarrow 2AlCl_3(aq) + 3H_2(g)$$

i.e.   mol Al(s) : mol H_2(g) = 2 : 3 = 1 : 3/2   ... A

g of Al(s) = 5.4

$\text{mol of Al(s)} = \dfrac{5.4}{27}$ (molar mass of Al is 27 g mol⁻¹)

$\text{mol of } H_2(g) = \dfrac{5.4}{27} \times \dfrac{3}{2}$ ...(from A)

$\text{dm}^3 \text{ of } H_2(g) \text{ at s.t.p.} = \dfrac{5.4}{27} \times \dfrac{3}{2} \times 22.4$

$$= \underline{6.72}$$

If you compare this method of setting out with that given on page 170 you will see that it differs only in the last line.

Summary

Remember that 1 mole of any gas
(a) contains 6.02×10^{23} molecules,
(b) has a mass in grams which is numerically equal to its relative molecular mass,
(c) has a volume of $22.4 \, dm^3$ at s.t.p.

Questions on Chapter 10

Note: in your answers to the following questions, assume that the molar volume of a gas under room conditions of temperature and pressure is $24 \, dm^3 \, mol^{-1}$, while at s.t.p. it is $22.4 \, dm^3 \, mol^{-1}$.

1   What volume of oxygen will be required for *complete* combustion of the following, and what volume of gaseous products will be formed? (All volumes measured at s.t.p.) (a) $10 \, cm^3$ of hydrogen sulphide, (b) $10 \, cm^3$ of carbon monoxide, (c) $10 \, cm^3$ of methane.

2   State Gay-Lussac's law of gaseous volumes. When $30 \, cm^3$ of carbon monoxide were exploded with $50 \, cm^3$ of oxygen and the products allowed to cool to the original temperature and pressure, the volume of the residual gas was $65 \, cm^3$, of which $35 \, cm^3$ was oxygen. Show that these figures are in accordance with Gay-Lussac's law.

3   A flask holds $0.02 \, g$ of hydrogen. At the same temperature and pressure it holds $1.42 \, g$ of gas X, $0.88 \, g$ of gas Y and $0.40 \, g$ of gas Z. Calculate the relative molecular masses of X, Y and Z.

4   Excess dilute hydrochloric acid is added to (a) $8.0 \, g$ of calcium carbonate, (b) $3.25 \, g$ of zinc. Name the gases given off and calculate the volume that each would occupy at s.t.p.

5   Calculate the relative molecular masses of the gases, X, Y and Z from the following information: (a) $5.6 \, dm^3$ of X has a mass of $16 \, g$, (b) $4.48 \, dm^3$ of Y has a mass of $8 \, g$, (c) $3.2 \, dm^3$ of Z has a mass of $4 \, g$. (All volumes are at s.t.p.)

6   A gaseous compound consists of 86% carbon and 14% hydrogen by mass. At s.t.p. $3.2 \, dm^3$ of the compound has a mass of $6 \, g$. Calculate (a) its empirical formula, (b) its relative molecular mass, (c) its molecular formula.

7   Calculate the volume at s.t.p. of (a) $35.5 \, g$ of chlorine molecules, (b) $5 \, g$ of hydrogen molecules, (c) $2.2 \, g$ of carbon dioxide, (d) $2.8 \, g$ of carbon monoxide.

8   What is the mass of (a) $11.2 \, dm^3$ of oxygen molecules, (b) $5.6 \, dm^3$ of sulphur dioxide, (c) $4.48 \, dm^3$ of ammonia, (d) $2.8 \, dm^3$ of hydrogen sulphide? (All volumes are at s.t.p.)

9   $60 \, cm^3$ of an oxide of nitrogen has a mass of $0.11 \, g$. When it is passed over heated copper, the metal is oxidised to copper(II) oxide and the gas reduced to nitrogen. The volume of this nitrogen, when cooled to the original temperature and pressure, is also $60 \, cm^3$. Calculate the molecular formula of the oxide of nitrogen. (1 mole of gas occupies $24 \, dm^3$ under the conditions of the experiment.)

10   $80 \, cm^3$ of ammonia gas, NH_3, were mixed with $30 \, cm^3$ of chlorine gas, Cl_2, at room temperature and pressure. The gases reacted completely so that no ammonia

or chlorine remained. A small quantity of solid ammonium chloride, NH_4Cl, was formed together with 10 cm^3 of nitrogen gas, N_2.

(a) Sketch the apparatus which could have been used to obtain this information about the volumes of ammonia, chlorine and nitrogen.

(b) Use the information about volumes of gases to write an equation for the reaction between ammonia and chlorine. You must show clearly how you have used the information. (L)

11   State (i) Gay-Lussac's law, (ii) Avogadro's law. Describe an experiment that illustrates Gay-Lussac's law.

X is a gaseous compound of silicon and hydrogen; its density was found to be 1.33 g dm^{-3}. When 100 cm^3 of X was completely decomposed, 200 cm^3 of hydrogen was obtained. (One mole of gas occupies 24 dm^3 at the conditions used in these determinations.)

(a) What is the relative molecular mass of X?

(b) Show how Avogadro's law enables you to deduce the number of hydrogen atoms in one molecule of X.

(c) What is the formula of X?

(d) X burns readily in air. What products would you expect to obtain under these conditions? (AEB)

12   (a) Calculate which of the following will give more hydrogen gas with excess of an acid, showing your calculation clearly: (i) 0.243 g magnesium, (ii) 0.327 g zinc.

(b) Devise an experiment to measure which of the above will produce more hydrogen gas. (SCE)

13   Calculate the minimum volume of hydrogen at s.t.p. required to convert 3.18 g of copper(II) oxide to copper metal. What mass of water would be formed? (SUJB)

14   (a) Calculate the mass of 1 dm^3 of oxygen, in g, under room conditions.

(b) The mass of 1 dm^3 of one gaseous member of the alkane series of hydrocarbons (general formula C_nH_{2n+2}) is $1\frac{5}{6} \text{ g}$ under room conditions. Deduce the formula of the gas.

(c) Calcium and phosphorus combine, under suitable conditions, to form a white solid compound. In an experiment, it was found that 9.10 g of this compound could be obtained from 6.00 g of calcium. Find the simplest formula for the compound. The white solid reacts with water to produce calcium hydroxide and the gas phosphine, PH_3. Write the equation for this reaction, employing the formula you have deduced above, and from your equation find the volume of phosphine, measured under room conditions, which should be produced from 0.1 mole of the solid compound. (W)

15   It is found that 4 g of oxygen, 7 g of the gaseous compound butene and 10.5 g of the gaseous element krypton all occupy the same volume at the same temperature and pressure.

(a) What fraction of a mole is 4 g of oxygen molecules?

(b) Calculate the relative molecular masses of butene and krypton.

(c) The empirical (simplest) formula of butene is CH_2. What is its molecular formula?

(d) The relative atomic mass of krypton is 84. What is the atomicity of krypton?

(e) Write a possible structural formula for butene.

(f) Give the name and structural formula of another compound having the empirical formula CH_2.

(g) Write the equation for the burning of butene in oxygen to give carbon dioxide and water.

(h) If 10 cm^3 of butene are burnt according to your equation, calculate the volume of oxygen used and the volume of carbon dioxide formed (all volumes at the same temperature and pressure). (C)

16 (a) When silver cyanide (formula AgCN) is heated in the absence of air, the only products are a colourless gaseous compound X and metallic silver.

(b) Compound X is found to have a density of $2.32\,g\,dm^{-3}$ at s.t.p.

(c) When a mixture of $20\,cm^3$ of X and $40\,cm^3$ of oxygen is ignited, a mixture of $40\,cm^3$ of carbon dioxide and $20\,cm^3$ of nitrogen is formed (all volumes at the same temperature and pressure).

Answer the following questions:

(i) What elements are present in the gaseous compound X?

(ii) Calculate the relative molecular mass of X.

(iii) How many molecules of carbon dioxide and of nitrogen are formed when one molecule of X reacts with oxygen?

(iv) What is the molecular formula of X?

(v) Write equations for the action of heat on silver cyanide and for the reaction between X and oxygen.

(vi) Suggest an experiment by which you could determine the percentage of carbon dioxide in a mixture of carbon dioxide and nitrogen. (C)

11 Electrochemistry

Many chemical reactions involve the transfer of electrons to form ions (8.1) and it seems likely that the ease with which atoms lose or gain electrons will have some bearing on the chemical activity of the element concerned. The experiments which follow are designed to test this idea and to investigate how electricity (which, like heat, is a form of energy) may bring about some chemical changes and be produced by others.

Experiments

11.1.a To determine the relative ease with which metals form positive ions in solution ★

If a rod of metal is dipped into water, some of its atoms give up electrons and the resulting positive ions go into solution. The electrons left behind on the undissolved metal give it an electric charge (positive or negative?). The size of this charge may be compared with that on a rod of some other element, in this case carbon, by connecting a voltmeter between the two and measuring the potential difference set up. The rods which connect the solution to the external circuit are referred to as *electrodes*.

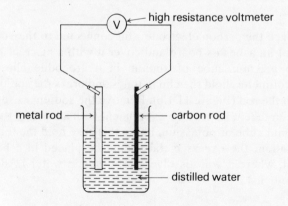

Figure 11.1.a

PROCEDURE Decide whether zinc is likely to become positively or negatively charged compared with carbon and then connect the electrodes to the correct terminals on the voltmeter. (This must be a high resistance instrument with range 0–3 volt.) Dip the electrodes into a beaker of distilled water and note the voltage registered. Repeat with pieces of magnesium, copper, iron, silver, lead and aluminium. You may have found earlier in your course that the reactivity of aluminium is greatly increased by dipping it in mercury(II) chloride solution for about 30 seconds. Do this, rinse and take another reading. (**Caution:** wash your hands thoroughly – mercury(II) chloride is poisonous.) Write a list of the metals in order of decreasing voltages.

Questions

1 Are the metals in your list in order of increasing or decreasing ease of forming positive ions in solution?
2 What is the relationship, if any, between the ease of forming ions and chemical reactivity?
3 Where would you expect potassium and sodium to come in the list?
4 Which of the voltage readings for aluminium have you used and why?

11.1.b   To place calcium, sodium and potassium in the list ★ ★

This experiment must not be performed by pupils.
Calcium, sodium and potassium cannot be used in Experiment 11.1.a (why not?) but readings of voltage may be made for these metals by modifying the experiment as follows.

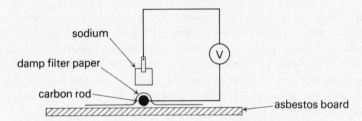

Figure 11.1.b

PROCEDURE   Place the carbon electrode, still connected to the voltmeter, in the centre of an asbestos board and cover it with a piece of damp filter paper. Grip a 3 mm cube of sodium in a crocodile clip, also connected to the voltmeter, hold the clip in tongs and press the sodium on to the paper above the rod (Figure 11.1.b). Remove the sodium as soon as the reading has been taken. Repeat the experiment with a calcium turning and then with a 3 mm cube of potassium but in this case hold the tongs at arm's length. (**Caution:** the excess metal should be placed in a beaker containing ethanol, where it will dissolve safely, *not* in water.) Add the three metals to your list.

Questions

1 Was your prediction about the positions of potassium and sodium in the list correct?
2 Can you see a connection between the position of a metal in your list and in the periodic table? Look for a general relationship concerning (a) the group number, and (b) the position of the metal in the group. Try to explain the relationships in terms of the number of electrons lost by each atom in forming an ion, and also in terms of the relative sizes of the atoms.

11.1.c Displacement of metals from solution

PROCEDURE Place a clean iron nail in a test tube containing 5 cm³ of copper(II) sulphate solution and leave it for about three minutes. Remove the nail, examine it carefully and try to answer questions 1 and 2 below. Repeat the experiment, using as many different combinations of metals and solutions as possible (e.g. copper and iron(II) sulphate solution, magnesium and copper(II) sulphate solution, copper and zinc sulphate solution) but do not, of course, use metallic potassium or sodium. Record your results as shown in Table 11.1.c.

Reactants	Reaction	Products
	$\checkmark$ or $\times$	

Table 11.1.c The reaction between metals and solutions

Questions

1 What do you think was deposited on the iron nail in the first experiment?
2 Copper(II) sulphate solution contains positively charged copper ions. What must have happened to some of them? Since electrons cannot just appear and disappear, what must have happened to some of the iron atoms of the nail?
3 Write an ionic equation for the reaction.
4 From the results of all the experiments, what general rule can you deduce to connect the displacement from solution of one metal by another with the relative positions of the two metals in the electrochemical series? (The electrochemical series is the name given to the list which you obtained in Experiments 11.1.a and 11.1.b).
5 Which ions seem to be the easiest to discharge to form atoms, those of metals at the top of the electrochemical series, or those at the bottom?
6 Write ionic equations for all the reactions which took place.

11.1.d Displacement of hydrogen from solution

You have probably already learned that dilute acids contain hydrogen ions and that some metals displace hydrogen when placed in an acidic solution. It should now be an easy matter for you to relate these facts to the relative positions of the elements in the electrochemical series.

PROCEDURE Add a 5 cm length of magnesium ribbon to a test tube containing about 5 cm³ of dilute sulphuric acid and collect any gas evolved over water. Identify the gas by a simple test and repeat the experiment using (a) a spatula measure of iron filings, (b) a piece of copper foil, in place of the magnesium ribbon.

Questions

1 What was the gas and how did you identify it?
2 Deduce the relative positions of the metals and hydrogen in the electrochemical series and explain the reaction, or lack of it, in terms of electron transfer between atoms and ions.
3 Write ionic equations for any reactions which took place.

11.1.e Displacement of non-metals from solution

So far, your experiments have been concerned with the formation of positive ions in solution. The halogens, chlorine, bromine and iodine, are used in this experiment to see if the rules and arguments developed up to now can be applied to elements which form negative ions.

PROCEDURE To separate test tubes containing 5 cm³ of solutions of (a) sodium bromide, (b) sodium iodide, add a little chlorine water. Shake each tube and observe the result. You will probably find it easier to identify the products if you add sufficient tetrachloromethane to just fill the rounded part of the tube, cork the tube, shake it vigorously and allow the contents to settle. (Halogens are more soluble in tetrachloromethane than in water, bromine giving a brown colour and iodine a violet one.) Repeat the experiment, but this time add 1 cm³ of bromine water to 5 cm³ of solutions of sodium chloride and sodium iodide in test tubes. Record your results in the same way as in Experiment 11.1.c.

Questions

1 Explain your results in terms of electron transfer between atoms and ions, remembering that the halogens form *negative* ions.
2 Write ionic equations for any reactions which took place.
3 The electrochemical series is a list of elements arranged in order of *decreasing* ease of forming *positive* ions in solution. However, as the tendency for atoms to donate electrons decreases, so their tendency to accept electrons increases. Thus, an alternative way of looking at the electrochemical series is to regard it as a list of elements arranged in order of *increasing* ease of forming *negative* ions in solution. Use similar arguments to those in Experiment 11.1.c to place the halogens in order of increasing ease of forming negative ions and add them to your original list.
4 What rule can you make connecting the displacement from solution of one non-metal by another with the relative positions of the two in the electrochemical series?
5 Which non-metal ions seem to be the easiest to discharge to form atoms, those higher in the electrochemical series, or those lower down?

11.2.a To investigate what happens to ions when an electric current passes through an electrolyte ★ ★

This experiment requires a *low current* 350 volt d.c. source. The Shandon Unikit Power Supply, which provides a maximum current of 10 mA, is suitable.

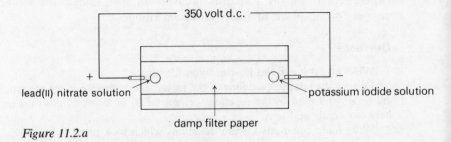

Figure 11.2.a

PROCEDURE   (a) Mix 1–2 cm³ of 1 M lead(II) nitrate solution with an equal volume of 1 M potassium iodide solution in a test tube and note the result.

(b) Moisten a 7 × 1 cm strip of filter paper with tap water and lay it on a microscope slide. Place a drop of the lead(II) nitrate solution near one end of the paper and a drop of the potassium iodide solution near the other end. Connect the 350 volt d.c. supply as shown in Figure 11.2.a and switch on the current (**care**). Watch the paper carefully.

(c) Repeat (b) with a fresh piece of filter paper but this time reverse the electrical connections. Look for signs of change around the crocodile clips.

Questions

1 What did you see in part (a)? Name the product.
2 What did you observe on the filter paper in part (b)? What do you think this was?
3 Can you explain the result in terms of movement of ions?
4 What did you observe in part (c)? What do you think the products were?
5 Explain the result in terms of ions, atoms and transfer of electrons.
6 Bearing in mind the results of other examples of electrolysis which you have seen (e.g. Experiment 8.1.a), what general conclusions can you draw concerning the behaviour of ions during electrolysis?

11.2.b   To investigate the relative conducting powers of some solutions ★

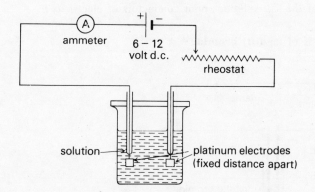

Figure 11.2.b

PROCEDURE   (a) Set up the circuit as shown in Figure 11.2.b and dip the platinum electrodes into 100 cm³ of 1 M sulphuric acid in a 250 cm³ beaker. Adjust the rheostat to give a current of 300 mA and then remove the electrodes from the acid and rinse them with distilled water. Without altering the rheostat, take a reading of current for the following aqueous solutions, all having the same concentration as the sulphuric acid: hydrogen chloride, nitric acid, ethanoic acid, potassium hydroxide, sodium hydroxide, ammonia, sodium chloride, potassium nitrate, ammonium ethanoate, sugar. You should also try using distilled water, trichloromethane, a solution of *dry* ammonia in trichloromethane, methylbenzene, and a solution of *dry* hydrogen chloride in methylbenzene. (**Caution:**

trichloromethane is an anaesthetic, methylbenzene is inflammable.) Record your results under the headings shown in Table 11.2.b.

Good/quite good conductors	Poor conductors	Non-conductors

Table 11.2.b   The conducting powers of some solutions

(b) The procedure for this part of the experiment is the same as for parts (a) and (b) of Experiment 5.5.a (page 70).

Questions

1   All the beakers in (a), except the last five mentioned, contain the same number of moles of dissolved substance. What can you say about the relative numbers of *ions* present in the solutions in the three groups in the table?

2   Did the solutions of ammonia and hydrogen chloride in non-aqueous solvents behave as if they contained ions?

3   Did aqueous solutions of these gases conduct electricity?

4   In which cases in (b) did the thermometer indicate that a reaction was taking place between gas and solvent?

5   Do you think that the reactions in question 4 produced ions? Give a reason for your answer.

6   Which salt is made from an acid and an alkali that both appear in the 'poor conductors' group? What type of conductor is this salt?

7   Were any of the salt solutions poor conductors of electricity?

11.2.c   The electrolysis of lead(II) bromide ★ ★

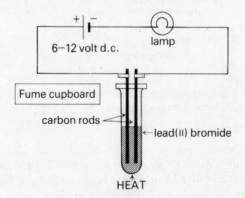

Figure 11.2.c

PROCEDURE   Half-fill a boiling tube with powdered lead(II) bromide and support it by means of a clamp in a fume cupboard (poisonous vapours will be formed during the experiment). Complete the rest of the apparatus as shown in Figure 11.2.c—the lamp will not light unless a current is flowing. Heat the tube *gently* to melt the solid and observe the lamp carefully. As soon as it lights, remove the lamp from the circuit to allow a larger current to flow. Look for bubbles around the electrodes and note the colour of any gas produced. After about five minutes, remove the electrodes and carefully pour the molten lead(II) bromide into another boiling tube. Watch out for another product in the bottom of the first tube and catch it in a spatula as it is poured out.

Questions

1 Did solid lead(II) bromide conduct electricity?
2 Explain the changes in conducting properties as the solid melted.
3 Which gas do you think was produced and at which electrode was it liberated?
4 What do you think the other product was and where was it liberated?
5 Explain the chemical changes which occurred at (a) the anode (positive electrode), (b) the cathode (negative electrode) in terms of ions, atoms, and transfer of electrons.

11.2.d   The electrolysis of dilute sulphuric acid using platinum electrodes ★

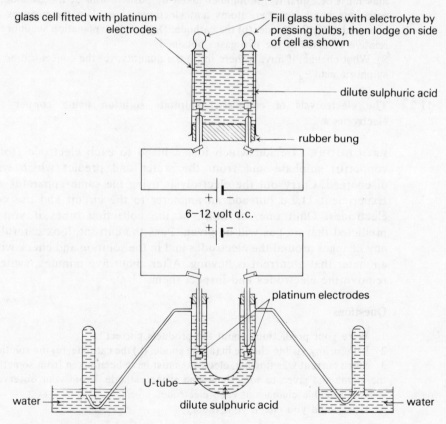

Figure 11.2.d

PROCEDURE   Use one of the voltameters shown in Figure 11.2.d to electrolyse dilute sulphuric acid. In the case of the U-tube version, allow the first few bubbles of gas to escape into the atmosphere before placing the collecting tubes in position (why?). Note the relative volumes of the gases in the tubes as the experiment proceeds and, when sufficient has been collected, test the one given off at the anode with a glowing splint and the one given off at the cathode with a flame.

Questions

1 Which ions do you think were attracted to the anode from (a) the sulphuric acid, (b) the water?

2 What was the gas liberated at the anode? How could you tell?

3 Which ions do you think were discharged at the anode? (Consult your answer to question 5, Experiment 11.1.e, and look up the positions of the ions you have selected in the electrochemical series on page 201.)

4 Something else, in addition to the gas which you tested, must have been formed at the anode when these ions were discharged. Which common compound is the most likely one?

5 Which ions would you expect to be attracted to the cathode from (a) the sulphuric acid, (b) the water?

6 What was the gas liberated at the cathode? How was it identified?

7 Bearing in mind that the number of electrons given up at the anode by negative ions must be equal to the number taken by positive ions at the cathode, write equations involving ions, atoms and electrons to explain the changes which occurred at (a) the anode, (b) the cathode. Does your explanation account for the relative volumes of the two gases obtained?

8 What change, if any, is there in (a) the quantity, (b) the concentration of the sulphuric acid?

11.2.e The electrolysis of copper(II) sulphate solution using copper electrodes ★

PROCEDURE   Decide which ions will go to each electrode from the copper(II) sulphate and from the water and predict which will be discharged. Carry out the electrolysis using the same apparatus as for Experiment 11.2.d but add an ammeter to the circuit and use copper electrodes. Omit one, or both, of the collecting tubes if you have predicted that no gas will be given. Pass the current, look carefully for any changes around the electrodes and in the solution and check with the ammeter that a current is flowing. After about five minutes, switch off, remove the electrodes and inspect them.

Questions

1 Were your predictions about the products correct?

2 Is there any visible change in (a) the anode, (b) the cathode, (c) the solution?

3 Since current was flowing, electrons must have been taken from something at the anode and given to something else at the cathode. From your observations, suggest possible changes at each electrode.

4 How could you check your theory?

11.2.f The electrolysis of copper(II) sulphate solution using copper electrodes – a quantitative experiment ★

PROCEDURE   Find the masses of two pieces of copper foil of size $12 \text{ cm} \times 3 \text{ cm}$. Place them on opposite sides of a 100 cm^3 beaker and fold the top of each one over the rim of the beaker to hold it in position (see Figure 11.2.f). Pour in copper(II) sulphate solution (mass concentration 12.5 g hydrated salt dm^{-3}) to a depth of 5 cm and then set up the circuit. Switch on,

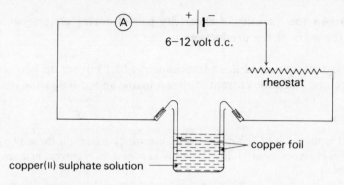

Figure 11.2.f

immediately adjust the current to 0.15 A and note the time. Keep this current steady for 45 minutes and then switch off the supply. (The time will not be needed in this experiment but will be useful in a calculation which you will make later on.) *Carefully* remove the electrodes from the beaker, immerse them in distilled water and then in propanone to assist drying. Find the mass of each one when completely dry.

N.B. If you handle the electrodes carelessly, you will dislodge the deposit of copper and ruin the results.

Questions

1 Which electrode showed a gain in mass and which one showed a loss?
2 How did the changes in mass of the two electrodes compare with one another?
3 Did the results fit the theory which you produced for the previous experiment? If they did not, suggest a new theory.
4 What change was there in (a) the quantity of copper(II) sulphate, (b) the concentration of the solution?
5 If you were given a large piece of impure copper and a small piece of pure copper, how could you obtain a large piece of the pure metal?

11.2.g   The electrolysis of sodium chloride solution using carbon electrodes ★

PROCEDURE   Predict what the products of this electrolysis will be, as you did for Experiment 11.2.e. Proceed as before but use carbon electrodes and a solution of sodium chloride which has been coloured with universal indicator. Look for a colour change around each of the electrodes and identify any gases by means of suitable tests. (The gas from the anode must be collected over a saturated solution of sodium chloride, not over water.)

Questions

1 What happened to the universal indicator at the anode?
2 Which ions must have been discharged?
3 Does the result agree with your prediction? If it does not, try to think of a reason for this. (Hint: are all the ions present in equal numbers?)
4 Which gas was given at the cathode and what happened to the universal indicator around this electrode?
5 Which ions must have caused this colour and where must they have come from?

11.3.a **To find how the quantity of electricity passed during electrolysis affects the mass of the products ★**

PROCEDURE   Proceed as in Experiment 11.2.f but set up two independent circuits. Keep the current at 50 mA in one and at 100 mA in the other.

Questions

1   What is the ratio of the quantities of electricity passed in these two experiments and in Experiment 11.2.f? (Quantity in coulombs = current in amps × time in seconds.)
2   What is the ratio of (a) the masses of copper deposited on the cathodes, (b) the losses in mass of the anodes?
3   Do these results indicate that the size of the charge on each copper ion is (a) constant, (b) variable? Is this in agreement with the theory of atoms and ions which we have been using so far?

11.3.b **To find the quantity of electricity required to deposit one mole of lead atoms from molten lead(II) bromide ★ ★**

In Experiment 11.2.c it was found that the electrolysis of lead(II) bromide gave lead at the cathode and bromine at the anode. That experiment is modified here so that both the quantity of electricity and the mass of lead may be measured.

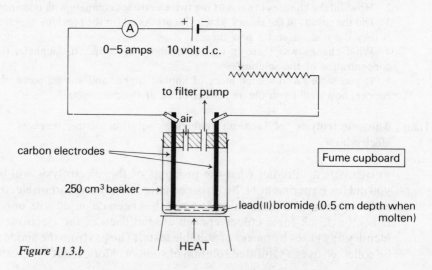

Figure 11.3.b

PROCEDURE   Set up the apparatus shown in Figure 11.3.b and heat the lead(II) bromide until it melts. Note the time and switch on the current. Immediately adjust the rheostat to give a reading of 3.0 A on the ammeter and turn on the pump to carry away the bromine vapour. Keep the current steady for 15 minutes and then switch off. Carefully pour off the molten lead(II) bromide into an evaporating basin, making sure that the lead remains in the original beaker. Hold the beaker at an angle so that a single bead of lead is formed. When cool, prise off the lead, remove any solid lead(II) bromide from it and find the mass of the clean metal.

Questions

1 What mass of lead was produced?
2 How many coulombs of electricity were passed?
3 Calculate the number of coulombs which would be required to deposit one mole of lead atoms (molar mass of lead = $207\,g\,mol^{-1}$).

11.3.c To find the quantity of electricity required to deposit one mole of silver atoms from an aqueous solution of silver nitrate ★

PROCEDURE Proceed as in Experiment 11.2.f but in this case use silver foil for the electrodes and a solution of silver nitrate (mass concentration $8.5\,g\,dm^{-3}$) as the electrolyte. Take great care not to dislodge the deposit of silver from the cathode at the end of the experiment.

Questions

1 Calculate the number of coulombs which would be required to deposit one mole of silver atoms (molar mass of silver = $108\,g\,mol^{-1}$).
2 If the loss in mass of the anode differs from the gain in mass of the cathode, which figure should you use? Give a reason for your answer.
3 From the results of Experiment 11.2.f, calculate the number of coulombs which would be required to deposit one mole of copper atoms (molar mass of copper = $63.5\,g\,mol^{-1}$).
4 Is there a simple ratio between the quantity of electricity required to deposit one mole of silver atoms and that required to deposit one mole of copper atoms?
5 How do these values compare with that for one mole of lead atoms in Experiment 11.3.b?
6 Can you explain the figures in terms of the relative charges on the various ions?
7 From your answers to questions 4, 5 and 6, what quantity of electricity would you expect to be required to liberate (a) one mole of hydrogen atoms, (b) one mole of chlorine atoms from solutions containing their ions?

11.3.d To find the quantities of electricity required to liberate one mole of hydrogen atoms and one mole of chlorine atoms from solutions containing their ions ★

The electrolysis of dilute hydrochloric acid gives hydrogen at the cathode and chlorine at the anode but much of the chlorine dissolves in the electrolyte. Errors from this source are minimised by adding sodium chloride to the acid and saturating it with chlorine before collecting any gas.

PROCEDURE Set up the apparatus as shown in Figure 11.3.d but without the syringes. Adjust the rheostat to give a current of 500 mA and allow it to pass for at least 5 minutes to saturate the electrolyte in the anode limb of the U-tube with chlorine. Switch off the supply, place the syringes in position, and then pass a steady current for 15 minutes. If necessary, adjust the syringe plungers to keep the surfaces of the electrolyte in the two limbs level; this will keep the gases at atmospheric pressure. Note the readings of gas volumes at 1-minute intervals.

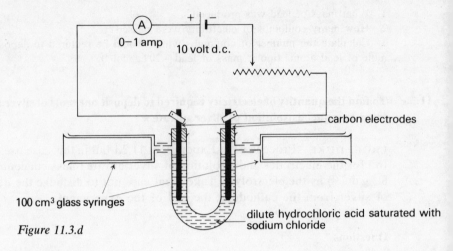

Figure 11.3.d

Questions

1 Plot a graph of volume of hydrogen against time. Do the results agree with Faraday's first law?
2 Calculate the number of coulombs of electricity passed during the whole 15-minute period.
3 The density of hydrogen is 0.000083 g cm^{-3} at room temperature and pressure. What mass of hydrogen was liberated during electrolysis?
4 How many coulombs would be required to liberate one mole of hydrogen atoms? (Molar mass of hydrogen atoms = 1 g mol^{-1}.)
5 The density of chlorine is 0.00296 g cm^{-3} at room temperature and pressure. What mass of chlorine was liberated during electrolysis?
6 How many coulombs would be required to liberate one mole of chlorine atoms? (Molar mass of chlorine atoms = 35.5 g mol^{-1}.)
7 Do your results agree with your predictions from the previous experiment?

11.4.a To obtain electrical energy from a chemical change ★

Electrolysis involves chemical changes brought about by the passage of an electric current. In this experiment we attempt to obtain an electric current from a chemical change.

PROCEDURE Decide whether zinc is likely to become more, or less, negatively charged than copper when dipped into a liquid and then connect a zinc rod and a piece of copper foil to the correct terminals of a high-resistance voltmeter (0–3 volt). Dip the electrodes into a beaker of dilute sulphuric acid so that about 6 cm × 5 cm of foil is immersed and note the voltage shown on the meter.

So far, you have only repeated Experiment 11.1.a using copper instead of carbon and acid instead of water. Now you should try to obtain an electric current from the apparatus. Disconnect the voltmeter (which takes very little current) and replace it with a 300 ohm resistor. Measure the voltage across the ends of the resistor as soon as it is connected and then make a note of the reading every 15 seconds for the first minute.

Take four more readings at intervals of 1 minute after this. Look carefully at both electrodes and if no change is seen on the copper foil, replace the 300 ohm resistor with a wire. This will mean that a much larger current is drawn (why?) and you should observe a change at the copper electrode. 'Switch off' the cell by disconnecting the external circuit and again look carefully at *both* electrodes.

Questions

1 Which metal had the greater negative charge?
2 What voltage was measured in the first instance?
3 Explain why an electric current flowed through the resistor. Was the current *steady*?
4 What did you observe at the copper electrode while the current was flowing?
5 Explain what happened at (a) the zinc electrode, (b) the copper electrode, in terms of electron transfer between atoms and ions. Write the overall equation for the reaction.
6 What disadvantage of this cell did you notice when the current was switched off?

11.4.b An improved cell (the Daniell cell) ★

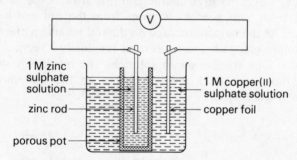

Figure 11.4.b

PROCEDURE Set up the cell shown in Figure 11.4.b and proceed as in the previous experiment.

Questions

1–4 As in the previous experiment.
5 Why are the solutions separated and why must the pot be porous?
6 What advantages has this cell over the one in Experiment 11.4.a?

11.4.c Galvanic couples and corrosion

A galvanic couple consists of two different metals in contact with one another and with an electrolyte.

PROCEDURE Coat a piece of zinc with copper by placing it in a solution of copper(II) sulphate for a few minutes. While you are waiting, add another piece of zinc to a test tube containing about 5 cm³ of dilute sulphuric acid and note the rate of evolution of hydrogen. Repeat this with the copper-coated zinc, collect the resultant gas over water and identify it.

Questions

1 What gas did you collect and how did you identify it?
2 Which of the two solids, zinc or zinc coated with copper, reacted faster with the acid?
3 Explain the second reaction in terms of electron transfer between atoms and ions. (The copper coating is porous so that essentially you have the same cell as in Experiment 11.4.a, but with no external circuit and with the electrodes in contact.)
4 If *any* two metals are placed in contact with one another in the presence of an electrolyte which one will dissolve, the one higher in the electrochemical series or the one lower down? (Consult your answer to question 3.)
5 Which metal will give the best protection to iron when plated on it, (a) zinc (galvanising), or (b) tin? (Tin is lower than iron in the electrochemical series.)
6 Why is it inadvisable to connect lead pipes to copper pipes when supplying drinking water in a house?

11.1 The electrochemical series

When a metal rod is dipped into water, some of its atoms go into solution as positive ions, leaving behind the electrons from their outermost shells. Thus the rod develops a negative charge and this attracts the ions back again; a few of these ions accept electrons from the rod and become atoms once more. As the negative charge on the rod builds up, the rate at which the ions return gradually increases until eventually it is equal to the rate at which new ions are going into solution. The result is a state of *dynamic equilibrium* (see 14.2) and the charge on the rod shows no further increase (Figure 11.1).

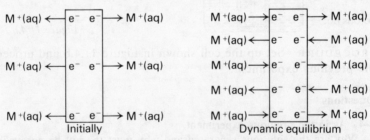

Figure 11.1

Metals do not all have the same tendency to form ions and thus different metals will develop different charges. Any attempt which is made to measure the potential difference between a metal and a solution will involve dipping a wire into the solution to connect it to the measuring instrument. This wire will, of course, develop a charge of its own. Thus it is necessary to choose a standard substance which will always produce the same charge, dip this into the solution and then compare the charge on it with that of the metal under investigation. In Experiment 11.1.a, a carbon rod is chosen as the standard electrode and the potential difference between it and various metals is measured with a high resistance voltmeter. The greater the tendency of a metal to throw off ions into solution, the higher will be the negative charge developed on it. Thus the potential differences measured will show the relative ease with which the metals form ions under these conditions. When the metals are listed in order of decreasing voltages the *electrochemical series* is obtained.

> The **electrochemical series** is a list of elements arranged in order of decreasing ease of forming positive ions in aqueous solution (or, alternatively, in order of increasing ease of forming negative ions in aqueous solution).

The order for some of the more commonly encountered elements is as shown below.

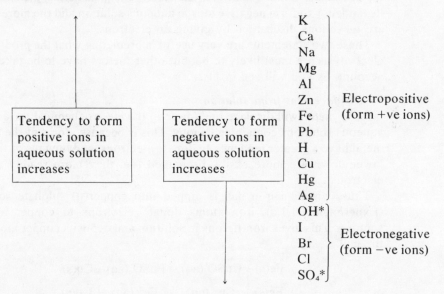

*OH and SO₄ radicals do not actually exist on their own but immediately decompose to form other products when electrons are removed from OH^- or SO_4^{2-} ions. However, the arguments concerning the discharge of ions, which are discussed a little later in this section, would place the radicals in the position shown.

Uses of the series

Several useful principles may be deduced from a consideration of the relative positions of the elements in the electrochemical series.

1 *Chemical activity*
(a) If you compare the electrochemical series with the reactivity series (20.1) you will see that, with the exception of calcium, the *metals* are in order of *decreasing* chemical activity. This shows that activity is connected with ease of forming ions.
(b) The *non-metals* in the electrochemical series are listed in order of *increasing* chemical activity, again showing that activity is connected with ease of forming ions.

In Chapters 19 and 20, ease of ionisation is related to position in the periodic table and many of the arguments given there may be applied here also. However, the electrochemical series was built up by finding the relative tendencies of atoms to form ions *in aqueous solution*. The fact that the positions of sodium and calcium in the reactivity series are reversed in the electrochemical series may be explained in terms of the interaction between water molecules and the ions.

2 Discharge of ions

Two general rules concerning the discharge of ions may be deduced from the definition of the electrochemical series.

(a) The *lower* a *metal* is in the electrochemical series, the smaller is its tendency to form positive ions in aqueous solution and the more easily are these ions discharged by accepting electrons.

(b) The *higher* a *non-metal* is in the electrochemical series, the smaller is its tendency to form negative ions in aqueous solution and the more easily are these ions discharged by giving up electrons.

These two statements are very useful in predicting what the products of electrolysis are most likely to be, but other factors have to be taken into account, as you will see in 11.2.

3 Displacement from solutions

(a) Any *metal* will displace one *lower* in the electrochemical series from aqueous solutions containing its ions. This is because atoms of the higher metal have a greater tendency to donate electrons, and the ions of the one lower down are more easily discharged (i.e. they more readily accept electrons).

Thus, when an iron nail is dipped into copper(II) sulphate solution (Experiment 11.1.c), iron atoms donate electrons to copper ions in solution. This gives iron(II) ions in solution and deposits copper atoms on the nail.

$$Fe(s) + CuSO_4(aq) \rightarrow FeSO_4(aq) + Cu(s)$$

or
$$Fe(s) + Cu^{2+}(aq) \rightarrow Fe^{2+}(aq) + Cu(s)$$

A similar argument explains why magnesium displaces hydrogen from dilute sulphuric acid in Experiment 11.1.d.

$$Mg(s) + H_2SO_4(aq) \rightarrow MgSO_4(aq) + H_2(g)$$

or
$$Mg(s) + 2H^+(aq) \rightarrow Mg^{2+}(aq) + H_2(g)$$

(b) Any *non-metal* will displace one *higher* in the electrochemical series from aqueous solutions containing its ions. This is because atoms of the lower non-metal have a greater tendency to accept electrons, and the ions of the one higher up are more easily discharged (i.e. more readily give up electrons).

Hence, in Experiment 11.1.e, where chlorine water is added to sodium iodide solution, chlorine atoms take electrons from iodide ions, giving chloride ions and iodine atoms. The iodine atoms then join together in pairs to form molecules.

$$Cl_2(aq) + 2NaI(aq) \rightarrow 2NaCl(aq) + I_2(aq)$$
or
$$Cl_2(aq) + 2I^-(aq) \rightarrow 2Cl^-(aq) + I_2(aq)$$

11.2   Electrolysis

Movement of ions

You have already learned in Chapter 8 that metals, hydrogen and the ammonium radical form positive ions while non-metals and most other radicals form negative ions. You should also be aware that the passage of an electric current through a solution causes it to decompose. Since electrons are flowing through the wires in the external circuit during electrolysis, it follows that they must be taken from the solution at one electrode and given back to it at the other. What, then, happens in the solution itself?

The answer to this question is provided by Experiment 11.2.a. In this experiment, a drop of lead(ɪɪ) nitrate solution and a drop of potassium iodide solution are placed at opposite ends of a strip of moist filter paper across which a potential difference is applied. Two different results are obtained, depending on which way round the electricity supply is connected; these are shown in Figure 11.2.1.

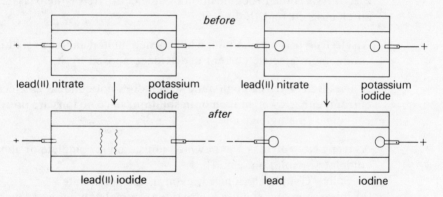

Figure 11.2.1

The results are easy to explain if we assume that the ions in the solution move towards the oppositely charged electrodes. Thus the yellow precipitate is lead(ɪɪ) iodide, formed when positively charged lead ions from the lead(ɪɪ) nitrate solution move towards the cathode at one end of the filter paper and meet negatively charged iodide ions from the potassium iodide solution moving towards the anode at the other end. If the electrodes are reversed, the ions move *apart*, giving a grey deposit of lead at the cathode and a brownish stain of iodine at the anode. The conclusion which we may draw from this is that during electrolysis, negative ions *give up* electrons at the anode and form atoms, while positive ions *take* electrons at the cathode, also forming atoms.

Strong and weak electrolytes

Experiment 11.2.b shows that the conducting powers of solutions differ widely, even when the concentrations of solute are the same. This variation can only be due to some of the solutions containing more ions than others.

When solids which have giant ionic lattices are dissolved in water, the ions separate and diffuse throughout the liquid, thus providing a solution which is a good conductor of electricity. Such substances are examples of *strong electrolytes*.

Covalent substances do not, in general, conduct electricity either when molten or when in solution because no ions are present. Thus when the covalent gas, hydrogen chloride, is dissolved in a covalent liquid like methylbenzene, the resulting solution is a non-conductor. However, since an *aqueous* solution of hydrogen chloride is a *good* conductor (i.e. a strong electrolyte), water must cause the gas molecules to form ions as they dissolve in it (see 12.6). Some covalent substances, such as ammonia and ethanoic acid, react with water in a similar manner but form far fewer ions. They are examples of *weak electrolytes*, their solutions being poor conductors of electricity.

Definitions

Electrolysis is the decomposition of an electrolyte by the passage of an electric current through it.

An **electrolyte** is a compound which, when molten and/or in solution, conducts an electric current and is decomposed by it.

A **non-electrolyte** is a compound which does not conduct an electric current, either when molten or in solution, since no ions are present, e.g. sugar, methylbenzene.

A **strong electrolyte** is one in which ionisation is complete or almost complete in solution, e.g.
(a) Salts – completely ionised, even in the solid state.
(b) Potassium and sodium hydroxides – completely ionised, even in the solid state.
(c) Sulphuric, nitric and hydrochloric acids – covalent in the absence of water.

A **weak electrolyte** is one in which ionisation in solution is slight, the solution containing mainly unionised molecules, e.g.
(a) Organic acids such as ethanoic acid.

$$CH_3COOH(aq) \rightleftharpoons CH_3COO^-(aq) + H^+(aq)$$

(b) Ammonia solution.

$$NH_3(g) + H_2O(l) \rightleftharpoons NH_4^+(aq) + OH^-(aq)$$

(c) Water is a *very* weak electrolyte; one molecule in about 550 million ionises as shown.

$$H_2O(l) \rightleftharpoons H^+(aq) + OH^-(aq)$$

or $$2H_2O(l) \rightleftharpoons H_3O^+(aq) + OH^-(aq)$$

Negatively charged ions are attracted to the **anode** (positive electrode) and are called **anions**; positively charged ions are attracted to the **cathode** (negative electrode) and are called **cations**.

Since electrons are removed at the *anode*, the process taking place is *oxidation*; at the *cathode*, electrons are added and so *reduction* is occurring. (If you have not studied oxidation and reduction in terms of electron transfer, you will find the topic in 8.6.)

The electrolysis of lead(II) bromide (Experiment 11.2.c)

An electric current will not flow through solid lead(II) bromide. This is because the ions in each crystal are only free to vibrate about fixed points in the lattice. When the solid is melted and a potential difference applied, the ions do not move with the random motion usually associated with liquids but travel towards the oppositely charged electrodes (Figure 11.2.2). At the anode, electrons are removed from the bromide ions giving bromine atoms, which combine in pairs to form molecules. At the cathode, electrons are added to the lead ions forming lead atoms.

Changes   At anode   At cathode

$$2Br^-(l) \rightarrow 2Br(g) + 2e^- \cdots \rightarrow 2e^- + Pb^{2+}(l) \rightarrow Pb(l)$$

$$\downarrow$$
$$Br_2(g)$$

(oxidation)   (reduction)

N.B. The arrow linking the electrons is not intended to show that the same two electrons taken from the bromide ions at the anode are given to the lead ion at the cathode; it simply indicates that the number of electrons transferred at the two electrodes must be the same. If you think of the circuit outside the voltameter as a diving board packed with people (the electrons), then if two extra people step on to one end of the board (the anode), two others must fall off at the other end (the cathode)!

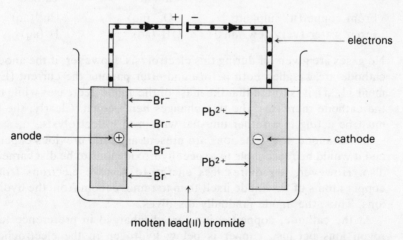

Figure 11.2.2 Movement of ions during the electrolysis of lead(II) bromide

The electrolysis of dilute sulphuric acid using platinum electrodes (Experiment 11.2.d)

	Ions attracted to anode	Ions attracted to cathode
From sulphuric acid	$SO_4^{2-}(aq)$	$2H^+(aq)$
From water (very few ions)	$OH^-(aq)$	$H^+(aq)$

When this experiment is carried out, one volume of oxygen is collected over the anode and two volumes of hydrogen over the cathode. We can explain these results as follows.

At the anode, hydroxide ions are discharged in preference to the sulphate ions because the former are higher in the electrochemical series. They give up electrons and form hydroxyl radicals, which are unstable and decompose into water and oxygen molecules as shown below.

At the cathode, hydrogen ions are discharged by accepting electrons. They become hydrogen atoms, which combine in pairs to form molecules.

Changes   At anode   At cathode

$$4OH^-(aq) \rightarrow 4OH(aq) + 4e^- \quad | \quad 4e^- + 4H^+(aq) \rightarrow 4H(g)$$
$$\downarrow \qquad\qquad\qquad\qquad\qquad\qquad \downarrow$$
$$2H_2O(l) + O_2(g) \qquad\qquad\qquad 2H_2(g)$$

(oxidation)   (reduction)

N.B. The ratio of molecules of oxygen to molecules of hydrogen is 1:2 so that, in effect, water is being electrolysed. As the experiment proceeds, more water molecules dissociate to replace the ions which have been discharged; thus, although the *quantity* of sulphuric acid is unchanged, its *concentration* increases as the water is used up.

The electrolysis of copper(II) sulphate solution using copper electrodes (Experiment 11.2.e)

	Ions attracted to anode	Ions attracted to cathode
From copper(II) sulphate	$SO_4^{2-}(aq)$	$Cu^{2+}(aq)$
From water (very few ions)	$OH^-(aq)$	$H^+(aq)$

No gases are given off during this electrolysis. However, if the anode and cathode are weighed both before and after passing the current (Experiment 11.2.f), it is found that the mass of the anode decreases while that of the cathode increases, the two changes being equal. Clearly, the anode must be acting in a rather unusual way in this electrolysis.

At the anode, the same ions are present as in the previous experiment and it would be reasonable to expect hydroxide ions to be discharged here also. However, it requires less energy to remove electrons from the copper atoms of the anode itself than to remove them from the hydroxide ions. Thus, the anode gradually dissolves.

At the cathode, copper(II) ions are discharged in preference to hydrogen ions because copper is below hydrogen in the electrochemical series, and so the cathode becomes coated with copper.

Changes	At anode	At cathode
	$Cu(s) \rightarrow Cu^{2+}(aq) + 2e^-$	$2e^- + Cu^{2+}(aq) \rightarrow Cu(s)$
	(oxidation)	(reduction)

N.B. The concentration of the solution remains unchanged but copper is transferred from the anode to the cathode. This process may be used to purify copper by making the impure metal the anode and using a thin sheet of pure metal as the cathode. As the anode dissolves and the cathode grows, the impurities dissolve or sink to the bottom of the voltameter.

The electrolysis of sodium chloride solution (coloured with universal indicator) using carbon electrodes (Experiment 11.2.g)

	Ions attracted to anode	Ions attracted to cathode
From sodium chloride	$Cl^-(aq)$	$Na^+(aq)$
From water (very few ions)	$OH^-(aq)$	$H^+(aq)$

Equal volumes of chlorine and hydrogen are produced by the electrolysis of a concentrated solution of sodium chloride, the chlorine being collected over the anode and the hydrogen over the cathode. The formation of chlorine in this experiment introduces another complication which must be considered when trying to predict what the products of a particular electrolysis will be.

At the anode, chloride ions are discharged in preference to hydroxide ions. You might well expect the reverse to take place because chlorine is *lower* in the electrochemical series than 'hydroxide'. However, the chloride ions are present in far greater numbers than the hydroxide ions and this results in the liberation of chlorine and the bleaching of the universal indicator. In very dilute solutions, oxygen may be given by the discharge of hydroxide ions as expected.

At the cathode, hydrogen ions are discharged in preference to sodium ions since hydrogen is much lower in the electrochemical series than sodium. The removal of hydrogen ions, which come from water molecules, leaves an excess of hydroxide ions and these turn the universal indicator blue.

Changes	At anode	At cathode
	$2Cl^-(aq) \rightarrow 2Cl(g) + 2e^-$	$2e^- + 2H^+(aq) \rightarrow 2H(g)$
	$\downarrow$	$\downarrow$
	$Cl_2(g)$	$H_2(g)$
	(oxidation)	(reduction)

N.B. The volumes of gases collected will probably not be in the ratio of 1:1 because chlorine dissolves to some extent in water.

Predicting the products of the electrolysis of aqueous solutions

You will have discovered by now that the products formed during the electrolysis of aqueous solutions depend upon three factors, (a) the relative positions of the ions in the electrochemical series, (b) the material of the electrodes and (c) the relative concentrations of the ions in solution. It is possible that a combination of (a) and (c) will result in more than one product being formed at a particular electrode; for example, when a fairly dilute aqueous solution of sodium chloride is electrolysed, both hydroxide *and* chloride ions are discharged, giving water, oxygen and chlorine.

The points which have been discussed in this chapter cover most of the examples of electrolysis which you are likely to encounter at this level. Using the knowledge which you have gained, decide which ions will go to each electrode and predict what the products are likely to be in each of the electrolyses listed below. Check your predictions by experiment where possible.

1 Molten sodium chloride with carbon electrodes.
2 Copper(II) sulphate solution with platinum electrodes.
3 Dilute sulphuric acid with copper electrodes.
4 Sodium sulphate solution with platinum electrodes.
5 Sodium hydroxide solution with platinum electrodes.

An alternative theory to explain the electrolysis of aqueous solutions

In the electrolysis of dilute sulphuric acid using platinum electrodes, we explained the formation of oxygen at the anode in terms of the discharge of hydroxide ions from the water. Similarly, the formation of hydrogen at the cathode in the electrolysis of sodium chloride solution was accounted for by the discharge of hydrogen ions from the water. However, only one water molecule in about 550 million actually dissociates into ions (which is why pure water is virtually a non-conductor of electricity). Why, then, should these few ions be discharged so much more quickly from solutions than they are from pure water?

There is an alternative theory which avoids this problem. It suggests that electrons can be taken or given up at the electrodes by water *molecules*. These molecules will, of course, be present in far greater numbers than any of the ions in the solution.

At the anode, the change would be

$$2H_2O(l) - 4e^- \rightarrow 4H^+(aq) + 2O(g)$$
$$\downarrow$$
$$O_2(g) \longrightarrow$$

volume of oxygen : volume of hydrogen $= 1:2$

and at the cathode,

$$4H_2O(l) + 4e^- \rightarrow 4OH^-(aq) + 4H(g)$$
$$\downarrow$$
$$2H_2(g) \longrightarrow$$

You may be wondering why it is that *pure* water does not conduct electricity very well, even though these molecules are in contact with the electrodes. The answer, as before, is that there are hardly any ions present to carry the current.

You will probably find it confusing to try to learn two theories but, since both explain the experimental results, which one should you choose? At present, opinion among chemists is divided as to which of the two is the more acceptable. Your teacher will advise you in your choice.

Applications of electrolysis

1 THE MANUFACTURE OF ELEMENTS A number of the more reactive metals are manufactured by electrolysis because chemical reduction of their compounds would be difficult or, in some cases, impossible. Sodium and magnesium, for example, are produced by electrolysing their molten chlorides, while aluminium is obtained from a solution of its oxide in molten cryolite (20.4). Chlorine is extracted from aqueous sodium chloride solution using a cell with a *mercury* cathode, sodium hydroxide being produced at the same time (20.2).

2 PURIFICATION OF METALS Copper is purified by electrolysing aqueous copper(II) sulphate solution with blocks of the impure metal as anodes and thin sheets of pure metal as cathodes. Pure copper is transferred from the anodes to the cathodes (see page 206). Zinc may be purified in a similar manner.

3 ELECTROPLATING A metal is often plated with another one to protect it from corrosion or to improve its appearance. You will no doubt be familiar with the chromium plating on the steel parts of cars and bicycles, which achieves both of these aims.

The principles of electroplating are quite simple. The article to be plated is made the cathode in a voltameter, the anode is made of the plating metal, and the electrolyte is a solution containing its ions. When a current passes, the plating metal is transferred from the anode to the cathode as in (2). However, careful control of conditions is necessary if successful results are to be obtained.

4 ANODISING OF ALUMINIUM If an aluminium article is made the anode during the electrolysis of dilute sulphuric acid, the oxygen produced will reinforce the coating of aluminium oxide already present on the surface of the metal. This coating resists corrosion and will adsorb dyes to give a decorative finish. You have probably seen such things as saucepan lids which have been coloured in this way.

Classification of substances by conduction of electricity

At various points in your study of chemistry you have seen how substances may be placed in groups according to certain properties, and you should have found that simple classifications help you to understand and remember the basic principles of the subject. You probably already know that the way in which a substance behaves when it is connected to a source of electric current is a useful clue to its structure (8.3). This short summary is a reminder of topics found in other parts of the book and you should refer to the relevant sections for a more detailed discussion.

Metallic elements conduct electricity when solid or molten but are not decomposed by it. The current is carried by loosely held electrons, which move from one atom to the next (8.5).

Non-metallic elements do not conduct electricity because their electrons cannot move from one atom to the next (carbon is a notable exception).

Electrolytes conduct electricity when molten and/or in solution (but not when solid) and are decomposed by it. The current is carried by ions, which are discharged at oppositely charged electrodes.

Non-electrolytes do not conduct electricity, either when molten or in solution, because no ions are present.

11.3 Faraday's laws of electrolysis

Faraday's first law

By using the ideas of atoms, ions and electrons it has been possible to explain the results of the experiments in this chapter. Thus the experiments provide evidence in favour of the theories proposed in Chapters 6 and 8. Further evidence is obtained if measurements are made of the masses of substances set free during electrolysis.

Suppose a certain number of electrons are passed through the external circuit during the electrolysis of copper(II) sulphate solution using copper electrodes, and the mass of copper liberated at the cathode is measured. Our simple ionic theory predicts that doubling the number of electrons will double the number of copper ions discharged, and hence double the mass of copper deposited. This result *is* obtained (Experiment 11.3.a), and so, once again, the theory seems to be correct.

The results of many experiments of this type were expressed in 1832 by Michael Faraday in his *first law of electrolysis*.

The mass of substance dissolved or liberated at an electrode during electrolysis is proportional to the quantity of electricity which passes through the electrolyte.

Quantity of electricity is not usually measured in numbers of electrons but in coulombs, one coulomb being passed when a current of one amp flows for one second. The quantity of electricity in coulombs which passes when a current I (in amps) flows for time t (in seconds) is then It, and if one coulomb deposits z g of a substance, then the mass in grams deposited by It is given by

$$m = zIt.$$

z is often referred to as the *electrochemical equivalent* of the element concerned.

Faraday's second law

By passing the *same* quantity of electricity through different electrolytes and measuring the masses of substances set free at the electrodes, Faraday in 1833 deduced his *second law of electrolysis*.

The masses of different elements liberated by the same quantity of electricity during electrolysis are proportional to the molar masses of their atoms divided by the number of charges per ion.

This too may be explained by using the simple ionic theory, as follows. If the electrolysis of silver nitrate solution is carried out using silver electrodes (Experiment 11.3.c) then eventually, if the anode is large enough, one mole of silver atoms will be deposited on the cathode. Since each Ag^+ ion requires 1 electron to discharge it, 1 mole of electrons will have to pass through the circuit. However, the same number of electrons will only discharge $\frac{1}{2}$ mole of copper ions from copper(II) sulphate solution because each Cu^{2+} ion has two positive charges. Thus the passage of one mole of electrons during *any* electrolysis will discharge *one mole divided by the number of charges per ion* of ions. This conclusion is in agreement with Faraday's second law.

The Faraday constant

The **Faraday constant** is the quantity of electric charge carried by one mole of electrons. Its value is $96\,500\,C\,mol^{-1}$ (coulombs per mole).

The number of moles of electrons which passes through the circuit during the discharge of one mole of any ions must be equal to the number of charges on each ion (i.e. must be numerically equal to the oxidation number of the ions). Thus, if we carry out experiments in which the masses of products liberated and the quantity of electricity which passes are measured, we can calculate the Faraday constant (see Experiments 11.3.b, 11.3.c and 11.3.d). Example 1 shows how this may be done.

Examples

1   One coulomb is found to liberate $0.001\,118$ g of silver, $0.000\,329$ g of copper and $0.000\,010\,45$ g of hydrogen. Use these results to calculate the Faraday constant. (Relative atomic masses: $Ag = 107.9$, $Cu = 63.54$, $H = 1.008$.)

	Ag	Cu	H
g coulomb^{-1}	0.001 118	0.000 329	0.000 010 45
mol of atoms coulomb^{-1}	$\dfrac{0.001\,118}{107.9}$	$\dfrac{0.000\,329}{63.54}$	$\dfrac{0.000\,010\,45}{1.008}$
coulomb (mol of atoms)$^{-1}$	$\dfrac{107.9}{0.001\,118}$	$\dfrac{63.54}{0.000\,329}$	$\dfrac{1.008}{0.000\,010\,45}$
	$= 96\,500$	$193\,000$	$96\,500$
mol of electrons (= number of charges per ion)	1	2	1

Thus the value of the Faraday constant is $96\,500\,C\,mol^{-1}$.

2   What mass of copper would be plated on the cathode from a solution of copper(II) sulphate by a current of 2 amps flowing for 15 minutes? (Relative atomic mass: $Cu = 63.5$; Faraday constant $= 96\,500\ C\ mol^{-1}$.)

$$Cu^{2+}(aq) + 2e^- \rightarrow Cu(s)$$

$$\text{mol } e^- : \text{mole } Cu(s) = 2:1 = 1:\tfrac{1}{2} \qquad \qquad \ldots A$$

$$\text{coulombs } = 2 \times 15 \times 60$$

$$\text{mol } e^- \quad = \frac{2 \times 15 \times 60}{96\,500}$$

$$\text{mol } Cu(s) = \frac{2 \times 15 \times 60}{96\,500} \times \frac{1}{2} \qquad \qquad \ldots \text{from A}$$

$$\text{g } Cu(s) \quad = \frac{2 \times 15 \times 60}{96\,500} \times \frac{1}{2} \times 63.5$$

$$= \underline{0.592}$$

Compare this method of setting out with that given on page 170.

3   A current was passed through two voltameters in series, one with platinum electrodes in dilute sulphuric acid and the other with copper electrodes in copper(II) sulphate solution. After a certain time, 0.009 g of hydrogen was collected in the first voltameter. What mass of copper was deposited on the cathode of the second voltameter? (Relative atomic masses: $H = 1$, $Cu = 63.5$.)

There are two steps in calculations of this type.
(a) Find the number of moles of electrons required to deposit the mass (or volume) given in the question.
(b) Use the answer in (a) to calculate the mass (or volume) of the second substance.

(a) $2H^+(aq) + 2e^- \quad \rightarrow H_2(g)$

$$\text{mol } H_2(g) : \text{mol } e^- = 1:2 \qquad \qquad \ldots A$$

$$\text{g } H_2(g) \qquad \qquad = 0.009$$

$$\text{mol } H_2(g) \qquad \quad = \frac{0.009}{2}$$

$$\text{mol } e^- \qquad \qquad = \frac{0.009}{2} \times 2 \qquad \ldots \text{from A}$$

$$= \underline{0.009}$$

(b) $Cu^{2+}(aq) + 2e^- \quad \rightarrow Cu(s)$

$$\text{mol } e^- : \text{mol } Cu(s) = 2:1 = 1:\tfrac{1}{2} \qquad \qquad \ldots B$$

$$\text{mol } e^- \qquad \qquad = 0.009$$

$$\text{mol } Cu(s) \qquad \quad = 0.009 \times \tfrac{1}{2} \qquad \ldots \text{from B}$$

$$\text{g } Cu(s) \qquad \qquad = 0.009 \times \tfrac{1}{2} \times 63.5$$

$$= \underline{0.286}$$

11.4 Obtaining electrical energy from chemical changes

The simple cell

In 1800, the Italian scientist, Alessandro Volta, invented an apparatus which produced the first continuous electric current. He experimented with several different arrangements but all of them consisted of pairs of metals in contact with an aqueous solution of an electrolyte. We can explain how the current was produced by considering the following example.

Suppose that a rod of zinc and one of copper are dipped into a beaker of water. Since zinc is higher than copper in the electrochemical series, it will throw off ions into the water more readily than copper and the electrons left behind will give it a greater negative charge (11.1). If the metals are now connected by a wire, electrons will flow from the zinc to the copper to equalise the charge on each rod.

A continuous flow of electrons (i.e. an electric current) can be obtained by replacing the water with a solution of an electrolyte, such as dilute sulphuric acid (Experiment 11.4.a). When this is done, hydrogen ions from the acid take the extra electrons from the copper, forming atoms, then molecules and finally bubbles of gas. Since hydrogen ions are being discharged by gaining electrons they are being reduced, and thus the copper must be acting as a *cathode.*

The reduction of the negative charge on the zinc means that more of it can dissolve as ions, and thus the supply of electrons to the copper is maintained. The zinc atoms are losing electrons (i.e. are being oxidised) and so the zinc rod is acting as an *anode* (Figure 11.4.1).

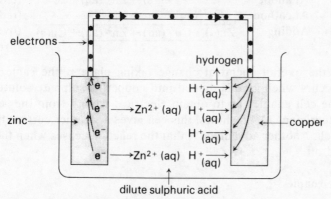

Figure 11.4.1 The simple cell

Changes	At anode	$Zn(s) \rightarrow Zn^{2+}(aq) + 2e^-$	(oxidation)
	At cathode	$2e^- + 2H^+(aq) \rightarrow 2H(g) \rightarrow H_2(g)$	(reduction)
	Adding	$Zn(s) + 2H^+(aq) \rightarrow Zn^{2+}(aq) + H_2(g)$	(overall reaction)

As you can see, the overall chemical change taking place is the same as that which occurs when zinc is placed in dilute acid on its own, but this arrangement enables us to obtain electrical energy from the reaction. The process is the reverse of electrolysis, where electrical energy is used up in bringing about a chemical change. Any apparatus in which a chemical reaction produces an electric current is referred to as a *voltaic cell*.

Experiment 11.4.a shows that Volta's cell has two major disadvantages. Firstly, a layer of hydrogen tends to collect on the copper foil and cut down the current (the cell is said to be 'polarised') and, secondly, the zinc continues to dissolve when the cell is switched off.

N.B. Though electrons flow from the zinc to the copper, by convention the *current* is said to flow from positive to negative, i.e. from copper to zinc.

The Daniell cell

This cell was invented by Professor John Daniell at King's College, London in 1836. It consists of a zinc rod dipping into a solution of zinc sulphate and a copper plate dipping into a solution of copper(II) sulphate. The two solutions are separated by a porous partition, which prevents direct reaction from taking place between the zinc rod and the copper(II) sulphate solution but allows ions to travel through it to complete the circuit (see Figure 11.4.b, page 199).

The explanation of the way in which the Daniell cell operates is almost the same as that for the simple cell but there is one important difference. At the copper cathode there are two types of ions present, hydrogen ions from the water and copper ions from the copper(II) sulphate. It is the latter which are discharged, because copper is below hydrogen in the electrochemical series.

Changes	At anode	$Zn(s) \rightarrow Zn^{2+}(aq) + 2e^-$	(oxidation)
	At cathode	$Cu^{2+}(aq) + 2e^- \rightarrow Cu(s)$	(reduction)
	Adding	$Zn(s) + Cu^{2+}(aq) \rightarrow Zn^{2+}(aq) + Cu(s)$	(overall reaction)

Clearly, the overall chemical change taking place is the same as that which occurs when zinc is dipped into copper(II) sulphate solution but, again, the cell enables us to obtain electrical energy from the reaction.

As Experiment 11.4.b shows, this cell gives a steadier current than the simple cell. Another advantage is that the reaction ceases when the cell is switched off.

Galvanic couples

Towards the end of the eighteenth century, the Italian scientist, Luigi Galvani, discovered that if a frog's leg was hung on an iron frame by means of a brass hook, it twitched whenever it came simultaneously into contact with both metals. We know now that this was because the fluid in the leg acted as an electrolyte and thus a simple cell was set up. The electric current produced by this cell caused the nerves to make the leg twitch. This discovery of Galvani's led his fellow countryman, Volta, to carry out further investigations which resulted in the invention of the first voltaic cell.

Nowadays we use the term *galvanic couple* to describe *any* pair of metals which are in contact with one another and with an electrolyte. A galvanic couple is, in effect, a simple cell where the electrodes are allowed to touch, so that electrons can flow directly from one metal to the other without passing round an external circuit.

If you have worked carefully through the section on the simple cell, you should have no difficulty in applying the same arguments and equations to a galvanic couple consisting of zinc and copper in dilute sulphuric acid. Remember that in *any* galvanic couple the metal which is *higher* in the electrochemical series is the one which acts as the anode and dissolves, and Experiment 11.4.c shows that it does so faster than when present alone.

The corrosion of metals

When iron rusts, it reacts with oxygen and water (see Chapter 3). The most obvious way of preventing corrosion is to keep these substances away from the surface of the metal by covering it with a layer of some inert material such as paint, grease or enamel. Another well-known method of rust prevention is to plate the iron with a metal such as zinc or tin. However, if this metal coating is scratched so that the iron is exposed, and then water containing an electrolyte enters the scratch, a galvanic couple is set up.

In the case of zinc plating (galvanising), the zinc, being higher in the electrochemical series than iron, acts as the anode and dissolves (Figure 11.4.2). The resulting zinc ions join with the hydroxide ions in the water to form a precipitate of zinc hydroxide, which fills up the scratch and helps to prevent further reaction.

Tin is *below* iron in the electrochemical series and in similar circumstances it will be the iron which acts as the anode and dissolves (Figure 11.4.3). As mentioned earlier, the presence of the tin will speed up the corrosion of the iron. Tin is only used to protect iron in food containers because zinc would poison the food.

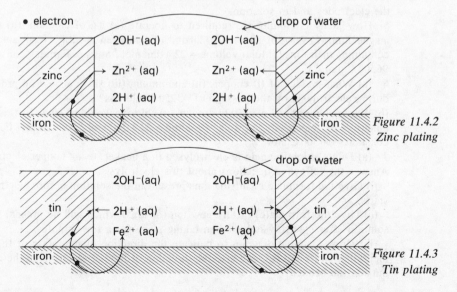

Figure 11.4.2 Zinc plating

Figure 11.4.3 Tin plating

Clearly, *any* metal which is above iron in the electrochemical series will act as an anode when the two are in contact with one another and with an electrolyte. This is made use of in the process known as *cathodic protection*. For example, magnesium bars are connected to underground pipes or to ships' hulls. The magnesium dissolves rather than the iron or steel and can be replaced from time to time as required.

You may well be wondering why it is that iron rusts even when no other metal is present. The explanation for this is that there are impurities in the metal and thus millions of tiny galvanic couples are set up over the whole surface when an electrolyte such as salt water comes into contact with it. If the iron is the anode in these couples it will dissolve. In all cases of rusting the first product is a solution containing iron(II) ions. Rust, which is hydrated iron(III) oxide, is formed as a result of further reaction with dissolved oxygen from the air.

Questions on Chapter 11

1   Explain why (a) sodium will conduct electricity both when solid and when molten, (b) chlorine will *not* conduct electricity either when solid or when molten, (c) sodium chloride will not conduct electricity when solid but *will* do so when molten.

2   Metal A comes between magnesium and aluminium in the electrochemical series and metal B comes between copper and silver. What, if anything, would happen if (a) samples of A and B were placed in dilute sulphuric acid, (b) a piece of zinc were placed in a solution of the sulphate of each metal? Explain your answers.

3   A ship has been made of copper plates joined together with steel rivets. Explain clearly why it would be inadvisable to take a long sea cruise in it.

4   Describe, with the aid of diagrams, (a) a chemical reaction which is brought about by the passage of an electric current, (b) a chemical reaction which produces an electric current. Give details of the various changes which occur at the electrodes and in solution.

5   How many coulombs are required to liberate (a) 8 g of calcium, (b) 6 g of magnesium, (c) 2.7 g of silver, (d) 2.8 dm^3 of hydrogen at s.t.p., from solutions containing their ions? (Molar volume = 22.4 dm^3 mol^{-1} at s.t.p. Faraday constant = 96 500 C mol^{-1}.)

6   (a) What masses of (i) copper, (ii) aluminium, (iii) silver would be produced during electrolysis by the passage of 48 250 coulombs?

(b) What volumes at s.t.p. of (i) hydrogen, (ii) chlorine would be given by this same quantity of electricity? (Molar volume = 22.4 dm^3 mol^{-1} at s.t.p., Faraday constant = 96 500 C mol^{-1}.)

7   (a) Dilute sulphuric acid is electrolysed in a beaker, using copper electrodes. Answer the following questions about this electrolysis.

(i) Give the formulae of *all* the ions present in the solution before electrolysis starts.

(ii) Give the formula(e) of any new ion (or ions) which will be present in the solution after electrolysis has been taking place for a few minutes.

(iii) Draw a simple diagram to indicate the direction of migration of the ions mentioned in (i) and (ii), and also the direction of flow of electrons in the circuit outside the beaker.

(iv) What changes, if any, would you *observe* at the anode, at the cathode, and in the solution?

(v) Explain the chemical change occurring at the anode (positive electrode.)

(b) Draw a labelled diagram of the apparatus you would use to electroplate a small object with a coating of a metal such as nickel or silver.

Do you think it would be possible to electroplate an object with magnesium? Give briefly the reasons for your answer. (C)

8 A metal M (of relative atomic mass 160) is very close to copper in the metal activity series (electrochemical series). You are provided with a solution of the sulphate of M, a stop-watch and the apparatus illustrated below. You are required to carry out an experiment to find the quantity of electricity required to liberate 160 g of M (one mole of M atoms).

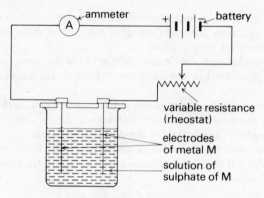

(a) What is the purpose of the variable resistance?

(b) Describe in outline how you would carry out the experiment, stating clearly the weighings and other measurements you would make.

(c) Indicate how you would calculate from your measurements the quantity of electricity required to deposit 160 g of M.

(d) Explain how the result of your calculation would enable you to decide whether the metal ion present in the sulphate has the formula M^+, M^{2+} or M^{3+}.

(e) Give a brief account of the chemical processes taking place at the anode (positive electrode) during the electrolysis and indicate how this process would differ if a platinum anode were used. (C)

9 Two experiments were carried out using the apparatus shown below.

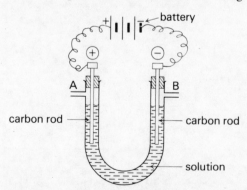

(a) A current was passed through copper chloride ($CuCl_2$) solution. Copper was deposited on electrode B and chlorine was evolved at A.

(b) A current was passed through a solution of sodium sulphate containing a little litmus solution. Oxygen was evolved at A and hydrogen at B. A red colour was seen near electrode A and a blue colour near electrode B.

Explain the changes taking place at each electrode during experiments (a) and (b) and account for the colours seen in experiment (b).

96 500 coulombs of electricity will liberate 1.00 g of hydrogen during electrolysis. Calculate the *mass* of copper and the *volumes* of oxygen and chlorine at s.t.p. formed during electrolysis by the passage of 96 500 coulombs. (Molar volume $= 22.4 \, dm^3 \, mol^{-1}$ at s.t.p.) (C)

10   A white crystalline salt B is melted in a test tube fitted with carbon electrodes. When an electric current is passed, small beads of a greyish metal are liberated at the cathode. This metal dissolves in dilute nitric acid, but has no reaction with dilute hydrochloric acid. Brown fumes are seen at the anode; these are identified as a halogen.

(a) Draw a labelled diagram of an apparatus which will show that a current is flowing during this process. How does the passage of this current differ from that of a current flowing through a copper wire?

(b) Suggest, giving reasons for your conclusion, the identity of the metal formed at the cathode. Describe one further chemical test which would confirm your suggestion.

(c) Some of the halogen was added to potassium iodide solution and, as the result of a chemical reaction, the mixture turned reddish-brown. Explain this reaction and give *one* test to confirm the identity of the element liberated.

(d) *Name* the salt B and give the *formulae* of the ions present in the melt.

(e) Would you expect the salt B to conduct an electric current in the solid state? Give reasons for your answer. (O & C)

12 Solvents and solubility

In Chapter 8 you discovered that in general ionic substances dissolve in water whereas covalent ones do not. With organic solvents the reverse is true; ionic solids are insoluble but covalent ones dissolve. Why is this? In order to find the answer to this question, we must look more closely at the whole process of dissolving.

Experiments

12.1.a To investigate the effect of temperature change on solubility

PROCEDURE The procedure and questions for this experiment are exactly the same as for Experiment 1.2.b (page 14).

12.1.b An experiment with hydrated sodium thiosulphate(VI)

PROCEDURE Half-fill a test tube with crystals of hydrated sodium thiosulphate(VI), $Na_2S_2O_3 \cdot 5H_2O$. Heat the tube *gently* until a solution is formed and then cool it under the tap. When it has cooled *completely* to room temperature, add a crystal of sodium thiosulphate(VI). Observe the tube carefully and feel it to see if there is any change in temperature.

Questions

1 What was the source of the water in which the crystals dissolved?
2 Did crystals form when the solution was cooled to room temperature? Were you expecting this result?
3 Before the crystal was added, was the solution (a) unsaturated, (b) saturated, (c) more than saturated? Explain your answer.
4 What did you see when the crystal was added? Was there any change in temperature?
5 From your answer to question 4, do you think that the contents of the tube were more stable before the crystal was added or afterwards? How did you reach your conclusion?

12.1.c To measure the solubility of a solid in water at various temperatures

PROCEDURE Find the mass of a boiling tube, add about 4.0 g of potassium chlorate(V) and find the mass again. Run in 10.0 cm³ of distilled water from a burette and heat the mixture gently, with shaking, until the crystals dissolve. (It may be necessary to boil the mixture carefully to achieve this, but you must avoid loss of the liquid by spitting.) Hold the tube up to the light and allow the solution to cool, stirring continuously with a 0–100 °C thermometer (care!). Note the temperature at which crystals *first* appear. Check the reading by reheating to *just* dissolve the crystals and allowing the solution to cool again. Add 2.5 cm³ of distilled water and repeat the experiment to find the new temperature of crystallisation. (Would you expect it to be higher or lower than the first temperature?) Continue in this way until at least five readings have been taken.

Questions

1 The solubility of a substance is usually expressed in grams of solute per 100 g of solvent. Work out the mass of potassium chlorate(V) which would have to be dissolved in 100 g of water to give solutions of the same concentrations as the ones you used. (Assume that 1 cm³ of water has a mass of 1 g.) Each calculation should be set out in a similar manner to that shown below.

10 g of water dissolves 4 g of potassium chlorate(V)

$\therefore$ 1 g of water dissolves $\dfrac{4}{10}$ g of potassium chlorate(V)

$\therefore$ 100 g of water dissolves $\dfrac{4}{10} \times 100$ g of potassium chlorate(V)

= 40 g of potassium chlorate(V)

2 The temperature at which crystals first appear is the one at which the solution becomes saturated, i.e. the temperature at which the solubility of potassium chlorate(V) is equal to the figures calculated in question 1. Tabulate your results as shown in Table 12.1.c.

Mass of potassium chlorate(V)/g	Volume of water/ cm³	Solubility/g (100 g water)⁻¹	Temperature/ °C
4	10	40	

Table 12.1.c The solubility of potassium chlorate(V) at various temperatures

3 Plot a graph of solubility of potassium chlorate(V) against temperature. Mark the axes as shown in Figure 12.1.c and draw a *smooth* curve through the points which you obtain.

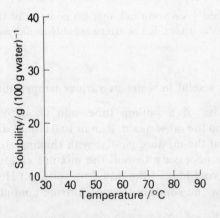

Figure 12.1.c Solubility curve for potassium chlorate(V)

4   Read from the graph the solubility of potassium chlorate(v) at (a) 50 °C, (b) 70 °C.

5   At which temperatures is the solubility of potassium chlorate(v) (a) 35 g (100 g water)$^{-1}$, (b) 20 g (100 g water)$^{-1}$?

6   Suppose a solution containing 25 g of potassium chlorate(v) in 100 g of water at 90 °C were slowly cooled to 50 °C. At what temperature would crystals first appear and what would be the total mass of crystals obtained?

12.1.d   Further information from solubility curves

PROCEDURE   Using the information given below, plot the solubility curves for aluminium potassium sulphate and copper(II) sulphate on the same piece of graph paper.

Temperature/°C	0	10	20	30	40	50	60
Solubility of KAl(SO$_4$)$_2$·12H$_2$O/ g (100 g water)$^{-1}$	4.8	6.9	9.6	14.3	20.2	30.3	46.9
Solubility of CuSO$_4$·5H$_2$O/ g (100 g water)$^{-1}$	22.4	27.2	32.4	37.9	44.9	52.9	62.6

Questions

1   A hot solution contains 25 g of CuSO$_4$·5H$_2$O and 25 g of KAl(SO$_4$)$_2$·12H$_2$O in 100 g of water at 60 °C. If the solution is allowed to cool, at what temperature will crystals first appear and what will be their composition?

2   Do you think that cooling the solution to room temperature (18 °C) would give a pure sample of either solid? Would separation of the two compounds be complete or only partial? Explain your answers.

12.1.e   To separate two solids by crystallisation

In this experiment the predictions which you made concerning the separation of copper(II) sulphate and aluminium potassium sulphate are put to the test.

PROCEDURE   Weigh out 10 g of a mixture made up of equal masses of hydrated copper(II) sulphate and hydrated aluminium potassium sulphate into a clean boiling tube. Add 20 cm^3 of distilled water and heat gently, with shaking, until both solids have dissolved. Allow the liquid to cool, scratch the inside of the tube with a glass rod if no crystals appear (why?), filter off the crystals and rinse them with a *little* distilled water.

Questions

1   Did you obtain the crystals you expected? How could you tell?

2   Do you think the crystals were pure? Give a reason for your answer.

3   Would you expect a purer crop of crystals to be obtained if the crystallisation were repeated? Explain your reasoning.

12.2.a   To find the number of molecules of water of crystallisation in hydrated magnesium sulphate crystals

PROCEDURE   The procedure for this experiment is exactly the same as that for Experiment 9.3.a. In addition to finding the number of molecules of water of crystallisation, find also the *percentage* by mass of water in the crystals.

12.3.a   To investigate what happens when certain substances are exposed to the atmosphere ★

PROCEDURE   Place one large, powder-free crystal of sodium carbonate-10-water, one of calcium chloride-6-water and a pellet of sodium hydroxide on separate watch glasses and find the mass of each. Inspect the solids after 15 minutes, after one hour and after a day or so. Check the masses at each inspection.

Questions

1   In what way did the appearance and mass of each solid change?
2   Can you suggest explanations for the changes which you observed?

12.4.a   To investigate the behaviour of various solvents in an electric field ★ ★

PROCEDURE   Suitable solvents for this experiment are water, ethanol, tetrachloromethane and methylbenzene. Set up four clean, *dry* burettes with beakers underneath them and place about 10 cm³ of one of the liquids in each. Comb your hair briskly with a nylon comb, turn on the tap of one of the burettes and hold the comb near the stream of liquid. Repeat this procedure with the other three solvents.

Questions

1   What happened to the comb when you combed your hair?
2   What effect, if any, did the comb have on each liquid?
3   All of the liquids are covalent compounds. Which, if any, behaved as if it might be made up of *polar* molecules, i.e. molecules in which one end is negatively charged and the other end positively charged?
4   Do the molecular structures of the liquids which behaved similarly have anything in common?

12.4.b   To investigate the dissolving of substances in water

PROCEDURE   The following solids are required for this experiment: sodium chloride, copper(II) sulphate-5-water, zinc oxide, citric acid, sodium nitrate, iodine, naphthalene and camphor. Place 5 cm³ of distilled water in a test tube and measure its temperature with a 0–100 °C thermometer. Add a spatula measure of sodium chloride and stir the mixture carefully with the thermometer. Note whether any of the sodium chloride dissolves and also whether there is a change in temperature. Repeat the experiment using each of the solids in turn. Record your results as shown in Table 12.4.b.

Substance	Bonding	Solubility	Temperature change
	Ionic or covalent	High, low or insoluble	$\checkmark$ or $\times$

Table 12.4.b The dissolving of solids in water

Questions

1 Which of the ionic substances (a) dissolved with a temperature change, (b) dissolved with no temperature change, (c) did not dissolve?
2 Which part of the water molecule would you expect to be attracted to (a) positive ions, (b) negative ions? (Hint: read about the structure of the water molecule on page 229.)
3 Answer question 1 for the covalent substances.
4 Which, if any, of the covalent substances do you think might have polar molecules? Give a reason for your answer.

12.4.c To investigate the dissolving of substances in methylbenzene

PROCEDURE Repeat the previous experiment but use methylbenzene instead of water. Record your results in a similar manner to that shown in Table 12.4.b. (**Care:** methylbenzene is inflammable.)

Questions

1 Can you link solubility in methylbenzene with the type of bonding found in the substances which dissolve?
2 Do you think that the mechanism of dissolving in this solvent is the same as, or different from that for water? Give a reason for your answer.

12.5.a To investigate the dissolving of one liquid in another ★

PROCEDURE The liquids required for this experiment are distilled water, ethanol, propanone, tetrachloromethane, methylbenzene and hexane. Place 1 cm³ portions of distilled water in five test tubes and add an equal volume of each of the other liquids to them, using a different liquid for each tube. Shake the tubes and note whether the liquids mix completely or form two distinct layers. Repeat this procedure until all the combinations of the liquids have been used. Record your results as shown in Table 12.5.a, marking a tick for pairs which mix and a cross for those which do not.

Questions

1 Which of the liquids, if any, (a) mixed with water but not with the other liquids, (b) mixed with all of the liquids except water, (c) mixed with all of the liquids, including water?

	Distilled water	Ethanol	Propanone	Tetrachloro-methane	Methyl-benzene	Hexane
Distilled water	√					
Ethanol		√				
Propanone			√			
Tetrachloro-methane				√		
Methyl-benzene					√	
Hexane						√

Table 12.5.a

2   All of the liquids are covalent compounds. Which ones seem likely to have polar molecules? Do you think that water molecules are more polar than the others? Explain your reasoning in both parts of your answer.

12.6.a   To investigate the solubility of gases in water ★ ★

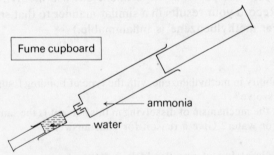

Fume cupboard

ammonia

water

Figure 12.6.a

PROCEDURE   Flush out a dry 100 cm³ syringe with dry ammonia and then fill it with a sample of the gas. (This operation should be carried out in a fume cupboard.) Fit a serum cap over the end of the syringe, clamp it at an angle of 45° and inject 2 cm³ of cold, freshly boiled distilled water into the ammonia from a small syringe. Observe the result and then repeat the experiment using gases such as oxygen, sulphur dioxide, hydrogen, hydrogen chloride, carbon dioxide, nitrogen and nitrogen dioxide in place of the ammonia.

Questions

1   Why did the distilled water have to be freshly boiled?
2   How could you tell whether a gas was (a) very soluble, (b) fairly soluble, (c) slightly soluble?
3   Record your results in a table, using the headings mentioned in question 2.

12.1　Solutions and suspensions

No doubt you know already that a *solution* consists of a *solute* dissolved in a *solvent* and that if the solvent is water then an *aqueous* solution is obtained.

If a spatula measure of hydrated copper(II) sulphate is added to a beaker of water and stirred well, the solid dissolves. However, if further portions of solid are added, a point will be reached when no more will go into the solution; we say that the solution has become *saturated*. The mass of solute required to saturate any solution varies with temperature (Experiment 12.1.a) and this must be taken into account in the definition.

> A **saturated solution** is one which contains as much solute as can be dissolved at the temperature concerned, *in the presence of undissolved solute.*

The importance of the words in italics is made clear by Experiment 12.1.b. In this experiment, hydrated sodium thiosulphate(VI) crystals are heated and found to dissolve in their own water of crystallisation. When the resultant solution is allowed to cool down to room temperature *no crystals form*. Obviously, since crystals were present originally they should have reformed on cooling. The solution must now be more than saturated; it is *supersaturated*.

> A **supersaturated solution** is one which contains *more* solute than a saturated solution at the same temperature.

A supersaturated solution is unstable. If a single crystal of solute is added to it, crystallisation occurs and the excess solid comes out of solution. The crystal which is added acts as a centre on which others of the same shape can form. Because crystals grow out from the one which is added, as a plant does from a seed, the technique is known as *seeding* the solution.

When crystallisation of a supersaturated solution occurs, there is a noticeable evolution of heat. This indicates again that a supersaturated solution is unstable since, in giving out energy to the surroundings, it must have moved to a lower energy state, i.e. it must have become more stable. N.B. Sodium thiosulphate(VI) easily forms a supersaturated solution, but with most other substances cooling must be slow, all dust on which crystals could grow must be excluded, and the container must be kept still. If these precautions are not taken crystallisation will occur.

Solubility

> The **solubility** of a solute in a solvent at a particular temperature is the mass of solute required to saturate 100 g of solvent at that temperature.

The solubilities of most substances in a given solvent increase as the temperature is raised, although for some substances, such as sodium chloride, the change is not very marked. By cooling solutions of differing concentrations and determining the temperatures at which they become saturated, a graph of solubility against temperature can be drawn (Experiment 12.1.c). Several examples of these graphs, or *solubility curves*, are shown in Figure 12.1.1.

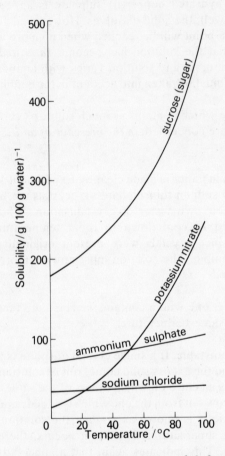

Figure 12.1.1 Some solubility curves

Solubility curves are useful for finding out how much solute will dissolve in a known mass of solvent at a given temperature, or for predicting the mass of crystals which will be deposited on cooling a solution of known concentration over a particular temperature range (Experiment 12.1.c). Another important application is in the separation of soluble substances by fractional crystallisation.

Fractional crystallisation

Fractional crystallisation is the process whereby substances are separated by the repeated partial crystallisation of a solution.

Suppose that we have a solution which contains 35 g of potassium chloride and 15 g of potassium sulphate dissolved in 100 g of water at 90°C. This situation is represented by points X and Y in Figure 12.1.2.

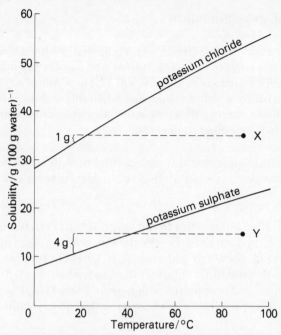

Figure 12.1.2 Solubility curves of potassium chloride and potassium sulphate

You can see that if the solution is allowed to cool, nothing happens until the temperature reaches 42°C, but at this point the solution becomes saturated with potassium sulphate and the solid begins to crystallise. Further cooling causes more potassium sulphate to be deposited until, at 21°C, potassium chloride also begins to separate out. The solid obtained at room temperature (18°C) will contain about 4 g of potassium sulphate and 1 g of potassium chloride. If this is filtered off, dissolved in water and crystallised again, the potassium sulphate will be obtained in a purer state. Repetition of the experiment will result eventually in the separation of pure potassium sulphate.

Suspensions

Some solids do not dissolve when placed in water. However, if the particles are small enough, stirring will cause them to float around, producing a cloudy liquid called a *suspension*. Suspensions may also be obtained using soluble substances if more solid is added than the solvent can dissolve at that temperature.

The differences between solutions and suspensions are due to the relative sizes of the particles. The particles in a solution are individual ions or molecules which cannot be filtered off and will not settle out. Those in a suspension are much bigger and are held back by filter paper; on standing, they settle to the bottom of the container, leaving clear liquid above.

A **suspension** is a mixture of a finely divided solid and a liquid in which it will not dissolve.

12.2 Water of crystallisation

Many crystals contain water chemically combined within them. This water is referred to as *water of crystallisation* and salts containing it are called *hydrated salts*. Often, as in Experiment 12.2.a, water of crystallisation can be driven off by gentle heating, and when this is done it is found that there is a definite number of water molecules associated with every formula unit of the compound concerned.

Water of crystallisation is that definite quantity of water with which some substances are associated on crystallising from an aqueous solution.

Water of crystallisation contributes to the shape of the crystals, since it is part of the lattice, and sometimes affects the colour too. Thus pink crystals of hydrated cobalt(II) chloride ($CoCl_2 \cdot 6H_2O$) turn to a blue powder, anhydrous cobalt(II) chloride ($CoCl_2$), when heated gently. Similarly, blue hydrated copper(II) sulphate ($CuSO_4 \cdot 5H_2O$) gives the white anhydrous form of the salt when its water of crystallisation is removed.

You will have noticed that in the formula of a hydrated salt, the water of crystallisation appears separately at the end. If the number of water molecules in the formula is definitely known, it should be included in the name of the compound. The word 'hydrated' should only be used where the formula is uncertain. Thus $CoCl_2 \cdot 6H_2O$ is cobalt(II) chloride-6-water, $CuSO_4 \cdot 5H_2O$ is copper(II) sulphate-5-water and $MgSO_4 \cdot 7H_2O$ is magnesium sulphate-7-water.

12.3 Efflorescence, deliquescence and hygroscopic substances

Efflorescence

When crystals of sodium carbonate-10-water are left exposed to the air for some time, they become powdery and their mass decreases (Experiment 12.3.a). The reason for these changes is that they *effloresce*, i.e. give up some of their water of crystallisation to the atmosphere. The white powder which forms is sodium carbonate-1-water.

$$Na_2CO_3 \cdot 10H_2O(s) \rightarrow Na_2CO_3 \cdot H_2O(s) + 9H_2O(g)$$

Efflorescence is the giving up of water of crystallisation to the atmosphere.

Deliquescence

Substances like calcium chloride-6-water and sodium hydroxide are *deliquescent*, i.e. they absorb water from the atmosphere and, if left long enough, dissolve in it. Obviously, bottles containing deliquescent substances should be opened as infrequently as possible but, even so, it is often found that such bottles contain solutions instead of solids.

Deliquescence is the absorption of water vapour from the atmosphere by a substance to such an extent that a solution results.

Hygroscopic substances

Hygroscopic substances absorb water vapour from the atmosphere but do not change in state (e.g. solids remain solid and do not form solutions).

Copper(II) oxide and concentrated sulphuric acid are both hygroscopic.

Because of their great affinity for water, both deliquescent and hygroscopic compounds find uses as drying agents.

12.4   How solids dissolve in liquids

Water and ice

Before we can begin to understand why water is such a useful solvent, we need to know something about the water molecule itself. Experiment 12.4.a, where a stream of water from a burette is deflected in an electric field produced by a charged comb, gives us a clue as to how water molecules differ from those of many other liquids.

Oxygen is very *electronegative*. This means that an oxygen atom attracts electrons and tends to pull them towards itself in any molecule in which it is found. Thus the electron pairs making up the covalent bonds between the oxygen atom and the two hydrogen atoms in a water molecule are not shared equally but are displaced towards the oxygen. As a result of this, the molecule is *polar*, with a small negative charge on the oxygen atom and small positive charges on the hydrogen atoms. This is why the stream of water is deflected in the electric field in Experiment 12.4.a. We normally use the signs $\delta +$ and $\delta -$ to show the charges on polar molecules; the 'δ' (delta) indicates that the charges are much smaller than on ions but it has *no specific value*. Thus we need not write $2\delta -$ on the oxygen atom to balance the two $\delta +$'s on the hydrogen atoms.

$$O^{\delta -}$$
$$^{\delta +}H \qquad H^{\delta +}$$

Obviously, there will be an attraction between the oxygen and hydrogen atoms of neighbouring water molecules and this will cause them to 'stick together' to some extent. These attractive forces are known as *hydrogen bonds* and act between any molecules which contain hydrogen atoms joined directly to oxygen atoms (or to atoms of the other very electronegative elements, fluorine and nitrogen). Hydrogen bonds are stronger than van der Waals' forces (8.5), but are much weaker than normal covalent or ionic bonds.

The presence of hydrogen bonds means that more energy than expected is required to overcome the forces of attraction acting between the individual water molecules. For this reason, the melting and boiling points of water are much higher than those of other group VI hydrides, H_2S, H_2Se and H_2Te. The figures given in Table 12.4 illustrate that the molar enthalpies of fusion and vaporisation are also exceptionally high for water. Just how 'out of step' these values are is even more strikingly shown if you plot graphs of the various quantities against the relative molecular masses of the compounds.

Compound	Relative molecular mass	m.p./ K (°C)	b.p./ K (°C)	Molar enthalpy of fusion/ kJ mol^{-1}	Molar enthalpy of vaporisation / kJ mol^{-1}
H_2O	18	273 (0)	373 (100)	6.0	41.1
H_2S	34	187 (−86)	212 (−61)	2.4	18.7
H_2Se	81	213 (−60)	231 (−42)	2.5	19.9
H_2Te	129	224 (−49)	271 (−2)	7.0	23.8

Table 12.4

When a liquid solidifies, the particles in it usually pack together as closely as possible. However, when water freezes this does not happen. Instead, the water molecules form a rather open molecular lattice in which each oxygen atom is surrounded tetrahedrally by four hydrogen atoms. Two of these hydrogen atoms are joined to the oxygen by covalent bonds and the other two by hydrogen bonds (see Figure 12.4.1). Thus it is the hydrogen bonds which are responsible for holding the molecules in the open structure.

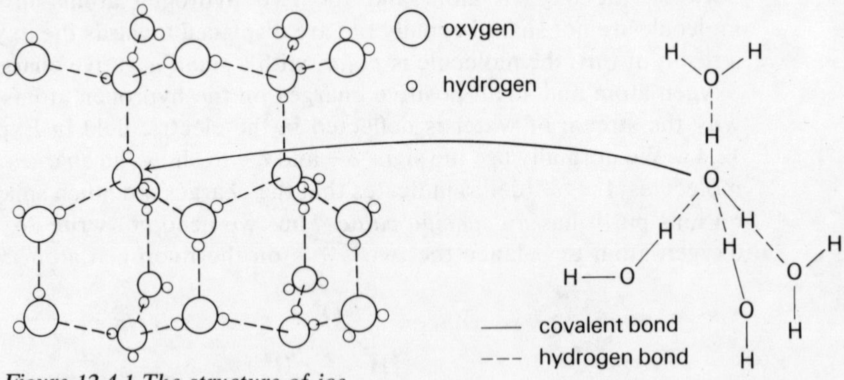

Figure 12.4.1 The structure of ice

Many, but not all, of the hydrogen bonds break when ice melts. Hence the open structure breaks up, the molecules pack closer together and the volume *decreases*. This, of course, is unusual; the particles in most solids are packed together as closely as possible and when melting occurs they move further apart. Can you explain why water has its maximum density at 4°C (i.e. why it contracts as it warms from 0°C to 4°C and then expands if the temperature is increased further)?

An iceberg – ice is less dense than water

Water as a solvent

The ability of water to dissolve many ionic crystals results from the polar nature of its molecules. When an ionic compound is placed in water the negative part of each water molecule (i.e. the oxygen atom) will be attracted to the positive ions on the surface of the solid. Similarly, the positively charged hydrogen atoms in the water molecules will be attracted to the negative ions on the surface of the solid. If the attraction between the ions and the water molecules is strong enough to overcome the forces holding the ions together in the crystal lattice, then the solid will dissolve and the ions will be carried off into solution surrounded by a 'jacket' of water molecules. This sequence of events is shown in Figure 12.4.2.

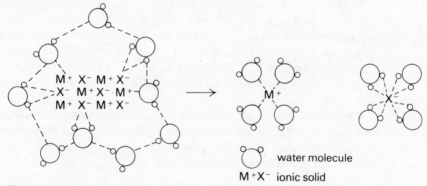

water molecule
M^+X^- ionic solid

Figure 12.4.2 How water dissolves an ionic solid

Although the forces of attraction holding non-polar covalent molecules together in a molecular lattice are much weaker than those which act in an ionic crystal, covalent substances do not usually dissolve in water. The reason for this is that the attraction of the water molecules for one another is far greater than their attraction for the molecules in the crystal.

Thus they will tend to stick together rather than carry off molecules from the solid.

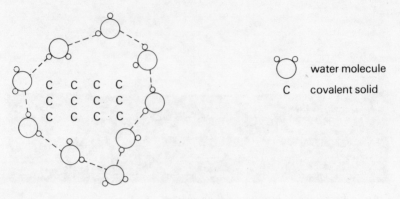

Figure 12.4.3 Non-polar solids do not attract water molecules

Nevertheless, there are some covalent solids, such as sugar, which *do* dissolve in water. This is because they themselves have polar molecules and the attraction between these and the molecules of water is great enough to break up the lattice.

Solvents other than water

Organic solvents, such as methylbenzene, do not generally dissolve ionic solids. The molecules of these solvents are non-polar, or only slightly polar, and hence the forces of attraction between them and the ions in a crystal are weak. There is little tendency for the solvent molecules to remove ions from the lattice, and so the solid does not dissolve.

In the case of a covalent solid, the forces of attraction between the individual molecules in the lattice are comparable with those between the solvent molecules. This means that the lattice is easily penetrated by the solvent molecules and a solution is formed. The particles of solute are not surrounded by a jacket of solvent molecules as happens when an ionic solid dissolves in water; they simply diffuse throughout the solution on their own.

The effect of various solvents on different types of solid can be summed up in the phrase 'like dissolves like'. A highly *polar* solvent such as water dissolves *polar* (or ionic) solids, whereas a *non-polar* solvent such as methylbenzene dissolves *non-polar* (covalent) solids.

12.5   The solubility of one liquid in another

The dissolving of one liquid in another may be explained in much the same way as the dissolving of a solid in a liquid. Water will mix with polar liquids, such as ethanol and propanone, because the oppositely charged ends of the different molecules attract one another (Figure 12.5.1). On the other hand, when water is added to a non-polar liquid, such as methylbenzene or tetrachloromethane, two layers separate out. The water molecules attract one another strongly but have little tendency to mix with the molecules of the second liquid (Figure 12.5.2). Ethanol and propanone *will* mix with non-polar liquids because their own molecules are far less polar than those of water and thus have a much smaller tendency to stick together. As before, remember that 'like dissolves like'.

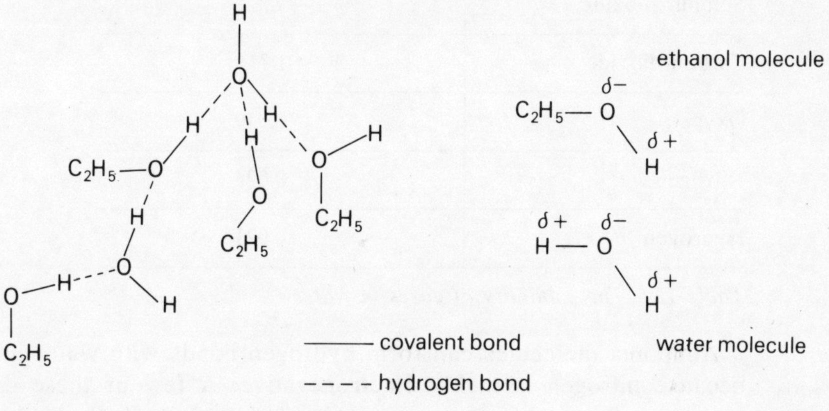

Figure 12.5.1 A mixture of ethanol and water

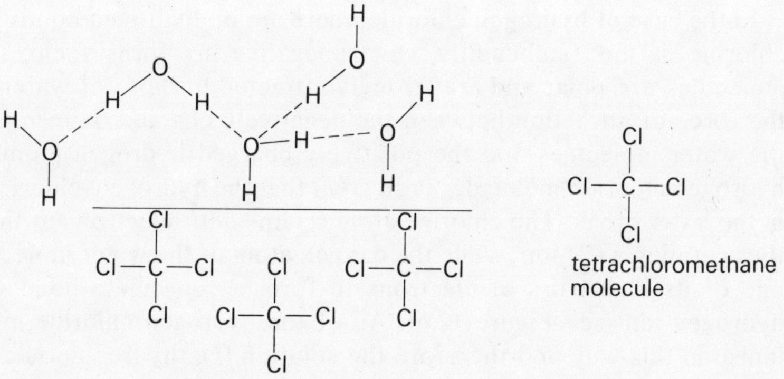

Figure 12.5.2 Water and tetrachloromethane will not mix

12.6 The solubility of gases in liquids

If you look at a glass of water which has been standing in a warm room for some time, you will see one way in which the solubility of gases differs from that of most solids. Bubbles form in the water as it warms up, showing that the air is becoming *less* soluble as the temperature rises.

The solubilities of different gases in water at a given temperature show a wide variation (Experiment 12.6.a). Table 12.6 shows that the gases which are most soluble are those which *react* with the water in some way.

Gas	Solubility/cm^3 (cm^3 water)$^{-1}$ at 0°C and total pressure of 101 325 Pa
Ammonia	1300
Hydrogen chloride	506
Sulphur dioxide	79.8
Carbon dioxide	1.71
Oxygen	0.049
Nitrogen	0.024
Hydrogen	0.021

Table 12.6 The solubility of gases in water

Ammonia molecules can form hydrogen bonds with water molecules because nitrogen is highly electronegative. A few of these dissolved molecules then split up into ions, giving a weakly alkaline solution.

$$NH_3(aq) + H_2O(l) \rightleftharpoons NH_4^+(aq) + OH^-(aq)$$

In the case of hydrogen chloride, there are no hydrogen bonds because chlorine is not sufficiently electronegative to form them, but the molecules *are* polar and are strongly attracted to those of water. In fact the force of attraction between the negatively charged oxygen atoms of the water molecules and the positively charged hydrogen atoms of the hydrogen chloride molecules is so great that the hydrogen–chlorine bonds in the latter break. The chlorine atom retains *both* electrons of this bond, thus forming a Cl$^-$ ion, while the oxygen atom of the water molecule uses one of its lone pairs of electrons to form a coordinate bond with the hydrogen ion (see Figure 12.6). All of the hydrogen chloride molecules ionise in this way and therefore the solution (i.e. hydrochloric acid) is a *strong* electrolyte.

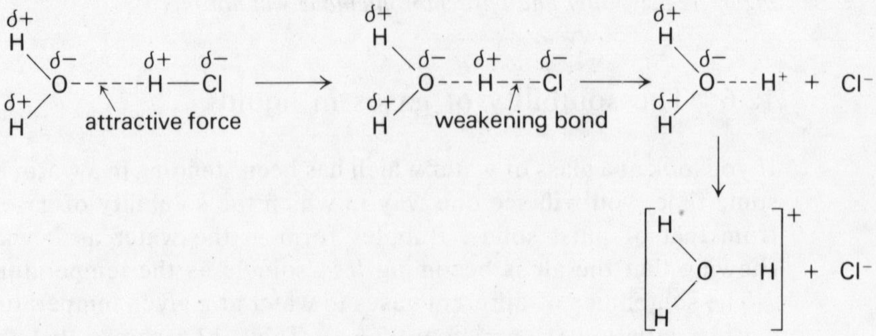

Figure 12.6 The reaction between water and hydrogen chloride

The reactions of the other very soluble gases with water are discussed with their general properties elsewhere in the book. They all involve the formation of ions to some extent.

Questions on Chapter 12

1 Explain why (a) potassium nitrate dissolves in water but not in tetrachloromethane, (b) paraffin wax dissolves in tetrachloromethane but not in water.

2 (a) Pure liquid hydrogen chloride and methylbenzene do not conduct electricity; pure water is a very poor conductor. (b) A solution of hydrogen chloride in methylbenzene does not conduct electricity but a solution of hydrogen chloride in water conducts it very well. Give an explanation for each of these statements.

3 Twenty-eight grams of a saturated solution of a salt at 25°C yielded 7 g of solid when evaporated to dryness. Find the solubility of the salt at 25°C.

Mention one precaution you would take if you did the above experiment.

(SUJB)

4 (a) What is meant by the following terms?
(i) a *saturated solution* of a solid
(ii) water of crystallisation
(b) The solubilities in grams of sodium nitrate in 100 g of water are given for various temperatures/°C.

Temperature	0	10	20	30	40	50	60	80	90	100
Solubility	73	80	88	96	104	114	124	148	162	180

(i) Construct on graph paper the solubility curve for sodium nitrate.
(ii) From your graph evaluate the solubility at 70°C.
(iii) 100 g of a solution of sodium nitrate is in the saturated condition at 10°C. How many grams of the salt would have to be added to make the solution just saturated at 80°C?
(c) In an experiment to determine the number of moles of water of crystallisation combined with one mole of anhydrous zinc sulphate, $ZnSO_4$, some zinc sulphate crystals were heated until all the water of crystallisation had been expelled. The following data were obtained:

$$\begin{aligned}
\text{mass of empty crucible} &= 20.54\,\text{g} \\
\text{mass of crucible and crystals} &= 22.99\,\text{g} \\
\text{mass of crucible and anhydrous salt} &= 21.92\,\text{g}
\end{aligned}$$

(i) How would the experimenter ensure that all of the water had been expelled?
(ii) Calculate x in the formula $ZnSO_4 \cdot xH_2O$ using the data and the relative atomic masses: $H = 1$, $O = 16$, $S = 32$, $Zn = 64$.
(iii) Name two sulphates, apart from zinc sulphate, which crystallise as hydrates and one sulphate which crystallises as the anhydrous salt from aqueous solution.

(NI)

5 Look at the graph (overleaf), then answer the following.
(a) Which compound is the more soluble in boiling water?

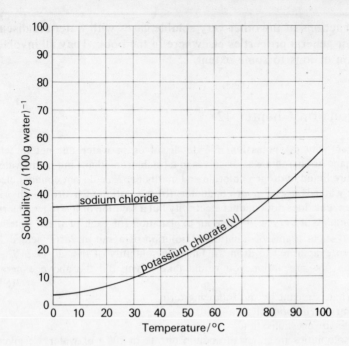

(b) At what temperature are sodium chloride and potassium chlorate(V) equally soluble in water?

(c) A mixture of 30 g sodium chloride, 30 g potassium chlorate(V), and 100 g of water, is stirred for a long time at 100°C. Will all the solid dissolve? If the mixture is now cooled to 20°C, what will happen?

(d) What is the mass of 6 moles of potassium chlorate(V) (KClO₃)?
Is it possible to make a 6 M solution of potassium chlorate(V) at 20°C? (Use the graph to justify your answer.)   (SCE)

6   Explain, with examples, the meaning of the terms deliquescence, efflorescence, supersaturated solution.

You are provided with a solution containing a mixture of two salts which can be separated by crystallisation. Indicate: (a) the nature of the solubility curves of the two salts; (b) how you would attempt to separate them.   (SUJB)

7   (a) The table below gives information about the solubility of sulphur dioxide in water at different temperatures. The masses of sulphur dioxide given in the table are those required to produce a saturated solution.

Temperature/°C	10	15	25	30	40	50	60
Solubility of sulphur dioxide/ g (dm³ of solution)⁻¹	154	126	90	77	58	42	32

Plot a graph of the solubility of sulphur dioxide against temperature, with temperature along the x-axis (the horizontal axis).
Use your graph to estimate:
(i) the mass of sulphur dioxide which would dissolve in one dm³ of saturated solution at 20°C, and

(ii) the temperature at which a saturated solution of sulphur dioxide would contain 64 g of sulphur dioxide in one dm^3 of solution. What is the concentration of this solution?

(b) (i) On boiling a solution of sulphur dioxide in water, *all* of the sulphur dioxide is set free. How would you prove that this actually happens?

(ii) On cooling a solution of sulphur dioxide in water, crystals separate out which are found to contain 37.2% by mass of sulphur dioxide and 62.8% by mass of water. It is thought that these crystals can be represented by a formula of the type $SO_2 \cdot xH_2O$. Calculate the value of x in this formula.   (L)

13 Rate of reaction

Different chemical reactions take place at different speeds. Many, particularly those involving solutions, seem to take place instantaneously. Others, such as those used to generate gases in the laboratory, take rather longer. If we look at processes in nature we can extend the time scale considerably; for example, the conversion of plant material to coal takes millions of years to complete.

How do we measure the rate of a reaction? As the reaction progresses, the concentration of the reactants decreases, and that of the products increases, but continuous measurement of concentration is not easy. Quantities which *can* be measured without too much difficulty as the experiment proceeds include such things as the volume of gas given off, the change in mass when a gas is given off, and the change in intensity of a colour.

The British chemist, Sir George Porter, shared a Nobel Prize in 1967 for his studies of very fast reactions. We shall be limited to somewhat slower changes, but even these can give some surprises. In 1975, the *Sunday Times* 'Young Scientists of the Year' award went to two girls who discovered that when magnesium is added to dilute acid, the rate of evolution of hydrogen is sometimes erratic. You may notice this effect yourself if you perform Experiment 13.2.a.

Experiments

13.1.a The reaction between calcium carbonate and dilute hydrochloric acid ★★

PROCEDURE   Place 20 g of large, *dust-free* marble chips in a 100 cm³ conical flask. Add 40 cm³ of 2 M hydrochloric acid, quickly insert a loose plug of cotton wool in the neck of the flask to prevent the escape of any acid spray and place the flask on a top-pan balance. (It may be necessary to remove the balance pan to prevent overloading. This will not affect the results, since it is mass *loss* which you will be measuring.) *Immediately* read the mass and start a stop clock. Note the mass at 30-second intervals for about 15 minutes and then plot a graph of total loss in mass/g on the vertical axis against time/s on the horizontal axis. Draw a *smooth* curve through as many of the points as possible.

Questions

1   What would be the shape of the graph if the rate were constant throughout the reaction?
2   Does your graph indicate a constant reaction rate? If it does not, explain how

its shape may be interpreted in terms of variation in the rate as the reaction proceeded.

3 If you found that the rate varied during the course of the reaction, can you explain why this variation occurred?

4 Is there a horizontal portion on the graph? If so, what is its significance?

13.1.b The reaction between magnesium and dilute hydrochloric acid

This is an alternative to Experiment 13.1.a.

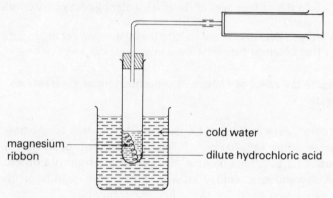

magnesium ribbon

cold water

dilute hydrochloric acid

Figure 13.1.b

PROCEDURE Clean some magnesium ribbon with emery paper, cut off a 7 cm length and find its mass. (This mass will be required for comparison in later experiments.) Place 30 cm³ of 0.5 M hydrochloric acid in a boiling tube. Connect a bung and delivery tube to a 100 cm³ glass syringe, make sure that the reading on the scale is zero, and then push the bung into the mouth of the boiling tube. Twist the syringe plunger and note the reading; this reading must be subtracted from all later ones. Remove the bung and push in the plunger once more. Coil the magnesium ribbon into a spiral and drop it into the acid. Quickly replace the bung and start a stop clock. Swirl the tube gently to ensure that the magnesium is immersed in the acid and then place it in a 250 cm³ beaker of cold water to keep the temperature steady. Record the volume of hydrogen at 30-second intervals for about 20 minutes, twisting the syringe plunger before each reading to check that it is not stuck. Plot a graph of volume of hydrogen/cm³ on the vertical axis against time/s on the horizontal axis. (Do not forget to correct the volume readings.) Draw a *smooth* curve through as many of the points as possible.

Questions

Answer the questions which follow the previous experiment.

13.2.a The effect of particle size on reaction rate

PROCEDURE (a)★★ Repeat Experiment 13.1.a but use 20 g of *small* marble chips in place of the large ones, or (b) repeat Experiment 13.1.b but use an equal mass of magnesium *turnings* in place of the ribbon and note the volume of hydrogen at 15-second intervals. Plot a graph of your results on the same piece of graph paper as before.

Questions

1   Was (a) the initial, (b) the average rate of reaction faster or slower in this case? How could you tell?
2   Since the mass of the solid is the same as before, how can you account for the difference in reaction rates? (Hint: if you had a cube of wood and you sawed it in half, how would the surface area of the two smaller pieces compare with that of the original cube?)
3   Did (a) the total decrease in mass, or (b) the total volume of hydrogen evolved, differ from that obtained before? How can you explain this?

13.3.a   To investigate the effect of change in concentration of reactants on reaction rate

PROCEDURE   (a)★★ Repeat Experiment 13.1.a, this time using 20 g of large marble chips but 40 cm³ of 1 M hydrochloric acid, or (b) repeat Experiment 13.1.b, using an equal mass of magnesium ribbon but 30 cm³ of 0.25 M hydrochloric acid. Plot a graph of your results on the same piece of paper you used for the other experiments.

Questions

1   Was the initial rate of reaction faster or slower than in the first experiment? How could you tell?
2   Can you offer an explanation for the difference?
3   In what way did the average rates differ?
4   Can you explain why the total decrease in mass in (a) differed from that in the previous experiment?

13.3.b   To find a relationship between concentration of reactants and rate of reaction

When sodium thiosulphate(VI) solution and a dilute acid are mixed together, they react to give a precipitate of sulphur. Suppose that the reaction is carried out in a beaker standing on a piece of paper marked with a cross. Provided that the same depth of liquid is used each time, it will always require the same mass of sulphur to make the liquid cloudy enough to obscure the cross. If this mass m is produced in time t, then the average rate of production of sulphur from the beginning of the reaction to time t will be m/t. Since m is constant, we can write

$$\text{average rate} \propto \frac{1}{t}$$

(e.g. if the rate doubles, the time is halved).

PROCEDURE   Using separate labelled measuring cylinders, measure out 40 cm³ of sodium thiosulphate(VI) solution (mass concentration 20 g dm⁻³) and 10 cm³ of 2 M hydrochloric acid. Place a 100 cm³ beaker on a piece of paper marked with a cross and pour in the thiosulphate(VI) solution. Add the acid and simultaneously start a stop clock. Stir the mixture well and then look down vertically through it at the cross. Stop the clock when the cross is obscured and record the time. Rinse the beaker and stirring rod *thoroughly* with tap water and then with a little distilled water and repeat the experiment, using in turn 30, 20 and 10 cm^3 of thiosulphate(VI) solution, but topping up to 40 cm^3 each time with distilled water so that the depth of the solution and the concentration of the acid remains constant.

The volume of thiosulphate(VI) solution used each time is proportional to the actual concentration of the salt (e.g. if the volume taken is halved and then made up with water, the concentration of the salt must be halved also). Therefore a graph of volume of thiosulphate(VI) solution against 1/time will show how variation in concentration of the thiosulphate(VI) affects the rate of this reaction. Plot volume of thiosulphate(VI) solution/cm^3 on the vertical axis against $(1/\text{time})/\text{s}^{-1}$ on the horizontal axis.

Questions

1   Did the average rate of this reaction (until the cross disappeared) increase, or decrease as the concentration of the sodium thiosulphate(VI) solution increased?
2   Would you expect the graph to go through the origin? Give a reason for your answer.
3   Is the graph a straight line or a curve?
4   Does the shape of the graph indicate a simple relationship between the concentration of the sodium thiosulphate(VI) and the average rate of reaction? If so, what is the relationship?

13.3.c   The 'iodine clock' reaction

Acidified hydrogen peroxide solution oxidises iodide ions to iodine; iodine reacts with starch to produce a deep blue compound. Thus when solution X (see below) is added to potassium iodide solution, a deep blue colour is immediately seen. However, if sodium thiosulphate(VI) is present it reduces the iodine back to iodide ions before it can react with the starch. This delays the appearance of the blue colour until sufficient iodine has been produced to react with all of the thiosulphate(VI). By using a *fixed* quantity of thiosulphate(VI) we can ensure that the blue colour always appears after the *same* quantity of iodine has been formed. This is similar to the last experiment where a fixed quantity of sulphur obscured a cross on a piece of paper, but generally the results of this experiment are more reliable.

PROCEDURE   Mix 70 cm^3 of 1 M sulphuric acid with 70 cm^3 of 0.1% starch solution and 7 cm^3 of 20 volumes hydrogen peroxide solution; this mixture will be referred to as 'solution X'. Add about 2 cm^3 of 0.005 M potassium iodide solution to an equal volume of solution X in a test tube and observe the colour; this is what you will be looking for during the rest of the experiment.

Using separate labelled measuring cylinders measure out 25 cm³ of solution X and 25 cm³ of the 0.005 M potassium iodide solution. *Carefully* run 1.0 cm³ of 0.005 M sodium thiosulphate(VI) solution from a burette into a 100 cm³ beaker and add the 25 cm³ of potassium iodide solution. Pour in the 25 cm³ portion of solution X and simultaneously start a stop clock. Stir well for a few seconds, place the beaker on a white tile and watch carefully. Record the time taken for the blue colour to appear. Repeat the experiment, using in turn 20, 15, 10 and 5 cm³ of potassium iodide solution, topping up to 25 cm³ with distilled water each time so that the concentrations of the other reagents remain constant. Remember to rinse the beaker and stirrer well with tap water and then distilled water between experiments. Pooling the class results can give a wider range of readings, but if this is done solution X must be made up in bulk to eliminate errors in mixing.

Plot a graph of volume of potassium iodide solution/cm³ on the vertical axis against $(1/\text{time})/\text{s}^{-1}$ on the horizontal axis. The volume of potassium iodide solution used is proportional to its concentration and the time measured is such a small fraction of the total reaction time that $1/t$ is proportional to the *initial* rate.

Questions

1 Did the initial rate of this reaction increase, or decrease as the concentration of the potassium iodide solution increased?
2 Would you expect the graph to go through the origin? Give a reason for your answer.
3 Is the graph a straight line or a curve?
4 What simple relationship between the concentration of the potassium iodide solution and the initial rate of reaction does the shape of the graph indicate?

13.4.a To investigate the effect of temperature change on the reaction between dilute nitric acid and sodium thiosulphate(VI) solution

The theory for this reaction is given at the beginning of Experiment 13.3.b. In this case the concentrations of both solutions are kept constant and the temperature is varied instead.

PROCEDURE Measure out 25 cm³ of 0.1 M sodium thiosulphate(VI) solution and 25 cm³ of 0.5 M nitric acid in separate labelled measuring cylinders. Pour the solutions into two boiling tubes and stand these in a 250 cm³ beaker of water. Warm the beaker gently and keep a check on the temperature of the acid. When the temperature reaches 20 °C, pour the acid into a 100 cm³ beaker standing on a piece of paper marked with a cross, add the thiosulphate(VI) solution and simultaneously start a stop clock. Stir with a thermometer and record both the time taken for the cross to become obscured and the average temperature during the reaction. Rinse the beaker and thermometer thoroughly and repeat the experiment at 10 °C intervals up to 60 °C; if ice is available, try cooling the solutions down to 10 °C as well. Plot a graph of temperature/°C on the vertical axis against $(1/\text{time})/\text{s}^{-1}$ on the horizontal axis.

Questions

1 Did the average rate of this reaction (up until the cross disappeared) increase, or decrease as the temperature was raised?
2 What would be the shape of the graph if the rate of reaction were directly proportional to the temperature? Is this shape obtained?
3 What are the values of $1/t$ at (a) 15 °C and 25 °C, (b) 20 °C and 30 °C, (c) 25 °C and 35 °C? Is there a rough relationship between the rate of this reaction at a particular temperature and the rate at a temperature 10 °C higher?

13.5.a To investigate the thermal decomposition of potassium chlorate(v) ★ ★

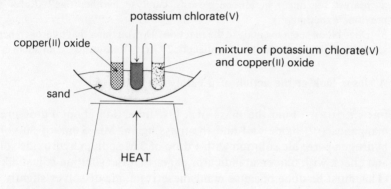

Figure 13.5.a

PROCEDURE Place a spatula measure of potassium chlorate(v), copper(II) oxide and a mixture of the two, in three separate ignition tubes. Push the tubes into sand in a sand tray as shown in Figure 13.5.a and heat the tray strongly. Test for the evolution of oxygen from the solids until the gas is detected at the mouths of two of the tubes. (**Caution:** make sure that fragments of charred wood do not drop on to the solids or a violent reaction may occur.)

Questions

1 Which of the three solids gave off oxygen first?
2 Which solid never gave oxygen at all?
3 Only one of the compounds present decomposed. Which one do you think it was?
4 From your earlier work in chemistry, can you suggest the purpose of the other compound?

13.5.b To find which substances catalyse the decomposition of hydrogen peroxide solution

Hydrogen peroxide solution decomposes slowly on standing, giving off a gas. The addition of a catalyst speeds up this reaction so that the gas may be collected and identified.

PROCEDURE (a) To separate 5 cm³ portions of 20 volumes hydrogen peroxide solution in a test tube add a little of the following powdered solids: copper(II) oxide, manganese(IV) oxide, aluminium oxide and lead(IV) oxide. If a gas is evolved, collect a tube-full over water and test it with a glowing splint.

(b) If you have a partly healed cut or scratch on your hand, use a dropping pipette to place one drop of the hydrogen peroxide solution on it. Look for bubbles of gas in the drop.

Questions

1   What was the gas? How could you tell?
2   Which of the solids in (a) catalysed the reaction and which had no effect? Does a catalyst for one reaction necessarily catalyse another one? (Refer to the previous experiment.)
3   Does blood seem to catalyse the reaction? Do you think that hydrogen peroxide solution might be useful as an antiseptic? Give a reason for your answer.

13.5.c   A closer look at the action of a catalyst ★

PROCEDURE   Find the mass of a test tube, add about 1 g of *granular* manganese(IV) oxide and find the mass again. Mix 5 cm³ of 20 volumes hydrogen peroxide solution with 1 drop of 2 M sodium hydroxide solution and check with universal indicator paper that the solution is just alkaline. (This must be done because manganese(IV) oxide dissolves slightly if the solution is acidic.) Add this mixture to the catalyst and then set up a control experiment using distilled water in place of the hydrogen peroxide solution. When the reaction has finished, inspect the two solids closely and filter the mixtures through separate *previously weighed* filter papers. Rinse out each tube with distilled water and filter the washings to make sure that all of the catalyst is transferred to the filter paper. Find the mass of the solids when completely dry.

Questions

1   Did the two solids differ in appearance (a) before the reaction, (b) after the reaction? Describe any differences which you observed.
2   Does your answer to question 1 indicate that (a) the manganese(IV) oxide definitely took part in the reaction, (b) the manganese(IV) oxide may have taken part in the reaction, or (c) the manganese(IV) oxide did not take part in the reaction?
3   Did the mass of either sample of manganese(IV) oxide change?
4   Do your answers to questions 2 and 3 seem to contradict one another? Can you think of a possible explanation of what happened to the manganese(IV) oxide?
5   What was the purpose of the control experiment?

13.5.d   Does a catalyst take part in a reaction?

In the previous experiment the catalyst changed in appearance, but not in mass. One possible explanation is that it reacted with the hydrogen peroxide and formed an unstable intermediate compound which rapidly decomposed to give the catalyst back again, together with the products of the reaction. We can use the following experiment, in which solutions of hydrogen peroxide and potassium sodium tartrate react to give oxygen and carbon dioxide, to test this theory of catalysis.

PROCEDURE   Place 50 cm³ of 2 volumes hydrogen peroxide solution and 1 g of potassium sodium tartrate in a beaker. Heat the beaker carefully, stirring with a thermometer, until a vigorous effervescence occurs. Note the temperature at which this happens.Repeat the experiment with a fresh mixture, but this time, when the temperature reaches 70–80 °C, remove the flame and add a little solid cobalt(II) chloride to the beaker. Stir and observe all that happens.

Questions

1   Did the cobalt(II) chloride catalyse the reaction? Give a reason for your answer.
2   Did the cobalt(II) chloride appear to take part in the reaction? Was it regenerated at the end of the reaction? What led you to draw your conclusions?
3   Do your results support the theory of the action of a catalyst given at the beginning of the experiment?

13.5.e   To investigate the effect of changing the mass of catalyst used in a reaction

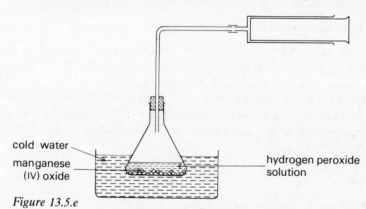

cold water

manganese (IV) oxide

hydrogen peroxide solution

Figure 13.5.e

PROCEDURE   Place 50 cm³ of 2 volumes hydrogen peroxide solution in a 100 cm³ conical flask. Connect a bung and delivery tube to a 100 cm³ glass syringe, make sure that the reading on the scale is zero and then push the bung into the neck of the flask. Twist the plunger and note the new reading; this 'zero error' must be subtracted from all readings taken later in the experiment. Remove the bung and push in the plunger once more.

Weigh out 0.05 g of precipitated manganese(IV) oxide on a watch glass. Tip the solid into the hydrogen peroxide solution, quickly connect the flask to the syringe and start a stop clock. Swirl the flask two or three times to disperse the catalyst and then place it in a trough of cold water to keep the temperature steady. Read the volume every 30 seconds until there is no further change, twisting the plunger before each reading is taken.

Repeat the experiment with fresh hydrogen peroxide solution but this time add 0.10 g of catalyst. Using the same piece of paper, plot graphs of volume of oxygen/cm^3 on the vertical axis against time/s on the horizontal axis. (Do not forget the zero error.)

Questions

1 What was the effect of the increased mass of catalyst on (a) the initial rate, (b) the total time of the reaction?
2 Can you account for this?

13.5.f To investigate the effect of changing the particle size of a catalyst

PROCEDURE Proceed as in the first part of the previous experiment but use 5 g of *granular* manganese(IV) oxide. Plot your graph on the same piece of paper as before.

Questions

1 What was the effect of increasing the particle size of the catalyst on (a) the initial rate, (b) the total time of the reaction?
2 How do you account for the differences?

13.6.a Can light affect the rate of a reaction?

PROCEDURE Add 5 cm^3 of 0.05 M sodium chloride solution to an equal volume of 0.05 M silver nitrate solution in a test tube, shake the mixture and place it in a dark cupboard. Now make a similar mixture but this time leave the tube in a rack on the bench. Repeat the whole procedure using 0.05 M sodium bromide solution and then 0.05 M sodium iodide solution in place of the sodium chloride solution. Inspect all of the tubes from time to time.

Questions

1 Which of the precipitates appeared to change?
2 What do you think was responsible for the change?
3 Do you know of any applications of this reaction?

13.1 Measuring the rate of a reaction

Why should we want to measure the rate of a reaction? Apart from simply satisfying our curiosity, there are practical gains to be made by studying reaction rates and finding out how to alter them. The speeding up of a reaction in industry will make a process more efficient and may save energy; alternatively, unstable substances may be handled and stored more safely if their rates of decomposition can be slowed down. In this chapter we are mainly concerned with reactions in which gases are evolved. This is because the courses of such reactions may be followed easily by measuring the changes in mass or volume which occur as they proceed. Such measurements can be made without disturbing the reacting mixture in any way.

Plotting a rate curve

If dilute hydrochloric acid is added to marble chips and the mass measured at regular intervals while the carbon dioxide is being given off (Experiment 13.1.a), a graph of decrease in mass against time may be drawn. The shape of this graph will be similar to that shown in Figure 13.1 and it illustrates a number of important points.

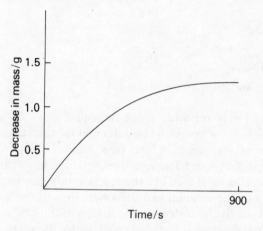

Figure 13.1 Rate curve for the reaction between 20 g of marble chips and 40 cm³ of 2 M hydrochloric acid

Clearly the reaction does not occur at the same rate throughout. The slope of the graph is steepest at the beginning, showing that the mass is decreasing at the greatest rate (i.e. the rate of evolution of carbon dioxide is greatest) when the acid is at its most concentrated. As the reaction progresses, the slope becomes less steep, indicating that the reaction is slowing down. This is because the acid is being used up and is becoming more dilute. Finally a point is reached where the graph becomes a horizontal straight line. Here the reaction has ceased; no further change in mass occurs because all of the acid has been used up.

Initial rate and average rate

If a tangent is drawn to any curve, its slope is equal to the slope of the curve at the point where the two lines touch. In Figure 13.1, the slope of the curve at any point is equal to the rate of evolution of carbon dioxide in $g\ s^{-1}$ at that particular moment. Thus a tangent drawn to the curve at its beginning will have a slope equal to the *initial rate* of the reaction. The total decrease in mass divided by the time taken for the reaction to go to completion (i.e. the time taken to reach the horizontal portion of the graph) will give a measure of the *average rate* of the reaction. In the same way, the average speed for a journey is the total distance travelled divided by the total time of travelling.

13.2 The effect of particle size on the rate of a reaction

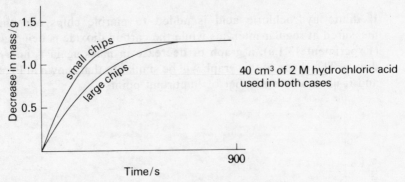

Figure 13.2 The effect of particle size on reaction rate

When Experiment 13.1.a is repeated using an equal mass of smaller marble chips, the initial rate of reaction is greater and the time taken for it to go to completion is less than before (see Figure 13.2). What is responsible for these differences? The only factor which has changed is the surface area of the chips and this, like the reaction rate, has increased. This means that more of the calcium and carbonate ions of the solid are brought into contact with the hydrated hydrogen ions of the acid and so it is not surprising that the reaction is speeded up. Because the same masses of reactants are used in each experiment, the overall decrease in mass is the same for both.

13.3 The effect of changes in concentration on the rate of a reaction

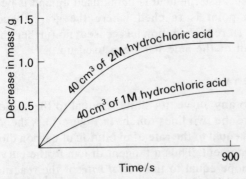

Figure 13.3 The effect of concentration change on reaction rate

The result obtained if the reaction between marble chips and dilute hydrochloric acid is carried out using different concentrations of acid is shown in Figure 13.3, and it is obvious that the initial rate is faster and that less time is taken to reach completion when the acid is more concentrated. The explanation for this is simply that by packing more hydrated hydrogen ions into the same volume of liquid, we increase the number of collisions per second between them and the surface of the solid and thus increase the rate at which the reaction proceeds. Doubling the concentration of the acid results in a doubling of the total mass of carbon dioxide produced because twice as much marble is consumed.

Are concentration and rate directly related?

In Experiments 13.3.b and 13.3.c it is found that the rate of a reaction is directly proportional to the concentration of one of the reactants. You must realise that this is not always the case; the relationship is often more complicated and a complete treatment of this topic must be left to a more advanced course.

Pressure changes

By increasing the pressure on gases we compress them into smaller volumes and thus increase their concentrations. This means that pressure changes have the same effect on the rates of reactions between gases as changes in concentration have on the rates of reactions in solution. Since liquids and solids are virtually incompressible, their reactions remain unaffected by variations in pressure.

13.4 The effect of change in temperature on the rate of a reaction

You must be aware from your own experiences in the laboratory that a rise in temperature causes the rate of a reaction to increase. It would seem reasonable to suggest that, because the reacting particles are speeded up when the temperature rises, more collisions per second must take place between them. However, although an increase in temperature of 10 K approximately *doubles* the rate of many reactions (Experiment 13.4.a), it may be shown that the number of collisions per second only increases by a *few per cent*. This is because relatively few of the particles are moving fast enough to have sufficient kinetic energy to react when they collide; in most collisions the particles simply bounce off one another and no reaction occurs. When the temperature rises by 10 K, the number of these high energy particles is roughly doubled and so the rate of reaction is doubled also. Remember that this doubling of reaction rate for every 10 K rise in temperature does not apply to *all* reactions at *all* temperatures. Nevertheless, it is a useful relationship to bear in mind.

13.5 The effect of a catalyst on the rate of a reaction

No doubt you have encountered catalysts earlier in your course and you probably know that they generally speed up chemical reactions in some way. For example, the thermal decomposition of potassium chlorate(v) into potassium chloride and oxygen takes place at a much faster rate if copper(II) oxide is present (Experiment 13.5.a). Similarly, the slow decomposition of hydrogen peroxide solution at room temperature is speeded up by the addition of metal oxides such as manganese(IV) oxide (Experiment 13.5.b). It is interesting to note that copper(II) oxide does *not* catalyse this second reaction; catalysts for one reaction do not necessarily affect the rate of another.

A closer look at the catalytic action of granular manganese(IV) oxide on the decomposition of hydrogen peroxide solution reveals that its appearance changes (the granules become powdery) but that its mass remains constant (Experiment 13.5.c). The difference in its appearance suggests that the catalyst has reacted in some way; on the other hand, the lack of change in its mass suggests that it has not. Which of these interpretations is correct? Chemists believe that catalysts *do* take part in reactions. They are thought to combine with one or more of the reactants to form an unstable intermediate compound which then reacts further to give the products of the reaction and regenerate the catalyst (see Figure 13.5). Some evidence in favour of this theory is obtained in Experiment 13.5.d where hydrogen peroxide solution oxidises a solution of potassium sodium tartrate to evolve a mixture of carbon dioxide and oxygen. The reaction is catalysed by cobalt(II) chloride which is pink in colour. As the experiment proceeds, the pink colour of the solution first changes to green and then, when the evolution of gases ceases, reverts to its original pink. Presumably the green colour is due to the catalyst forming an intermediate compound.

We can now explain why a catalyst alters the rate of a reaction. The formation of the intermediate compound requires less energetic collisions between particles than does the direct combination of the reactants. Thus more particles have sufficient energy to react at any given temperature and the reaction rate is increased. Obviously, if the surface area of the catalyst is made greater, the chances of fruitful collisions are improved (Experiments 13.5.e and 13.5.f).

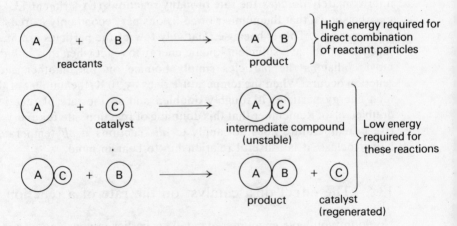

Figure 13.5 The action of a catalyst

So far we have only considered catalysts which speed up chemical reactions; these are known as *positive catalysts*. In some cases it may be desirable to slow down a reaction which is occurring too quickly. *Negative catalysts* (*inhibitors*) are used for this purpose, one well-known example being the 'anti-knock' compound, tetraethyl-lead(IV), $Pb(C_2H_5)_4$. This is added to petrol to prevent premature ignition of the mixture of petrol and air in the cylinder of an engine. A definition of a catalyst which covers both positive catalysts and inhibitors is as follows.

A **catalyst** is a substance which alters the rate of a chemical reaction but may be recovered unchanged in mass at the end.

13.6   The effect of light on the rate of a reaction

There are a few reactions for which light provides sufficient energy to noticeably affect the rate. Such reactions are said to be *photosensitive* and several examples are mentioned elsewhere in the book. They include the reactions between hydrogen and chlorine (19.2), hexane and bromine (Experiment 21.2.a), and perhaps the most important chemical reaction of all, photosynthesis (3.3).

Another photosensitive reaction is the decomposition of silver halides to give metallic silver (Experiment 13.6.a). This reaction is widely used in photography. Light falls on to a film which is coated with a mixture containing a silver halide and, on developing and fixing, a 'negative' is obtained. The bright parts of the picture are black on the negative because the greater light intensity has produced more silver there. By shining light through the negative on to a piece of paper which has also been coated with a silver halide, a 'negative of the negative', i.e. a positive print, is obtained.

Summary

The rate of a chemical reaction may be increased by
(a) increasing the surface area of the reactants (i.e. making them more finely divided),
(b) increasing the concentrations of the reactants (or increasing the pressure if gases are involved),
(c) increasing the temperature,
(d) adding a (positive) catalyst,
(e) increasing the light intensity (applies to relatively few reactions only).

Questions on Chapter 13

1   A crystal of calcite (pure calcium carbonate) weighing 7.50 g was placed in a flask with $50 \, cm^3$ of dilute hydrochloric acid. The flask was kept at constant temperature and the carbon dioxide evolved was collected in a graduated vessel. The volume of carbon dioxide was recorded at 20-minute intervals and corrected

to s.t.p. Some of the calcite remained undissolved at the end of the experiment. The results of the experiment are given in the table below.

Time from start of reaction/min	Volume of carbon dioxide formed at s.t.p./cm^3
20	655
40	910
60	1065
80	1100
100	1120
120	1120

(a) Write the equation for the reaction between calcium carbonate and hydrochloric acid.

(b) 655 cm^3 of gas were evolved during the first 20-minute interval. Write down the volumes of gas evolved:
   (i) during the second 20-minute interval (20 to 40 minutes),
   (ii) during the third 20-minute interval (40 to 60 minutes),
   (iii) during the fourth 20-minute interval (60 to 80 minutes).
Explain the variation in the volumes of gas formed during these intervals.

(c) Why is there no increase in volume of gas after 100 minutes?

(d) What is the mass of 1120 cm^3 of carbon dioxide at s.t.p.? (Molar volume = 22.4 dm^3 mol^{-1} at s.t.p.)

(e) What mass of calcite will have reacted with the acid after 120 minutes?

(f) Calculate the original concentration of the acid used in grams of hydrogen chloride per dm^3.

(g) What will be the least volume of hydrochloric acid of this concentration needed to dissolve the calcite remaining at the end of the reaction?   (C)

2   In a series of experiments in which magnesium ribbon of uniform width reacted with excess dilute hydrochloric acid, the rate of evolution of hydrogen was found to be as follows:

Length of ribbon/cm	1.0	2.0	3.0	4.0	5.0	6.0	7.0
Rate of evolution of hydrogen/cm^3 min^{-1}	1.1	1.8	2.7	3.6	4.6	5.4	6.4

Draw a graph and state what conclusion you reach. Calculate the rate of evolution of hydrogen from a piece of magnesium ribbon 12 cm long under the same conditions.
    Describe, with experimental details, how these measurements could be made.
                                                                              (L)

3   (a) Describe an experiment that you have either seen or performed to show how temperature affects the rate of *either* the reaction of a metal with a dilute acid *or* the decomposition of hydrogen peroxide solution.
    Include in your description the apparatus used, the experimental procedure and the measurements made.

(b) An excess of dilute hydrochloric acid is added to 10 g of marble (calcium carbonate) and left at constant temperature until the reaction has stopped.

(i) Calculate the maximum volume of carbon dioxide at s.t.p. formed when the reaction has stopped. (Molar volume = $22.4\ dm^3\ mol^{-1}$ at s.t.p.)

(ii) Draw a graph to show approximately how you would expect the volume of carbon dioxide formed to vary with *time* during the experiment. (C)

4   What is a *catalyst*? Write equations for *two* catalysed reactions of industrial importance and for *each* reaction name the catalyst used.

Describe experiments you would carry out to compare the effectiveness of two given metallic oxides when used as catalysts in the decomposition of hydrogen peroxide solution. (C)

5   Manganese(IV) oxide is a *catalyst* in the decomposition of aqueous hydrogen peroxide to water and oxygen. Using this as an example, explain the term *catalyst*.

With the aid of a diagram, describe an apparatus which you could use to measure the volume of oxygen evolved in a given time. How would you show that the manganese(IV) oxide had acted as a catalyst?

Describe *one* reaction, apart from the decomposition of hydrogen peroxide, in which a metal is used as a catalyst. (O & C)

6   A chemist attempts to prepare some hydrogen by allowing sulphuric acid to react with zinc. Using lumps of zinc of about 0.5 cm diameter and 0.2 M acid he finds that gas is given off but the reaction is too slow. State, and explain as far as you can, three ways in which he could alter the conditions of reaction so that the gas would be formed more rapidly. Which of these would you recommend as the most suitable, and why? If the chemist wanted to produce $36\ dm^3$ of hydrogen, at room temperature and pressure, how much zinc (to the nearest 5 g) would he need? (The volume of 1 mole of gas at 101325 Pa and room temperature is approximately $24\ dm^3$. Relative atomic masses: Zn = 65.4, H = 1.00.) (L)

14   Reversible reactions

When magnesium is added to dilute sulphuric acid, hydrogen is evolved and magnesium sulphate solution is formed. However, if hydrogen is bubbled through the magnesium sulphate solution in an attempt to reform the original materials, no change is detected. The reaction between magnesium and dilute sulphuric acid is *irreversible* and *goes to completion* (i.e. it continues until all of the metal or all of the acid is used up). There are, however, some reactions which do not behave in this way; they do *not* go to completion and they *can* be reversed.

Experiments

14.1.a   To study the effect of acid and alkali on bromine water

PROCEDURE   Place 1 cm depth of bromine water in a test tube. Add two drops of sodium hydroxide solution, shake the tube and then add two drops of dilute hydrochloric acid. Repeat the additions of acid and alkali several times.

Questions

1   What colour was the bromine water originally?
2   What happened on adding (a) the sodium hydroxide solution, (b) the dilute hydrochloric acid?
3   Did the reaction between the bromine water and the alkali appear to be reversible?

14.1.b   The action of water on bismuth chloride ★

PROCEDURE   Fill the rounded part of a test tube with bismuth chloride, dissolve the solid in 1 cm³ of concentrated hydrochloric acid and then fill the tube with distilled water. Tip this mixture into a 250 cm³ beaker and add more of the concentrated hydrochloric acid until a clear solution is obtained. Repeat the addition of water and acid several times.

Questions

1   What did you see when distilled water was added to the solution of bismuth chloride in hydrochloric acid?
2   What happened when concentrated hydrochloric acid was added to the product in question 1?
3   Did the reaction between water and bismuth chloride appear to be reversible?

14.1.c The effect of heat on ammonium chloride – a closer look

You should be familiar with the idea that ammonium chloride sublimes (see Experiment 1.2.h). When sublimation occurs does the ammonium chloride simply vaporise, or does a more complicated change take place?

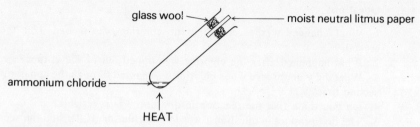

glass wool ⟶ ⟵ moist neutral litmus paper

ammonium chloride ⟶

↑ HEAT

Figure 14.1.c

PROCEDURE Place a small quantity of ammonium chloride (sufficient to just cover the end of a spatula) in the bottom of a test tube and set up the apparatus as shown in Figure 14.1.c. (**Caution:** do not handle the glass wool unnecessarily.) Heat the ammonium chloride gently and observe the litmus paper on both sides of the glass wool. Be careful not to char the paper. When no solid remains at the base of the tube, allow the apparatus to cool and then remove the litmus paper and examine it.

Questions

1 In what way did the colour of the litmus paper change during the experiment? Which gases could have been formed from the ammonium chloride, NH_4Cl, to cause these changes?
2 Which of these gases diffused through the glass wool at the greater rate? Why did it do this?
3 Did the gases seem to recombine to form ammonium chloride again on cooling?
4 Write an equation for the reaction which took place.

14.2.a To study the reaction between iodine and chlorine ★ ★

This experiment must be carried out in a fume cupboard.

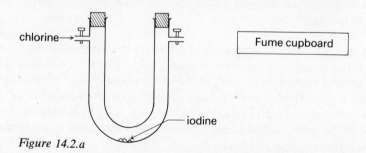

chlorine ⟶

Fume cupboard

iodine

Figure 14.2.a

PROCEDURE Place one or two crystals of iodine (**care**) in the bottom of a U-tube and stand it in front of a white screen. Pass dry chlorine over the iodine (Figure 14.2.a) until no further change is seen. Disconnect the chlorine supply, invert the U-tube for a moment, so that the chlorine is tipped out, and then look for evidence of another reaction. When this reaction appears to have finished, pass the chlorine through the apparatus as before but this time close the taps when the gas supply is disconnected.

Questions

1 What changes were seen when the chlorine was passed over the iodine crystals?

2 What happened after the chlorine was tipped out of the U-tube?

3 What did you observe when chlorine was passed through the apparatus for the second time?

4 Do you think that the reaction in question 3 is reversible?

5 The first product in question 1 was iodine chloride, $ICl(l)$, and the second was diiodine hexachloride, $I_2Cl_6(s)$. Write an equation for the reaction between iodine chloride and chlorine to form diiodine hexachloride.

6 Bearing in mind the changes which you observed and your answer to question 4, which of the following do you think is the better explanation of what happened when the chlorine supply was disconnected and the taps closed? (a) The diiodine hexachloride partly decomposed to form iodine chloride and chlorine and then the reaction stopped. (b) The diiodine hexachloride was continually decomposing into iodine chloride and chlorine, but these products were recombining to give the diiodine hexachloride back again.

14.2.b To investigate the equilibrium between a solute and its saturated solution ★ ★

This experiment must not be performed by pupils.

When solid lead(II) chloride is placed in a beaker of distilled water it dissolves until a saturated solution is formed. Once this point has been reached, the concentration of the solution and the mass of the solid in the bottom of the beaker remain constant, provided that the temperature does not change – the solid and solution are in *equilibrium* with one another. Does the lead(II) chloride stop dissolving, or is there a state of *dynamic equilibrium*, where ions are still going into solution but others are being deposited on the solid at the same rate, so that the two effects cancel one another out? To answer this question we must be able to follow the movements of the various particles. One way of doing this is to use radioactive lead ions.

PROCEDURE   Place 5 cm³ of distilled water in a Geiger-Müller tube and count the radiation for five minutes. This gives a measure of the *background radiation* present in the atmosphere of the laboratory.

Add a spatula measure of non-radioactive lead(II) chloride to about 50 cm³ of distilled water in a 100 cm³ beaker, heat to boiling to dissolve some of the solid and then cool again so that a saturated solution at room temperature is obtained. Suck off some of the clear liquid with a dropping pipette, centrifuge it if necessary, and transfer 5 cm³ to the Geiger-Müller tube. Count the radiation for five minutes.

Place 5 cm³ of 0.2 M lead(II) nitrate solution in a boiling tube together with 10 cm³ of 0.1 M thorium nitrate solution (this always contains some

radioactive lead-212 ions owing to decay of the radioactive thorium ions).
Add dilute hydrochloric acid to precipitate lead(II) chloride. This sample
of lead(II) chloride will, of course, contain radioactive lead ions as well as
the ordinary non-radioactive ones. Centrifuge and add the precipitate to
about 30 cm^3 of the clear saturated solution of lead(II) chloride prepared
earlier. Stir the mixture with a magnetic stirrer for about twenty minutes,
centrifuge, and then count the radiation from a 5 cm^3 portion of the
solution for five minutes.

Questions

1 What was the background count per minute?
2 Compare the count per minute obtained from (a) the first lead(II) chloride
solution, (b) the second lead(II) chloride solution with that obtained for the
background radiation.
3 Did any of the radioactive lead ions from the solid enter the saturated
solution? If so, what must have happened to some of the lead ions already in the
solution? (Remember that the concentration of a saturated solution in contact
with undissolved solute remains constant, provided that the temperature does not
change.)
4 Do your results indicate that the equilibrium between a solute and its saturated
solution is static, or dynamic?

14.3.a **The effect of a change in concentration on the position of equilibrium in a
reversible reaction ★**

Solutions containing iron(III) ions and thiocyanate ions react to produce a
blood-red colour, according to the following equation.

$$Fe^{3+}(aq) + CNS^-(aq) \rightleftharpoons [Fe(CNS)]^{2+}(aq)$$
$$\text{yellow} \qquad \text{colourless} \qquad \text{blood red}$$

Use this equation to interpret the results of the experiment described
below.

PROCEDURE Mix together one drop of 0.5 M iron(III) chloride solution
and one drop of 0.5 M potassium thiocyanate solution in a test tube and
add sufficient distilled water to form a pale pink solution. Divide this
solution into three equal parts and add one drop of the original iron(III)
chloride solution to one of them and one drop of the potassium thiocyanate solution to another. Compare the colours of these solutions with that
of the untouched third sample. Now add a little solid sodium fluoride to
the two solutions, stir well and again compare the colours with that of the
original solution. (The sodium fluoride removes iron(III) ions from the
equilibrium.)

Questions

1 In what way did the colour change on adding (a) iron(III) chloride solution, (b)
potassium thiocyanate solution, (c) solid sodium fluoride, to the equilibrium
mixture?
2 Does an increase in the concentration of one of the reactants (a) cause more

product to form, (b) cause less product to form, (c) have no effect on the equilibrium position?

3 Explain your answer to question 2 in terms of the effect of a change in concentration of the reactants on the rates of the forward and backward reactions.

4 What is the effect of a decrease in concentration of one of the reactants on the position of equilibrium? Again, explain your answer in terms of variation in the rates of the forward and backward reactions.

14.3.b The effect of a temperature change on the position of equilibrium in a reversible reaction ★ ★

At room temperature 'nitrogen dioxide' actually consists of a mixture of dinitrogen tetraoxide molecules, N_2O_4, and nitrogen dioxide molecules, NO_2, in dynamic equilibrium.

$$N_2O_4(g) \underset{\text{exothermic}}{\overset{\text{endothermic}}{\rightleftharpoons}} 2NO_2(g)$$

pale yellow $\qquad\qquad$ dark brown

Any movement in the position of equilibrium will be indicated by a corresponding change in colour and volume (why?).

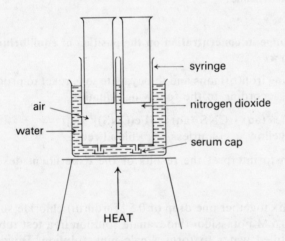

Figure 14.3.b

PROCEDURE Flush out a $100\ cm^3$ glass syringe with nitrogen dioxide, collect a $50\ cm^3$ sample of the gas in it and close the end with a serum cap. Repeat this procedure with a second syringe. Keep one sample for reference and place the other, together with a similar syringe containing $50\ cm^3$ of air, in a $1\ dm^3$ beaker of water (see Figure 14.3.b). Heat the beaker until the water boils, twist the syringe plungers, and note the volumes of the gases. Compare the colour of the hot nitrogen dioxide with that of the reference sample. Allow the syringes to cool to room temperature and again read the volumes and compare the colours. Finally, place the syringes in a beaker containing a mixture of ice and water. Once more compare the volumes of air and nitrogen dioxide and the colours of the nitrogen dioxide samples.

Questions

1 Since the volumes of nitrogen dioxide and air were equal at room temperature, what can you say about the numbers of molecules of gas in the two syringes at the beginning of the experiment?

2 Were the volumes equal at 100°C? What must have happened to the number of molecules in the nitrogen dioxide sample as it was heated?

3 How did the colour of the hot nitrogen dioxide compare with that of the reference sample at room temperature?

4 Refer to the equation given in the introduction to the experiment and explain your results in terms of changes in the rates of the forward and backward reactions and in the position of equilibrium.

5 In a similar way, explain the results obtained when the air and nitrogen dioxide were cooled in the beaker of ice and water.

6 Show how your results could have been predicted by using Le Chatelier's principle.

14.3.c   Can light affect the equilibrium position in a reversible reaction?★★

In Chapter 13 you learned that the energy in light is great enough to affect the rates of a few reactions. If a reversible photosensitive reaction has reached equilibrium, we might expect a change in the intensity of light to alter the position of this equilibrium, just as a change in temperature did in the last experiment.

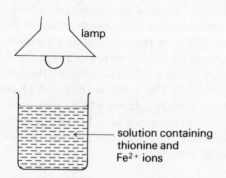

Figure 14.3.c

PROCEDURE   Add about 0.01 g of thionine to 10 cm³ of distilled water in a test tube, stir well and leave overnight to obtain a saturated solution. Filter off the excess solid and place 5 cm³ of the filtrate in a 250 cm³ measuring cylinder. Add 10 cm³ of 1 M sulphuric acid, make up to the 250 cm³ mark with distilled water and pour the mixture into a 500 cm³ beaker containing 1 g of iron(II) sulphate-7-water. Stir well, so that the crystals dissolve and the following equilibrium is set up.

$$Fe^{2+}(aq) + thionine(aq) \rightleftharpoons Fe^{3+}(aq) + reduced\ thionine(aq)$$
pale green   violet   yellow   colourless

Place a white screen behind the beaker and illuminate the solution from above by means of a 100 watt lamp, as shown in Figure 14.3.c. When no further change occurs, switch off the lamp and observe the solution for a minute or so. Repeat the illumination several times.

Question

1 In what way were the rates of the forward and backward reactions and the position of equilibrium affected by the intensity of the light?

14.3.d The effect of a change in pressure on the equilibrium position in a reversible reaction ★★

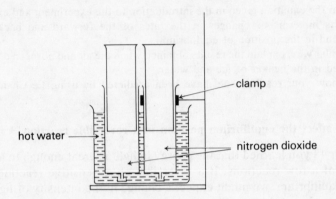

Figure 14.3.d

PROCEDURE Place two 100 cm³ glass syringes, each containing 50 cm³ of nitrogen dioxide and previously greased to ensure gas-tightness, in a 1 dm³ beaker of water. Heat the water to a temperature of about 90°C and then place the beaker on the bench in front of a white screen (Figure 14.3.d). Clamp one of the syringes so that its serum cap is pressed firmly against the bottom of the beaker and press the plunger until the volume of the gas is approximately halved. Hold this position for about 15 seconds, comparing the colour of the gas with that in the other syringe, and then release the plunger. Repeat the experiment in a beaker of water at a temperature of about 5°C but this time pull the plunger *out* until the volume of gas is approximately doubled.

Questions

1 What happened to the colour of the hot gas as its pressure was increased? Explain why this change occurred.
2 Did the colour change while the increased pressure was held constant? If so, did it darken or lighten?
3 In what way did the colour of the cold gas alter as the pressure was decreased? How can this be explained?
4 Did the colour darken, lighten, or stay the same, while the reduction in pressure was maintained?
5 Refer to the equation given in Experiment 14.3.b and explain the observations made in questions 2 and 4 in terms of variation in the rates of the forward and backward reactions and in the position of equilibrium between the molecules of dinitrogen tetraoxide and nitrogen dioxide.
6 Do the results agree with the predictions of Le Chatelier's principle? Explain your answer.

14.1   Reactions which can go both ways

By now you must have come across several reactions which can be made to go backwards to regenerate the original reactants. For example, if an alkaline solution of litmus is acidified, the colour of the litmus changes from blue to red; the addition of alkali to this red solution reverses the change and gives the blue colour back again. Likewise, if blue copper(II) sulphate crystals are heated, water is given off and the white anhydrous salt remains. When water is added to this solid, blue hydrated copper(II) sulphate is formed once more.

$$CuSO_4 \cdot 5H_2O(s) \rightleftharpoons CuSO_4(s) + 5H_2O(l)$$

(The sign $\rightleftharpoons$ shows that the reaction can proceed in both directions.)

A further example of this type of change is provided by Experiment 14.1.a where the reaction between bromine water and sodium hydroxide solution causes the colour of the bromine to be discharged. Acidification of the resultant colourless liquid drives the reaction back again and the orange colour of the bromine reappears.

Similarly, the addition of water to a solution of bismuth chloride in hydrochloric acid produces a white precipitate of bismuth chloride oxide (Experiment14.1.b). This disappears when more hydrochloric acid is added, only to be reformed upon further dilution with water.

All of these changes are examples of *reversible reactions*.

> A **reversible reaction** is one which can proceed in either direction, depending on the conditions under which it is carried out.

Thermal dissociation and thermal decomposition

When solid ammonium chloride is heated it sublimes, that is, it changes directly to a vapour which reforms the solid on cooling. At first sight, this appears to be nothing more than a rather specialised change of state but further investigation shows that something more complicated than ordinary vaporisation is taking place (see Experiment 14.1.c). The ammonium chloride decomposes into ammonia and hydrogen chloride as it is heated and these gases recombine on cooling. Thus a reversible chemical reaction is occurring, according to the following equation.

$$NH_4Cl(s) \rightleftharpoons NH_3(g) + HCl(g)$$

The reaction proceeds to the right on heating and to the left as the products cool down; it is an example of *thermal dissociation*.

> **Thermal dissociation** is the splitting up of a compound by the action of heat to give products which recombine on cooling.

Other examples of substances which dissociate on heating include calcium carbonate and nitrogen dioxide. Can you think of any more?

$$CaCO_3(s) \rightleftharpoons CaO(s) + CO_2(g)$$
$$2NO_2(g) \rightleftharpoons 2NO(g) + O_2(g)$$

Of course, many substances decompose on heating to give products which do *not* recombine on cooling. They are said to undergo *thermal decomposition*.

e.g.

$$NH_4NO_3(s) \rightarrow N_2O(g) + 2H_2O(g)$$
$$2KClO_3(s) \rightarrow 2KCl(s) + 3O_2(g)$$

Thermal decomposition is the splitting up of a compound by the action of heat to give products which do not recombine on cooling.

14.2   Dynamic equilibrium

A chemical reaction usually continues until one of the reactants is completely used up. Thus, if we were to add a little dilute hydrochloric acid to some marble chips and wait until the evolution of carbon dioxide ceased, we would expect the residue to contain *either* unreacted acid *or* excess marble. It is possible that neither would be present (why?), but we would be very surprised to find *both*. Bear this in mind as you read through the following section.

The reaction between iodine chloride and chlorine

Iodine chloride is formed as a brown liquid when chlorine is passed over iodine (Experiment 14.2.a).

$$I_2(s) + Cl_2(g) \rightarrow 2ICl(l)$$

The passage of more chlorine causes a further reaction to take place, producing yellow crystals of diiodine hexachloride. This reaction is reversible, as is easily demonstrated by leaving the crystals exposed to the atmosphere. When this is done, iodine chloride and chlorine are reformed. The equation is thus:

$$2ICl(l) + 2Cl_2(g) \rightleftharpoons I_2Cl_6(s).$$

If a sample of diiodine hexachloride is left in a *sealed* tube, some of it decomposes into iodine chloride and chlorine but then the reaction seems to stop. Why should this happen? Alternatively, since iodine chloride and chlorine are *both* present, why do they not combine? There seem to be no obvious answers to these questions, so could it be that the decomposition of the diiodine hexachloride does *not* stop and that the iodine chloride and chlorine *do* combine?

Let us consider the most likely course of events if both reactions take place simultaneously. At the beginning of the experiment the diiodine hexachloride starts to decompose and small amounts of iodine chloride and chlorine are produced. These substances immediately begin to react together but, since their concentrations are low, their rate of combination is slow. As more diiodine hexachloride decomposes, the concentrations of iodine chloride and chlorine build up and the rate of the backward reaction increases. Eventually, a point is reached where the rate of decomposition of the diiodine hexachloride is equal to the rate of its reformation and, when this happens, the concentrations of all the substances in the sealed tube remain constant and there is no *apparent* change.

There is no reason why the forward and backward reactions should *not* take place at the same time and, by assuming that they do, we have produced an explanation of what was observed during the experiment. How can we obtain proof of our theory? One way is to label individual atoms and follow their paths throughout the course of the reaction. This may sound impossible but chemists *can* label atoms; they make them radioactive and then check on their movements as the reaction proceeds by means of a Geiger-Müller counter. It would be difficult for us to apply this technique to the reaction between iodine chloride and chlorine, but we can use it to investigate another process which appears to stop before reaching completion, namely the dissolving of a sparingly soluble salt in water.

Can a solid dissolve in its own saturated solution?

If a point is reached in a reversible reaction where the proportions of the reactants and products become constant and no change appears to be taking place, we say that the various substances present are *in equilibrium* with one another. This situation occurs in a sealed tube containing iodine chloride, chlorine and diiodine hexachloride. Another mixture in equilibrium is obtained if a sparingly soluble salt, such as lead(II) chloride, is placed in water. At first the salt dissolves but then the solution becomes saturated and the dissolving seems to stop. However, if solid lead(II) chloride containing radioactive lead ions is added to the saturated solution, it is found that the solution becomes radioactive (Experiment 14.2.b). Thus some of the radioactive solid must dissolve in the already saturated solution. Since the concentration of dissolved lead(II) chloride is fixed, provided that the temperature does not change, some of the ions which are already in solution must be precipitated on the surface of the solid. The rates of these two processes must be equal, so that their effects cancel out.

$$PbCl_2(s) + water \rightleftharpoons Pb^{2+}(aq) + 2Cl^-(aq)$$

An equilibrium of this kind, in which two opposing reactions are continuing at the same rate is called a *dynamic equilibrium*. It is believed that all chemical equilibria are of this type.

> **Dynamic equilibrium** is reached in a reversible reaction when the rate of the forward reaction is equal to the rate of the backward reaction.

The way in which *any* dynamic equilibrium is set up is similar to that suggested for the diiodine hexachloride/iodine chloride/chlorine mixture in the previous section. At the beginning of the reaction, the reactants combine to form the products, but as time passes their concentrations decrease and the rate of the forward reaction falls. At the same time, since the concentrations of the products are building up, the rate of the reverse reaction increases. Eventually the two rates become equal and cancel out.

In many reversible reactions, one or more of the products is allowed to escape. If this happens it is impossible for an equilibrium to be set up because the backward reaction cannot take place.

14.3   Varying the position of equilibrium – Le Chatelier's principle

The effect of a change in concentration of reactants on the position of equilibrium

When solutions of iron(III) chloride and potassium thiocyanate are mixed, the following equilibrium is set up:

$$Fe^{3+}(aq) + CNS^-(aq) \rightleftharpoons [Fe(CNS)]^{2+}(aq)$$
$$\text{yellow} \quad\quad \text{colourless} \quad\quad\quad \text{blood red}$$

If more iron(III) chloride or potassium thiocyanate is added, the colour of the solution darkens (see Experiment 14.3.a). Thus a new equilibrium mixture must be formed which contains more of the $[Fe(CNS)]^{2+}(aq)$ ions. The reason for this is that an increase in the concentration of the reactants speeds up the forward reaction and disturbs the dynamic equilibrium. More product forms, slowing down the forward reaction and speeding up the backward reaction until the two rates become equal once more. When this happens there are more of the blood-red ions than there were before – the position of equilibrium has moved to the right.

The removal of iron(III) ions (e.g. by adding sodium fluoride) has the reverse effect; the rate of the forward reaction is decreased. Thus more $[Fe(CNS)]^{2+}$ ions dissociate, slowing down the backward reaction and speeding up the forward one until dynamic equilibrium is re-established. There are now fewer blood-red ions than before and the colour is noticeably paler; the equilibrium position has been displaced to the left.

We can explain the results of other experiments in a similar way. For example, when bismuth chloride is dissolved in hydrochloric acid, (Experiment 14.1.b), an equilibrium is established:

$$BiCl_3(aq) + H_2O(l) \rightleftharpoons BiOCl(s) + 2HCl(aq).$$

The position of equilibrium is well to the left and the solution is clear. However, the addition of water produces a white precipitate of bismuth chloride oxide; thus the equilibrium must be displaced to the right. If concentrated hydrochloric acid is added, the position of equilibrium moves back to the left and the solution becomes clear again. You should make sure that you can interpret these results in terms of changes in the rates of the two opposing reactions and the effects of these changes on the dynamic equilibrium.

You should have noticed by now that the addition of one or more of the substances which appear on one side of the equation always displaces the position of equilibrium *away from* that side. On the other hand, removing one or more substances from the mixture has the opposite effect; the equilibrium position is displaced *towards* that side of the equation. The ways in which changes of concentration, pressure and temperature alter the position of an equilibrium were investigated by the French scientist, Henry Le Chatelier who, in 1888, put forward the rule known as *Le Chatelier's principle*.

> If a system in equilibrium is subjected to a change, the position of equilibrium will move in such a way as to tend to eliminate the change.

Thus, when one of the substances on the left hand side of the equation is added to a system in equilibrium, the position of equilibrium moves to the right so that the concentration of the added substance is reduced.

The effect of a change in temperature on the position of equilibrium

If equal volumes of air and nitrogen dioxide at room temperature are heated at constant pressure, the colour of the nitrogen dioxide changes from light brown to dark brown and its volume increases much more than that of the air does (Experiment 14.3.b). The process reverses as the gases cool.

The reason for these changes is that, at room temperature, 'nitrogen dioxide' really consists of an equilibrium mixture of two substances, dinitrogen tetraoxide and nitrogen dioxide.

$$N_2O_4(g) \underset{\text{exothermic}}{\overset{\text{endothermic}}{\rightleftharpoons}} 2NO_2(g)$$

On heating, the dinitrogen tetraoxide undergoes thermal dissociation and the equilibrium shifts to the right. Thus the colour of the mixture darkens and the volume increases by more than would be expected from simple thermal expansion. The temperature rise speeds up both the forward and the backward reaction but, clearly, has a greater effect on the former.

These changes could have been predicted by using Le Chatelier's principle as follows. If the temperature of the equilibrium mixture were increased, the system should adjust itself in such a way as to lower the temperature again. Since the forward reaction is endothermic, a shift in the position of equilibrium to the right would have the desired effect.

The effect of a change in pressure on the position of equilibrium

Consider again the equilibrium between dinitrogen tetraoxide and nitrogen dioxide.

$$N_2O_4(g) \quad \rightleftharpoons \quad 2NO_2(g)$$
$$\text{pale yellow} \quad \text{dark brown}$$

According to Le Chatelier's principle, an increase in the pressure exerted on the equilibrium mixture will result in changes occurring to reduce the pressure again. For this to happen the position of equilibrium must move to the left, thereby decreasing the number of molecules of gas and hence the pressure exerted by them. Thus as the external pressure rises, the colour of the mixture should lighten. In practice the colour *darkens* at first, due to the increased concentration of gas. After a short time the expected effect is seen (see Experiment 14.3.d).

When the pressure on the system is reduced, the reverse change takes place. The equilibrium position shifts to the right to produce more molecules of gas and increase the pressure. As a result of this, the colour of the mixture deepens. (Again there is an initial *fading* of colour due to the decrease in gas concentration.)

The effect of a catalyst on the position of equilibrium

As you learnt in the previous chapter, the function of a catalyst is to alter the rate of a chemical reaction. However, in a reversible reaction, a catalyst changes the rates of the forward and backward reactions to the same extent and so the position of equilibrium is unchanged. Nevertheless, the addition of a positive catalyst will enable equilibrium to be established more rapidly and therefore the yield will be obtained in a shorter time.

14.4   Reversible reactions in industry

Many industrial processes involve reversible reactions and, since cost is an extremely important consideration in industry, a great deal of care is taken in determining the conditions of concentration, temperature and pressure which will produce the product as economically as possible. We will now consider the equilibria set up in some very important manufacturing processes and see which factors govern the choice of conditions under which the reactions are carried out.

The Haber process

In the Haber process (17.3) nitrogen and hydrogen are combined to give ammonia.

$$N_2(g) + 3H_2(g) \rightleftharpoons 2NH_3(g) + heat$$

Which conditions will give the best yield of ammonia? Since the forward reaction is exothermic and leads to a decrease in the number of molecules of gas, Le Chatelier's principle predicts that low temperature and high pressure should be used in order that the position of equilibrium should be established as far to the right as possible. Table 14.4, which gives the

Temperature/ K (°C)	Yield of ammonia at 250 atm	Pressure/atm	Yield of ammonia at 823 K (550 °C)
1273 (1000)	negligible	1	negligible
823 (550)	15%	100	7%
473 (200)	88%	1000	41%

Table 14.4 Yields of ammonia at various temperatures and pressures

percentages of ammonia in the equilibrium mixture at various temperatures and pressures, shows that these predictions are correct.

Unfortunately, at low temperatures the *rate* at which equilibrium is reached is very slow; also, if very high pressures are used, expensive industrial plant will be required. Thus, in practice, optimum conditions of 773 K (500 °C) and 250 atmospheres are chosen so that both the rate of reaction and the yield are reasonable. An iron catalyst ensures that the reaction proceeds as quickly as possible under the conditions used but, even so, the position of equilibrium is not usually reached by the time the gases leave the catalyst chamber. The net result is a yield of about 10%, the unchanged gases being recirculated until, eventually, they combine.

The Contact process

Sulphuric acid is manufactured by the Contact process (18.8), which involves the combination of sulphur dioxide and oxygen to give sulphur trioxide.

$$2SO_2(g) + O_2(g) \rightleftharpoons 2SO_3(g) + heat$$

Just as in the Haber process, the forward reaction is exothermic and is accompanied by a decrease in the number of molecules of gas. Thus low temperature and high pressure will favour the production of sulphur trioxide. These conditions have the same drawbacks as before and once again a compromise has to be made. In this case a temperature of 723 K (450 °C) and a complex vanadium catalyst (see page 360) are used to give a satisfactory rate of reaction. Even at a pressure of only 1 atmosphere the yield is still very good (about 98%). Therefore the cost of compressors to produce higher pressures is not justified and the plant is operated at atmospheric pressure.

The limekiln

In the limekiln, calcium oxide (quicklime) is produced by heating calcium carbonate to a temperature of about 1273 K (1000 °C).

$$CaCO_3(s) \rightleftharpoons CaO(s) + CO_2(g) - heat$$

Although the reaction is reversible, the backward part does not take place because the carbon dioxide is allowed to escape from the top of the kiln. This means that equilibrium cannot be established and the reaction goes to completion. Obviously there is no problem in obtaining a high yield in any reversible reaction where one or more of the products can be removed.

14.5 Further examples of reversible reactions

The purpose of this short section is to ensure that you understand the various principles which have been discussed earlier in the chapter. The examples in Table 14.5 have been taken from industrial processes and laboratory experiments which you will find described in detail elsewhere in the book. In each case the conditions under which they are carried out have been chosen in order to favour the forward reaction. Study the table carefully and check that you understand why these conditions have been selected.

Reaction	Conditions
Reaction between iron and steam (20.1) $3Fe(s) + 4H_2O(g) \rightleftharpoons Fe_3O_4(s) + 4H_2(g)$	Steam sweeps hydrogen out of apparatus; high temperature and excess steam used.
Blast furnace (20.6) $Fe_2O_3(s) + 3CO(g) \rightleftharpoons 2Fe(s) + 3CO_2(g)$	High temperature and excess CO used; CO and CO_2 escape from top of furnace.
Esterification (21.10) $CH_3COOH(l) + C_2H_5OH(l) \rightleftharpoons CH_3COOC_2H_5(l) + H_2O(l)$	Mixture boiled; concentrated H_2SO_4 added to catalyse reaction and absorb water.
Hydrolysis of esters (21.10) $CH_3COOC_2H_5(l) + H_2O(l) \rightleftharpoons CH_3COOH(l) + C_2H_5OH(l)$	Mixture boiled; sodium hydroxide solution added to catalyse reaction and neutralise acid.
Effect of heat on red lead oxide (20.7) $2Pb_3O_4(s) \rightleftharpoons 6PbO(s) + O_2(g)$	Solid heated; oxygen allowed to escape.

Table 14.5 Some reversible reactions

Summary

For a reversible reaction:
1 An increase in the concentration of any component will
(a) increase the rate at which equilibrium is attained
(b) cause the position of equilibrium to shift in such a way as to reduce the concentration of that component.
2 An increase in temperature will
(a) increase the rate at which equilibrium is attained
(b) cause the position of equilibrium to shift in the direction of the endothermic reaction.
3 An increase in pressure on a system containing gases will
(a) increase the rate at which equilibrium is attained
(b) cause the equilibrium to shift in the direction which results in the production of fewer gas molecules. (Where the number of molecules of gas is the same on both sides of the equation, altering the pressure will have no effect on the position of equilibrium.)
4 The presence of a catalyst will
(a) alter the rate at which equilibrium is attained
(b) have no effect on the position of equilibrium.
5 Allowing one or more of the components to escape from the mixture will
(a) prevent the attainment of equilibrium

(b) cause the reaction to go to completion.

A decrease in concentration, temperature or pressure will, of course, reverse the effects described in 1, 2 and 3.

Questions on Chapter 14

1 Explain what happens when a sparingly soluble solid is placed in water.
2 In the Haber process, nitrogen and hydrogen are converted to ammonia.
(a) State the conditions used in the conversion to ammonia and write the equation for the reaction that takes place.
(b) The conversion is exothermic. What would be the effect of increasing (a) the temperature, (b) the pressure, at which the reaction is carried out? (W)
3 (a) $3Fe(s) + 4H_2O(g) \rightleftharpoons Fe_3O_4(s) + 4H_2(g)$ ... 1

$Br_2(aq) + H_2O(l) \rightleftharpoons H^+(aq) + Br^-(aq) + HOBr(aq)$ ... 2

Use the information provided by the above equations to answer the following questions.

(i) What do you understand by the terms 'reversible reaction' and 'chemical equilibrium'?

(ii) Draw and label a diagram of an apparatus suitable for converting some iron filings as *completely as possible* to the oxide, Fe_3O_4, and explain why the reaction is non-reversible under the conditions you have chosen.

(iii) Equation 2 represents the chemical changes occurring in bromine water. Free bromine, $Br_2(aq)$, is the only substance which makes the solution brown. Explain why the addition of drops of aqueous sodium hydroxide to well-stirred bromine water causes the brown colour to fade and finally disappear, whilst the addition of an aqueous acid to the now colourless solution causes the brown colour to reappear.

(b) Describe and explain what is seen to occur when a little solid ammonium chloride is heated in a test tube. (L)
4 (a) What is meant by the term 'reversible reaction'? Give two examples of reversible reactions, at least one of which is used in industry.
(b) What is meant by the term 'catalysed reaction'? Give one example of a catalysed reaction which can be investigated in the laboratory and describe briefly how it is decided that a catalyst does influence this reaction. (L)

15   Energy changes in chemistry

The energy produced by chemical changes in our bodies keeps us alive. We also rely on chemical reactions to supply us with energy for cooking, heating, travelling and so on. Since chemical energy plays such an important part in our lives, an investigation into its sources should be of considerable interest to everyone.

Experiments

15.1.a   Do all spontaneous chemical reactions give out heat? ★

PROCEDURE   (a) Pour 25 cm³ of 2 M hydrochloric acid from a measuring cylinder into a 100 cm³ beaker and measure its temperature. Rinse the thermometer and then find the temperature of 25 cm³ of 2 M sodium hydroxide solution in a second measuring cylinder. Tip the alkali into the acid, stir gently with the thermometer and note any temperature change which may occur.
(b) Measure the temperature of a fresh 25 cm³ portion of 2 M hydrochloric acid in the beaker. Add four spatula measures of anhydrous sodium carbonate, stir gently with the thermometer and note any variation in temperature.
(c) Repeat (b) using pure ethanoic acid (**care:** corrosive liquid, harmful vapour) in place of the dilute hydrochloric acid and solid ammonium carbonate in place of the anhydrous sodium carbonate.
(d) Repeat (b) using distilled water in place of the dilute hydrochloric acid and solid ammonium nitrate in place of the anhydrous sodium carbonate.
(e) Repeat (a) using 1 M solutions of lead(II) nitrate and potassium iodide in place of the acid and alkali.

Questions

1   Which reactions, if any, gave out heat energy and which, if any, absorbed it?
2   Write an equation for each reaction, adding '+heat' or '−heat' to the right-hand side as required.

15.3.a   To find the enthalpies of combustion of a series of alcohols

Because of the considerable heat losses which occur during this experiment, the heat capacity of the apparatus (i.e. the amount of heat required to raise its temperature by 1 K) must be found by using an alcohol of known enthalpy of combustion. If the experiment is then repeated under exactly the same conditions but using a different alcohol, the heat losses should be the same.

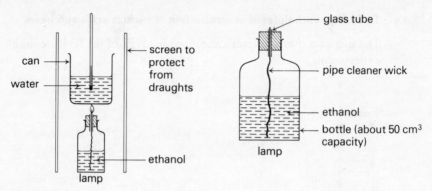

Figure 15.3.a

PROCEDURE   Pour 250 cm³ of tap water from a measuring cylinder into a can (capacity about 500 cm³) and then clamp the can so that its base is about 1 cm above the wick of the lamp (see Figure 15.3.a). Measure the temperature of the water and when it becomes steady, half-fill the lamp with ethanol and find its mass. Light the lamp, quickly place it in position and stir the water in the can *gently* with the thermometer. As soon as a temperature rise of 20 °C has been obtained, blow out the flame and find the mass of the lamp and unburned ethanol.

Repeat the whole procedure, using fresh water in the can and (a) methanol, (b) propan-1-ol, (c) butan-1-ol in the lamp.

Questions

1   What are the major sources of error in the experiment?
2   Work out the heat capacity of the apparatus in the following way.

Initial mass of lamp + ethanol	$= a$ g
Final mass of lamp + ethanol	$= b$ g
∴ Mass of ethanol used	$= (a-b)$ g
Temperature rise	$= 20$ °C
Heat evolved by burning ethanol	$=$ heat gained by apparatus
∴ Heat evolved by burning $(a-b)$ g ethanol	$=$ heat capacity of apparatus $\times$ temperature rise

∴   Heat evolved by burning 1 mol (46 g) ethanol/kJ (i.e. enthalpy of combustion of ethanol/kJ mol⁻¹) $= \dfrac{46}{(a-b)} \times$ heat capacity of apparatus $\times 20$

But enthalpy of combustion of ethanol $= 1371$ kJ mol⁻¹

∴   Heat capacity of apparatus $= 1371 \times \dfrac{(a-b)}{46} \times \dfrac{1}{20}$ kJ K⁻¹

3   Calculate the enthalpies of combustion of the other alcohols, as in question 2, but substitute the value you have just found for the heat capacity of the apparatus.
4   In what way does the formula of each alcohol differ from that of the next one in the homologous series?
5   Do the values of enthalpy of combustion follow a pattern?
6   Does there seem to be a connection between the structures of the alcohols and their enthalpies of combustion? If so, can you explain it in terms of bond breaking and bond making?

15.4.a To find the enthalpies of neutralisation of various acids and bases

The use of a thermometer calibrated in steps of 0.1 °C is desirable in this experiment.

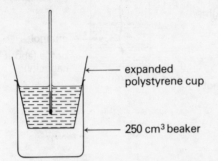

expanded
polystyrene cup

250 cm³ beaker

Figure 15.4.a

PROCEDURE   Pour 50 cm³ of 2 M hydrochloric acid from a measuring cylinder into an expanded polystyrene cup, supported in a 250 cm³ beaker as shown in Figure 15.4.a. Measure the temperature of the acid, rinse and dry the thermometer and then find the temperature of 50 cm³ of 2 M sodium hydroxide solution in a second measuring cylinder. If the two temperatures differ, take an average. Tip the alkali into the acid and stir *gently* with the thermometer. (Vigorous stirring can produce a measurable temperature rise of its own.) Note the maximum temperature reached.

Rinse and dry all the apparatus thoroughly and repeat the experiment, using in turn, 50 cm³ portions of 2 M nitric acid and 1 M sulphuric acid in place of the hydrochloric acid. Next, investigate the effect of neutralising each of the acids with 50 cm³ of 2 M potassium hydroxide solution. Finally, repeat the experiment using 50 cm³ of 2 M ethanoic acid and 50 cm³ of 2 M ammonia solution.

Questions

1   Why was the polystyrene cup used instead of a glass container and why was it placed in the beaker?
2   Work out the enthalpy of neutralisation of each acid, i.e. the heat which would be evolved on neutralising 1 mole of $H^+(aq)$ ions from it. Set out your calculation as shown below; you may assume that the density of each solution is $1\ g\ cm^{-3}$ and that its specific heat capacity is the same as that of water, i.e. $4.2\ J\ g^{-1}\ K^{-1}$.

No. of moles of HCl(aq) in 50 cm³ of $= \dfrac{50}{1000} \times 2 = 0.1$
2 M hydrochloric acid

Total mass of solution $= 100\ g$

Initial temperature $= T_1$

Final temperature $= T_2$

Temperature rise $= (T_2 - T_1)$

∴.   Heat produced when 0.1 mol of HCl(aq)   = mass of solution × specific
   is neutralised by NaOH(aq)   heat capacity × temp. rise
          $= 100 \times 4.2 \times (T_2 - T_1)\ J$

∴.   Heat produced when 1 mol of HCl(aq) $= 10 \times 100 \times 4.2 \times (T_2 - T_1)\ J$
   is neutralised by NaOH(aq)

i.e. Heat of neutralisation of hydrochloric $= \dfrac{10 \times 100}{1000} \times 4.2 \times (T_2 - T_1)\ kJ\ mol^{-1}$
   acid by sodium hydroxide solution

3 Express this result in an equation and in an energy level diagram.

4 Within the limits of experimental error, is there any connection between the results obtained for the strong acids when neutralised by sodium hydroxide solution? Try to explain any pattern which emerges. (Hint: write an ionic equation for each reaction.)

5 Do the results differ noticeably if the acids are neutralised by potassium hydroxide solution instead of sodium hydroxide solution? Does this fit in with your explanation in question 4?

6 How does the enthalpy of neutralisation of the weak acid, ethanoic acid, with the weak alkali, ammonia solution, compare with that of the others? Can you suggest an explanation for the value obtained in this case?

15.5.a To compare the enthalpies of solution of a salt in its anhydrous and hydrated states

PROCEDURE   Set up the same apparatus as in the previous experiment, pour 50 cm^3 of distilled water into the cup and measure its temperature. Accurately weigh out about 1.6 g of anhydrous copper(II) sulphate on a watch glass, tip the solid into the water and stir *gently* with the thermometer until dissolution is complete. Note the maximum temperature change. (Again, the use of a thermometer calibrated in steps of 0.1 °C is desirable.)

Repeat the experiment, using about 2.5 g of copper(II) sulphate-5-water in place of the anhydrous salt.

Questions

1 What are the possible sources of error in this experiment?

2 Calculate the enthalpies of solution of the two forms of the salt as shown below. Assume that the densities and specific heat capacities of the solutions are the same as those of water and that the salt and water were both at the same temperature at the beginning of the experiment.

Mass of copper(II) sulphate	$= m$	
Temperature change	$= T$	
Heat produced when mass m of copper(II) sulphate dissolves in water	$=$ mass of solution $\times$ specific heat capacity $\times$ temp. rise $= 50 \times 4.2 \times T$ J	
$\therefore$ Heat produced when 1 mol (159.5 g) of copper(II) sulphate dissolves in water	$= \dfrac{159.5}{m} \times 50 \times 4.2 \times T$ J	
i.e. Enthalpy of solution of copper(II) sulphate	$= \dfrac{159.5}{m} \times \dfrac{50}{1000} \times 4.2 \times T$ kJ mol^{-1}	

3 Express your results in the form of energy level diagrams.

4 When these solids dissolve, the lattice breaks up and the *hydrated* ions diffuse throughout the solution. Can you explain the difference in the values of enthalpy of solution obtained for the two forms of the salt? (Remember that hydration energy is *evolved* when ions become hydrated and that energy must be *supplied* to break up a crystal lattice.)

15.6.a To find the enthalpy of precipitation of silver chloride

PROCEDURE The procedure for this experiment is exactly the same as that for Experiment 15.4.a. In this case, 25 cm³ of 0.5 M silver nitrate solution is placed in the expanded polystyrene cup and an equal volume of 0.5 M sodium chloride solution is added from the measuring cylinder to precipitate the silver chloride. The experiment is then repeated using separate 25 cm³ portions of 0.5 M solutions of potassium chloride and ammonium chloride in place of the sodium chloride solution.

Questions

1 Calculate the enthalpy of precipitation of silver chloride (i.e. the heat change which occurs when 1 mole of the salt is precipitated) for each reaction. Set out your results in a similar fashion to that shown in Experiment 15.4.a.
2 What connection is there between the three values obtained, within the limits of experimental error? Can you explain this connection? (Hint: write an ionic equation for each reaction.)
3 Express your results on an energy level diagram.

15.7.a To measure the enthalpy change for a displacement reaction

You should be familiar with the fact that any metal will displace another one lower down in the electrochemical series from aqueous solutions containing its ions (11.1). In this experiment the enthalpy changes for two such displacement reactions are measured and compared.

PROCEDURE Carry out the same procedure as in Experiment 15.5.a but use 50 cm³ of 0.2 M copper(II) sulphate solution in the cup and add 0.8–0.9 g of zinc powder (an excess) from the watch glass.

Repeat the experiment with 50 cm³ of 0.2 M silver nitrate solution and 0.4–0.5 g of copper powder (again an excess).

Questions

1 Why was excess metal used in each reaction?
2 The equations for the reactions are

$$Zn(s) + CuSO_4(aq) \rightarrow ZnSO_4(aq) + Cu(s)$$
$$Cu(s) + 2AgNO_3(aq) \rightarrow Cu(NO_3)_2(aq) + 2Ag(s).$$

Calculate the enthalpies of reaction in a similar way to that shown in Experiment 15.4.a. You must work out the heat changes which occur when 1 mol of copper(II) sulphate reacts in the first case and 2 mol of silver nitrate react in the second.
3 Draw energy level diagrams for the two reactions.
4 Can you suggest a reason for the difference in the two values obtained? (Hint: compare the positions of the metals in the electrochemical series on page 201.)

ALTERNATIVE PROCEDURE FOR EXPERIMENTS 15.4.a TO 15.7.a

As an alternative to expanded polystyrene cups, the apparatus shown in Figure 15.7.a may be used for these experiments. The experimental procedure is much the same as that used with the cups but the following additional points should be noted.

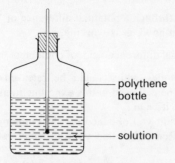

Figure 15.7.a

(a) The bottle may have to be inverted to bring the bulb of the thermometer into contact with the liquid.

(b) The bottle should be held by the neck to minimise heat transfer from the hand.

(c) The bottle should be swirled *gently* until the maximum temperature change has been obtained.

15.7.b To compare the thermal energy and electrical energy obtained from a chemical reaction ★ ★

The first reaction in Experiment 15.7.a is the same as the one which occurs in the Daniell cell (11.4). The second reaction can also be made to produce an electric current and gives a clearer result, as far as this experiment is concerned.

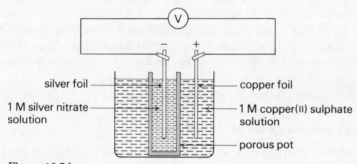

silver foil

1 M silver nitrate solution

copper foil

1 M copper(II) sulphate solution

porous pot

Figure 15.7.b

PROCEDURE Set up the cell shown in Figure 15.7.b and measure its e.m.f. with a *high resistance* voltmeter or a potentiometer.

Questions

1 What was the e.m.f. of the cell?
2 The reaction occurring in the cell was

$$Cu(s) + 2Ag^+(aq) \rightarrow Cu^{2+}(aq) + 2Ag(s).$$

(a) How many mol of electrons must flow through the circuit for 1 mol of copper atoms to go into solution from the anode and 2 mol of silver atoms to be deposited on the cathode?

(b) What is the charge carried by this number of electrons? (The Faraday constant is 96 500 coulomb mol^{-1}.)

3   When a charge of 1 coulomb passes through a potential difference of 1 volt, 1 joule of energy is released. Thus, when any cell is working,

no. of joules released = potential difference × no. of coulombs.

Calculate the maximum number of joules which could be released by the dissolving of 1 mol of copper atoms in the cell used in this experiment, assuming that its working voltage is the same as its e.m.f.

4   How does this compare with the result you obtained for the enthalpy of the same reaction in Experiment 15.7.a? (Remember that your enthalpy value was probably *low* because of heat losses.)

5   Does this cell seem very efficient at converting the energy available from the chemical reaction into electrical energy?

15.1   Energy changes in chemistry

What is energy?

> **Energy** is usually defined as the capacity for doing work.

There are many different forms of energy but all of them are interconvertible. Thus the 'chemical energy' stored up in petrol can be changed to heat energy, kinetic energy and electrical energy in a car engine linked to an alternator. The electrical energy can be changed to light energy in the headlamps or into heat energy to demist the rear window; if the car climbs a hill some of the kinetic energy associated with its movement is transformed into potential energy due to its position. It is important to remember that in all these changes the total amount of energy remains *constant*; none is created and none is destroyed. This is true of all energy conversions and is summed up in the *law of conservation of energy*.

> Energy can neither be created nor destroyed.*

Energy supplies – what of the future?

The phrase 'energy crisis' looms large in newspaper headlines from time to time and no doubt you are aware that we are fast using up our sources of energy. Stocks of fossil fuels are not limitless, and oil in particular seems certain to run out early in the next century. When you consider how much our everyday life depends on oil for heating, travelling, and producing electricity, not to mention making clothing, plastics and so on, you will see that the problem is not one which can be ignored. Other sources of energy *must* be found.

Nuclear power is one possibility and solar energy (i.e. energy from the sun) another. The trouble with solar energy is that though more than enough is supplied in the summer (even in Britain!), we have no way of storing up the surplus for the winter. However, it has been suggested that heat from the summer sun might be used to drive steam turbines and generate electricity to electrolyse water. The hydrogen produced by this process could be stored and used as a fuel when required.

*The interconversion of mass and energy is mentioned briefly on page 99.

Solar heating panels on the roof of a house

Ideas like this may or may not be feasible, but the increasing worldwide demand for energy must be satisfied in some way if life as we know it is to continue. Clearly, scientists have a formidable task ahead of them.

Exothermic and endothermic reactions – the ΔH convention

As you know, during an exothermic reaction energy is evolved. Where does this energy come from? Since it cannot be created or destroyed it must have been stored up somehow in the reactants and then released during the reaction. As a result of this, the products must contain less stored energy than the reactants did. On the other hand, in an endothermic reaction, where energy is absorbed from the surroundings, there must be more energy stored up in the products than was originally present in the reactants. Endothermic reactions are rarely spontaneous at room temperature but two such examples are given in Experiment 15.1.a.

It is impossible for us to measure the *total* energy stored up in a particular substance but we can measure the *change* in it which occurs during a chemical reaction. The symbol used for such a change (at constant pressure) is ΔH where Δ (delta) means 'change of' and H is the 'heat content' or *enthalpy* of the system. ΔH is *negative* for an exothermic change because the system *loses* energy to the surroundings; in an endothermic reaction, ΔH is *positive* because the system *gains* energy from the surroundings. These changes in enthalpy may be represented on *energy level diagrams* as shown in Figure 15.1.1.

When we talk about the enthalpy change which occurs during a reaction we are referring to the total quantity of heat which must be transferred between the products and the surroundings in order for the products to

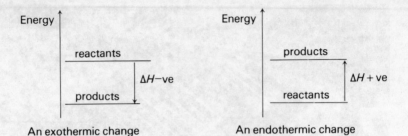

Figure 15.1.1

end up at the original temperature of the reactants. The unit of energy is the *joule* (J) or, more conveniently, the kilojoule (kJ), i.e. 1000 joules.

One **joule** is the energy required to raise the temperature of 1 g of water by $\frac{1}{4.18}$ K; thus 4.18 joules will raise the temperature of 1 g of water by 1 K.

Although 1 kilojoule *sounds* as if it is a large unit of energy, it is really quite small. For example, 1 kJ will
(a) heat about 3 cm^3 of water from room temperature to boiling point,
(b) power a 1 kW electric fire for 1 second,
(c) be produced when 0.04 cm^3 (about 1 drop) of methylated spirit burns,
(d) be produced when 10 cm^3 of dilute (2 M) hydrochloric acid is neutralised by sodium hydroxide solution.
Obviously the amount of heat energy which is evolved or absorbed during a reaction will depend on how much of each reactant is present. Thus it is necessary to produce a definition of *enthalpy of reaction* which clearly states the amounts of reactants involved. This definition is given below and may be applied to *all* reactions.

The **enthalpy of reaction** (ΔH) is the heat change which occurs when the numbers of moles of reactants indicated by the equation react together.

The value of the enthalpy of reaction is often included with the chemical equation. Thus

$$C(s) + O_2(g) \rightarrow CO_2(g) \qquad \Delta H = -394 \text{ kJ}$$

means that when 1 mol of carbon burns in excess oxygen, the heat *evolved* is 394 kJ.

Why is there an enthalpy change during a reaction?

Any chemical reaction involves a change in the bonding between atoms or ions; bonds in the reactants are broken and then new ones are formed to give the products. Because energy has to be supplied to overcome the strong electrostatic forces which constitute a chemical bond, the *breaking* of bonds is an *endothermic* change. The reverse process, that is the *formation* of these bonds, must therefore be an *exothermic* change. Thus bond-breaking absorbs energy but bond-making releases it.

The energy which is required to break one mole of bonds is known as the *bond energy* and it gives us a measure of the strength of a particular bond. For example, the bond energy of the H–H bond in a hydrogen molecule is $435 \, kJ \, mol^{-1}$ while that of the Cl–Cl bond in a chlorine molecule is $240 \, kJ \, mol^{-1}$. This means that the H–H bond is stronger than the Cl–Cl bond.

Consider the reaction between hydrogen and chlorine to give hydrogen chloride.

$$H_2(g) + Cl_2(g) \rightarrow 2HCl(g)$$

The energy needed to split one mole of hydrogen molecules into atoms is 435 kJ and that needed to split one mole of chlorine molecules into atoms is 240 kJ. Therefore the total energy which must be *supplied* for the formation of two moles of hydrogen chloride from its elements is $(435 + 240) = 675 \, kJ$. However, the energy *released* by making the new H–Cl bonds is 860 kJ. Thus the reaction is exothermic, the value of ΔH being $(675 - 860) = -185 \, kJ$. These energy changes, together with a pictorial representation of the various processes occurring, are shown on an energy level diagram in Figure 15.1.2.

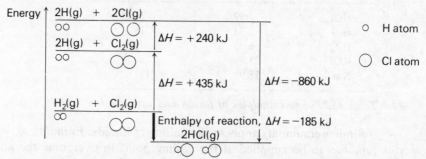

Figure 15.1.2 *Energy level diagram for the reaction between hydrogen and chlorine*

The reason why enthalpy changes occur during chemical reactions is now clear. It may be summarised in the following important statement.

> The enthalpy change which accompanies a chemical reaction results from the differing amounts of energy involved in breaking bonds in the reactants and in making new ones in the products.

We now proceed to apply these general ideas to specific types of change.

15.2 Enthalpy of fusion and enthalpy of vaporisation

It was mentioned in Chapter 6 that while a solid is melting it continually absorbs heat energy, and yet its temperature does not rise until it has all been liquefied. This heat energy is used in overcoming the forces which hold the atoms, molecules or ions in position in the lattice and it is referred to as *enthalpy of fusion*. By measuring the enthalpies of fusion for the *same number* of particles of different substances we can obtain a comparison of the strengths of the forces holding the lattices together. The number of particles chosen for these measurements is, as usual, 6.02×10^{23}. Thus we are dealing with 1 mole of the substance concerned.

The **molar enthalpy of fusion** (ΔH_f) is the quantity of heat required to change one mole of solid into liquid at its melting point.

The values of some molar enthalpies of fusion are shown in Table 15.2. You can see that the intermolecular forces in solid hydrogen chloride and solid hydrogen sulphide must be considerably less than those in ice. This is explained by the existence of hydrogen bonds in ice (12.4). The separate particles in the lattices of sodium chloride and potassium chloride are oppositely charged ions. These are held together by strong ionic bonds and so the enthalpies of fusion of these two compounds are very much higher than those of the other compounds in the table.

Compound	ΔH_f/kJ mol^{-1}	ΔH_v/kJ mol^{-1}
HCl	1.99	16.2
H$_2$S	2.38	18.7
H$_2$O	6.02	41.1
KCl	25.5	162
NaCl	28.9	171

Table 15.2 Some enthalpies of fusion and vaporisation

Similar comments apply to the boiling of liquids. *Enthalpy of vaporisation* has to be supplied at the boiling point to overcome the attractive forces between the particles and separate them to form a gas.

The **molar enthalpy of vaporisation** (ΔH_v) is the quantity of heat required to change one mole of liquid into vapour at its boiling point.

Because the particles have to be separated to a much greater extent when changing from a liquid to a gas than when changing from a solid to a liquid, the molar enthalpy of vaporisation for a particular substance is considerably more than its molar enthalpy of fusion. This is shown by the figures in Table 15.2.

15.3 Enthalpy of combustion

The **enthalpy of combustion** of a substance is the heat change which occurs when 1 mole of it is completely burned in excess oxygen.

The enthalpy of combustion of a liquid such as ethanol can be measured by burning a known mass of it and using the heat so produced to raise the temperature of a known mass of water. This may be done by means of the simple apparatus of Experiment 15.3.a or the more elaborate version shown in Figure 15.3.1.

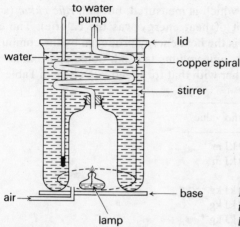

Figure 15.3.1 Apparatus for finding the enthalpy of combustion of a liquid

Example

When 0.7 g of ethanol was completely burned in excess oxygen, the heat produced raised the temperature of 250 g of water by 20 °C. Assuming that there were no heat losses in the apparatus, calculate the enthalpy of combustion of ethanol. Also, write an equation for the reaction and draw an energy level diagram to represent it. (Molar mass of ethanol = 46 g mol^{-1}, specific heat capacity of water = 4.18 J g^{-1} K^{-1}.)

Heat produced by combustion of ethanol = heat gained by water.

∴ Heat produced by burning 0.7 g ethanol = mass of water × specific heat capacity × temperature rise

$$= 250 \times 4.18 \times 20 \text{ J}$$

∴ Heat produced by burning 1 mol (46 g) of ethanol

$$= \frac{250 \times 4.18 \times 20}{1000} \times \frac{46}{0.7} \text{ kJ}$$

$$= 1370 \text{ kJ}$$

i.e. Enthalpy of combustion of ethanol = 1370 kJ mol^{-1}

The equation for the reaction is thus

$$C_2H_5OH(l) + 3O_2(g) \rightarrow 2CO_2(g) + 3H_2O(l) \quad \Delta H = -1370 \text{ kJ (mol } C_2H_5OH)^{-1}$$

and the energy level diagram is as shown in Figure 15.3.2.

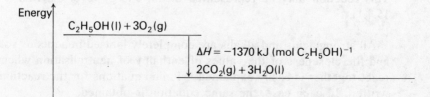

Figure 15.3.2 Energy level diagram for the combustion of ethanol

It is important to know how much energy is available from the burning of a given amount of a substance when considering the relative advantages and disadvantages of different fuels or foods. The units chosen for this are generally larger than those used in the laboratory and it is not enthalpy of combustion which is measured, but *calorific value* (so called because the original unit of heat energy was the calorie). The calorific value of a solid or liquid is the heat energy produced by the combustion of unit mass of it; for a gas, unit volume is used. Some approximate values for domestic fuels, together with that for bread, are given in Table 15.3.

Fuel or food	Calorific value
Town gas	$20\ 000\ \text{kJ m}^{-3}$
North Sea Gas (methane)	$37\ 000\ \text{kJ m}^{-3}$
Coke	$27\ 000\ \text{kJ kg}^{-1}$
Anthracite	$33\ 000\ \text{kJ kg}^{-1}$
Fuel oil	$45\ 000\ \text{kJ kg}^{-1}$
Bread	$11\ 000\ \text{kJ kg}^{-1}$

Table 15.3 Some calorific values

Of course, the choice of fuel for heating a house depends on a number of factors in addition to calorific value. A householder must consider such things as cost, convenience and, if he lives in a smokeless zone, pollution, before deciding on a particular fuel to heat his home.

15.4 Enthalpy of neutralisation

The **enthalpy of neutralisation** is the heat change which occurs when one mole of hydrated hydrogen ions from an acid reacts with one mole of hydrated hydroxide ions from an alkali.

By measuring the temperature rise produced when a known volume of acid is neutralised by a known volume of alkali it is possible to calculate the enthalpy change which accompanies the reaction (see Experiment 15.4.a). It is found that for strong acids and alkalis the value is always about $57\ \text{kJ mol}^{-1}$.

e.g. $HCl(aq) + NaOH(aq) \rightarrow NaCl(aq) + H_2O(l)$ $\Delta H = -57.4\ \text{kJ mol}^{-1}$

This reaction may be represented on an energy level diagram (Figure 15.4.1).

All strong acids and alkalis are completely ionised in aqueous solution and the closeness of the values of enthalpy of neutralisation when they react together is easily explained if ionic equations for the reactions are written. In each case, the same equation is obtained,

i.e. $H^+(aq) + OH^-(aq) \rightarrow H_2O(l)$

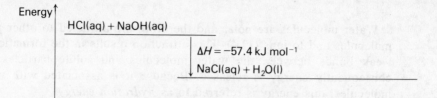

Figure 15.4.1 Energy level diagram for the reaction between hydrochloric acid and sodium hydroxide solution

and so it is not surprising that the enthalpy of neutralisation should always be approximately the same. (Remember that if a dibasic acid such as sulphuric acid is used, each mole of it contains *two* moles of $H^+(aq)$. The enthalpy of neutralisation in this case is *half* the enthalpy of reaction per mole of acid, as should be clear from the definition.)

When a *weak* acid is neutralised by a *weak* alkali the enthalpy of neutralisation is considerably less than 57 kJ mol^{-1}.

e.g. $\quad CH_3COOH(aq) + \text{‘}NH_4OH(aq)\text{’} \rightarrow CH_3COONH_4(aq) + H_2O(l)$

$$\Delta H = -51.5 \text{ kJ mol}^{-1}$$

The reason for this is that solutions of weak acids and alkalis contain mainly unionised molecules. Thus energy must be supplied to dissociate the molecules into ions before neutralisation can take place. As a result of this less energy is evolved than is the case with strong acids and alkalis (see Figure 15.4.2).

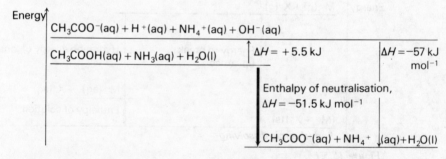

Figure 15.4.2 Energy level diagram for the reaction between ethanoic acid and ammonia solution

15.5 Enthalpy of solution and enthalpy of hydration

When a substance dissolves in water to give a concentrated solution, the process is usually accompanied by a heat change. If this concentrated solution is diluted, a further heat change often takes place. This has to be taken into account when defining *enthalpy of solution*.

The **enthalpy of solution** of a substance is the heat change which occurs when one mole of it dissolves in so much water that further dilution results in no detectable change in temperature.

Water molecules are polar and therefore are attracted to other polar molecules and to ions (12.4). This attraction results in the formation of weak bonds between the water molecules and solute particles and consequently energy is released. Because it is associated with water molecules, this energy is referred to as *hydration energy*.

When a gas dissolves in water, hydration energy is evolved. There are no intermolecular forces to overcome and thus no energy has to be supplied. The dissolving of any gas in water is therefore an exothermic process. Obviously, the more polar gases will have higher hydration energies than those of the non-polar ones and so their enthalpies of solution will be greater. However, if the substance which is dissolving is a liquid or a solid, energy will have to be supplied to overcome the forces of attraction acting between its atoms, molecules or ions. If the hydration energy is greater than that needed to separate the particles, the dissolving

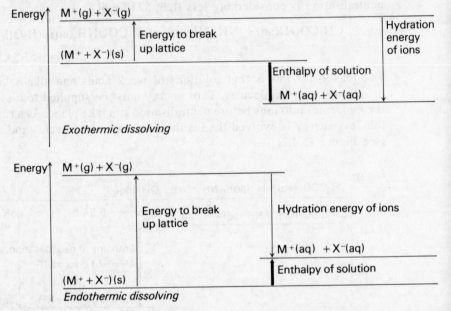

Figure 15.5.1

Compound	Enthalpy of solution/kJ mol⁻¹
$CO_2(g)$	− 19.9
$HCl(g)$	− 72.8
$NH_3(g)$	− 35.4
$SO_2(g)$	− 35.7
$SO_3(s)$	− 205.9
$H_2SO_4(l)$	− 74.2
$NH_4NO_3(s)$	+ 26.5
$NaCl(s)$	+ 5.4
$NaOH(s)$	− 41.4

Table 15.5 *Some enthalpies of solution*

will be exothermic; if it is smaller, then the dissolving will be endothermic (see Figure 15.5.1). Of course, if the hydration energy is *very* small compared with that required to break up the lattice, the substance will not dissolve. This is why some ionic solids are insoluble in water but it is *not* the reason for the insolubility of most covalent substances (see 12.4).

The dissolving of anhydrous copper(II) sulphate in water releases heat energy but the dissolving of the hydrated salt absorbs it (Experiment 15.5.a). Why should this be? The answer is fairly obvious if you have understood the discussion so far. The hydrated salt contains water, therefore some of the hydration energy of the ions must already have been released. The energy evolved when it dissolves in water and the ions become fully hydrated is insufficient to completely overcome the attractive forces holding the lattice together. The extra energy needed has to be supplied by the surroundings.

The equations for the two processes are:

$$CuSO_4(s) \quad + water \rightarrow Cu^{2+}(aq) + SO_4^{2-}(aq) \qquad \Delta H = -57 \text{ kJ mol}^{-1}$$
$$CuSO_4 \cdot 5H_2O(s) + water \rightarrow Cu^{2+}(aq) + SO_4^{2-}(aq) \qquad \Delta H = +10 \text{ kJ mol}^{-1}$$

These may be represented on a single energy level diagram, as shown in Figure 15.5.2.

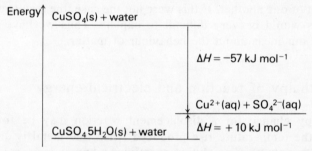

Figure 15.5.2 Energy level diagrams for the dissolving of copper(II) sulphate in water

Use the diagram to work out the enthalpy change for the following reaction.

$$CuSO_4(s) + 5H_2O(1) \rightarrow CuSO_4 \cdot 5H_2O(s)$$

15.6 Enthalpy of precipitation

The **enthalpy of precipitation** of a substance is the heat change which occurs when one mole of it is precipitated from aqueous solution.

Since ions move independently of one another in dilute solution we would expect the enthalpy of precipitation of a particular substance to be the same no matter what type of spectator ions were present. This is found to be so in Experiment 15.6.a, where silver chloride is precipitated by adding solutions of various chlorides to one of silver nitrate and measuring the temperature rise produced.

The reaction in each case is the same, i.e.

$$Ag^+(aq) + Cl^-(aq) \rightarrow AgCl(s) \qquad \Delta H = -58.5 \text{ kJ mol}^{-1}$$

and hence only one energy level diagram (Figure 15.6) need be drawn.

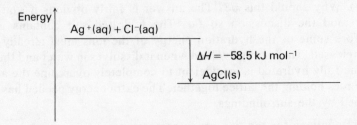

Figure 15.6 Energy level diagram for the precipitation of silver chloride

It is worth remembering that the constancy of enthalpy of neutralisation for strong acids and alkalis and the constancy of enthalpy of precipitation for a given substance have both been explained by the idea that in a dilute solution of a strong electrolyte, the ions are completely independent of one another. If this were not the case, the results of these experiments would be very difficult to explain. Thus we have further support for our ideas about the behaviour of matter.

15.7 Enthalpy of reaction and electrical energy

The enthalpy change for a displacement reaction may be found by measuring the temperature rise produced when one metal is displaced from a known volume of a solution containing a known concentration of its ions by another one which is higher in the electrochemical series (see Experiment 15.7.a). When this is done, it is found that more energy is released when the metals are far apart in the electrochemical series than when they are close together. This is not surprising, since metals close to one another have similar tendencies to form ions in aqueous solution. The 'driving force' for the reaction will be greater if the metal which is added forms ions very readily and the one which is being displaced does not.

e.g.
$$Zn(s) + Cu^{2+}(aq) \rightarrow Zn^{2+}(aq) + Cu(s) \qquad \Delta H = -217 \text{ kJ}$$
$$Cu(s) + 2Ag^+(aq) \rightarrow Cu^{2+}(aq) + 2Ag(s) \qquad \Delta H = -147 \text{ kJ}$$

(The energy level diagrams are shown in Figure 15.7.)

If the displacement is carried out in an electrolytic cell, the maximum energy available from the reaction may be calculated from the relationship

maximum electrical = e.m.f. of cell × no. of
energy released   coulombs passed.

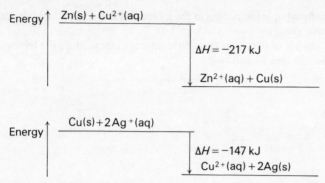

Figure 15.7 Energy level diagrams for displacement reactions

In the second reaction above, the dissolving of 1 mol of copper atoms and deposition of 2 mol of silver atoms would be accompanied by the passage of 2 mol of electrons (i.e. $2 \times 96\,500$ coulombs) through the circuit. The e.m.f. of the cell is found to be 0.46 volt (Experiment 15.7.b) and therefore

$$\text{maximum electrical energy released} = 0.46 \times 2 \times 96\,500 \text{ J (mol Cu)}^{-1}$$

$$= 89 \text{ kJ (mol Cu)}^{-1}.$$

You will notice that this cell is not a very efficient energy converter – it 'loses' 58 kJ for every mole of copper which dissolves. What happens to the lost energy? If you were to measure the temperatures of the solutions in the cell you would find that the liquids warm up slightly as the current is passed. Thus the thermal energy released by the reaction is only partly converted into electrical energy.

Not all cells are as inefficient as this one; some even produce *more* electrical energy than expected because they cool down while they are working and hence absorb energy from the surroundings. However, detailed treatment of this topic must be left until a more advanced course.

Questions on Chapter 15

1 . The heat produced by the complete combustion of 2 g of methanol, CH_3OH, was found to raise the temperature of 500 g of water by 21.4°C. Assuming that there were no heat losses, calculate the enthalpy of combustion of methanol. (Specific heat capacity of water = $4.2 \text{ J g}^{-1} \text{ K}^{-1}$)

2   (a) When 50 cm³ of 2 M ethanoic acid was neutralised by an equal volume of 2 M sodium hydroxide solution, both initially at 20.0°C, the temperature of the mixture rose to a maximum of 33.2°C. What is the enthalpy of neutralisation for this acid and alkali?

(b) In a similar experiment, using 50 cm³ of 2 M hydrochloric acid in place of the ethanoic acid, a temperature rise of 13.7°C was obtained. Calculate the enthalpy of neutralisation in this case. Comment on the significance of the two values obtained. (Specific heat capacity of each solution = $4.2 \text{ J g}^{-1} \text{ K}^{-1}$)

3   (a) Explain, with one example of each, the terms *exothermic* and *endothermic* reaction.

Hydrogen gas burns in chlorine to produce hydrogen chloride and evolve heat.

Indicate briefly what is happening to the molecules and what you think may be the source of this energy.

(b) When a sample of benzene is cooled in an ice/water mixture the temperature of the benzene changes as follows.

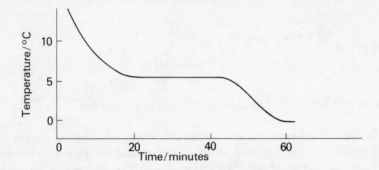

Explain these changes and what is happening to the benzene molecules.

Sketch and explain the shape of the graph you would expect if the benzene were withdrawn from the ice/water mixture after 1 hour and left in the laboratory.

(SUJB)

4 (a) Describe and write equations for
(i) *one* reaction of industrial importance in which heat is absorbed,
(ii) *one* reaction of industrial importance in which heat is produced,
(iii) *two* reactions in which light energy is absorbed.

(b) What is meant by the *calorific value* of a fuel?

A fuel such as coal and a food such as bread are used in very different ways. Explain why the calorific value is an important characteristic of both the fuel and the food.

(c) Give the names and approximate composition (by volume) of *two* gaseous fuels. (C)

5 Give examples of exothermic reactions occurring between (a) a gas and a solid, (b) a liquid and a solid, (c) two gases.

State the conditions under which the reactions occur and, in each case, give the equation for the reaction. Suggest a reason for the exothermic nature of each reaction. (O & C)

6 20 cm³ of a solution containing 1 mole of HCl per dm³, i.e. 1 M hydrochloric acid, was placed in a beaker and 1 M sodium hydroxide was added from a burette 5 cm³ at a time. After each addition of alkali the liquid in the beaker was stirred with a thermometer and its temperature recorded (Experiment 1).

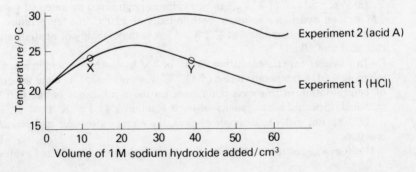

The experiment was then repeated using 20 cm³ of a solution containing 1 mole per dm³ of another strong acid, A, in place of the 1 M hydrochloric acid (Experiment 2).

The results of the two experiments are represented in the graph at the base of the previous page.

(a) Explain why, in each experiment, the temperature rises to a maximum and then falls.

(b) What is the maximum *rise* in temperature (i) in Experiment 1, (ii) in Experiment 2?

(c) What approximate volume of 1 M sodium hydroxide is needed to cause the maximum rise in temperature in Experiment 2?

(d) What conclusion about acid A can you draw from the results of Experiment 2? Explain your answer.

(e) If a little indicator had been added to the acid at the start of each experiment, what would have been the least volume of 1 M sodium hydroxide needed to cause a colour change (i) in Experiment 1, (ii) in Experiment 2?

(f) Give *two* disadvantages of the use of a thermometer instead of an indicator in the titration of an acid with an alkali.

(g) What changes, if any, would you expect to see if a little copper(II) sulphate solution were added to each of the liquids represented by points X and Y on the graph? (C)

16 Carbon and silicon

When you look around, most of what you see contains either carbon or silicon. Although carbon is found in all living matter, it makes up only about 0.03% by mass of the earth's crust. Silicon, on the other hand, is the second most abundant element on earth, being present to the extent of 27.72%. It is found in enormous quantities in sand and rocks and in man-made materials derived from them, such as bricks, cement, concrete and glass.

Experiments

16.1.a To compare diamond and graphite★

PROCEDURE (a) Test pieces of graphite and diamond for hardness by trying to mark paper and glass with them. Notice how each of the substances feels.
(b) Using the apparatus in Figure 2.1.d (page 31), investigate the electrical conductivity of each solid.

Questions

1 How did the hardness of the piece of graphite compare with that of the diamond? Can you suggest uses of each substance which result from their relative hardnesses?
2 How did the electrical conductivities of the two substances compare with one another?
3 In spite of the differences in their physical properties, diamond and graphite are both pure carbon. How might this be proved?

16.1.b To observe what happens when wood is heated in the absence of air★★

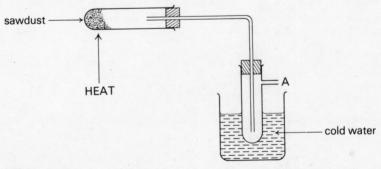

sawdust

HEAT

A

cold water

Figure 16.1.b

PROCEDURE   Place sawdust or small pieces of wood to a depth of 5 cm in a Pyrex boiling tube and assemble the apparatus shown in Figure 16.1.b. Heat the wood and attempt to ignite the gas which emerges from tube A. When no further change is observed, allow the apparatus to cool, test the collected liquid with universal indicator paper and inspect the solid residue.

Questions

1   Did the gas burn?
2   What happened to the universal indicator paper? What did this show?
3   Describe the appearance of the solid residue. What do you think this residue was?
4   If the wood had been heated in a plentiful supply of air, would the same product have been obtained? Explain your answer.

16.2.a   To investigate the reaction between wood charcoal and metal oxides

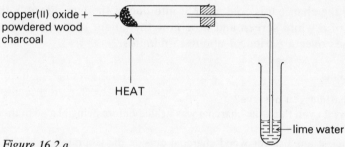

copper(II) oxide +
powdered wood
charcoal

HEAT

lime water

Figure 16.2.a

PROCEDURE   Thoroughly mix equal volumes of copper(II) oxide and powdered wood charcoal. Place 2 cm depth of the mixture in a test tube and heat it strongly as shown in Figure 16.2.a until no further change takes place. Remove the delivery tube from the lime water *before* you stop heating (why?). Repeat the experiment using zinc oxide, lead(II) oxide and magnesium oxide in place of the copper(II) oxide. Replace the lime water each time.

Questions

1   What happened to the lime water in the first experiment? What must have been formed?
2   What must the copper(II) oxide have changed into? What type of reaction is this?
3   Which of the oxides did not react?
4   Write equations for those reactions which did occur.
5   What connection is there between the position of a metal in the reactivity series and the ease with which its oxide reacts with carbon?

16.2.b   The effect of animal charcoal on a solution of magenta

PROCEDURE   Half-fill a 100 cm³ beaker with a dilute solution of magenta (a dye) and add a spatula measure of powdered animal charcoal.

Boil the mixture gently for two minutes, filter, and note the colour of the filtrate. Finally, pour 5 cm³ of ethanol on to the residue on the filter paper and collect the filtrate in a clean test tube.

Questions

1 How did the colour of the two filtrates compare with that of the original magenta solution?
2 Do you think that the magenta solution underwent a chemical change or a physical change? Give a reason for your answer.
3 Suggest one use for animal charcoal outside the laboratory.

16.2.c The effect of wood charcoal on bromine vapour ★ ★

PROCEDURE Place a small piece of wood charcoal in a crucible, cover it with sand and heat it strongly for a few minutes. Meanwhile, add 2 drops of bromine (**care**) to each of two gas jars, close the jars with cover slips and invert them several times to distribute the bromine vapour evenly. When the charcoal is cold, drop it into one of the jars. Place both jars in front of a white screen and note how the colour of the gas in each one changes over a period of about 15 minutes.

Questions

1 What did you observe?
2 Why do you think the charcoal was heated before being placed in the bromine vapour?
3 Suggest one use for wood charcoal outside the laboratory.

16.4.a To prepare carbon dioxide

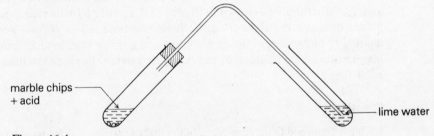

marble chips + acid

lime water

Figure 16.4.a

PROCEDURE Add two or three marble chips to 5 cm³ of dilute hydrochloric acid in a test tube. Pass the resultant carbon dioxide through lime water (see Figure 16.4.a) for about three minutes. Repeat the experiment, using in turn dilute ethanoic acid, dilute nitric acid and dilute sulphuric acid in place of the dilute hydrochloric acid. Use fresh lime water each time.

Questions

1 What did you see in the first case?
2 What happened when the carbon dioxide was passed through lime water for (a)

a few seconds, (b) several minutes? Name the substances formed in each case.

3 Which of the acids did not react with the marble chips? Suggest a reason for this.

4 Write equations for the reactions between (a) the marble chips and dilute hydrochloric acid, (b) the carbon dioxide and lime water.

16.4.b Some properties of carbon dioxide ★★

The most convenient method of preparing carbon dioxide in the laboratory is to react dilute hydrochloric acid with marble chips, using the apparatus shown in Figure 16.4.1 (page 300) or a Kipp's apparatus.

PROCEDURE (a) Collect a gas jar of carbon dioxide by downward delivery and close it with a cover slip. Sprinkle a few drops of methylbenzene on to a ball of cotton wool and place it in the centre of a small glass trough. Ignite the cotton wool and pour the jar of carbon dioxide on top of it (see Figure 16.4.b).

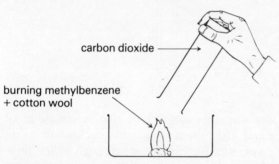

carbon dioxide

burning methylbenzene
+ cotton wool

Figure 16.4.b

(b) Ignite a 5 cm length of magnesium ribbon in a bunsen flame and then lower it into a jar of carbon dioxide. (**Caution:** do not look directly at the flame; use tinted glass if available.) When the reaction has finished, inspect the residue carefully.

(c) Bubble some carbon dioxide through distilled water containing a few drops of universal indicator.

(d) Pour 2 cm³ of distilled water into a test tube full of carbon dioxide. Quickly stopper the tube, shake it for about a minute and then open it under water contained in a trough. Note how far the water rises in the tube.

(e) Repeat (d) using 2 cm³ of sodium hydroxide solution in place of the distilled water.

Questions

1 How did you check that the gas jars were full of carbon dioxide?

2 What can you say about the density of carbon dioxide relative to that of air? Give *two* reasons for your answer.

3 What happened to the burning methylbenzene when the carbon dioxide was poured on it? Can you suggest a use for carbon dioxide outside the laboratory? Which *two* properties make it particularly useful for this?

4 What happened to the burning magnesium in (b)? What did the residue appear to consist of? Write a possible equation for the reaction.

5 What effect did the carbon dioxide solution have on the universal indicator? What did this tell you?

6 How did the solubility of carbon dioxide in water compare with that in sodium hydroxide solution? Explain how you reached your conclusion.

7 Bearing in mind how carbon dioxide reacts with lime water (calcium hydroxide solution), what do you think was formed when it reacted with the sodium hydroxide solution in (e)? Write a possible equation for the reaction.

8 What would you expect to be formed if *excess* carbon dioxide were bubbled into sodium hydroxide solution? (Again, think what happens with lime water.)

16.4.c The effect of heat on carbonates

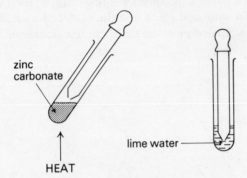

HEAT

Figure 16.4.c

PROCEDURE Place a spatula measure of zinc carbonate in an ignition tube. Heat the tube and, after about 10 seconds, suck some gas out of it by means of a dropping pipette. Bubble this gas through a *little* lime water in a second ignition tube (see Figure 16.4.c). If the test is negative, repeat it a few times but make sure that you dry the outside of the pipette before reinserting it in the hot tube (why?).

Repeat the experiment using anhydrous sodium carbonate, copper(II) carbonate, powdered calcium carbonate (precipitated chalk) and potassium carbonate in place of the zinc carbonate. If little or no carbon dioxide is given, allow the residue to cool, tip it into a test tube containing 2 cm³ of dilute hydrochloric acid and test for carbon dioxide again. Use fresh lime water each time.

(N.B. As an alternative, the procedure described in Experiment 16.2.a may be used here. Two spatula measures of each of the solids mentioned above should be used in place of the mixture of wood charcoal and metallic oxide.)

Questions

1 Which solids did not decompose, or only decomposed slightly, on heating? How did you reach your conclusion?

2 Which solids readily gave carbon dioxide on heating? What do you think the residues were in these cases? (The colours should give you a clue.)

3 Write an equation for each decomposition.

4 What connection is there between the thermal stability of a metallic carbonate and the position of the metal in the reactivity series?

16.4.d The effect of heat on hydrogencarbonates

PROCEDURE (a) Heat 1 cm depth of sodium hydrogencarbonate in a test tube and check for the evolution of carbon dioxide. When no further change occurs, allow the residue to cool, add 2 cm³ of dilute hydrochloric acid and test any gas evolved with a fresh sample of lime water.
(b) Place a little sodium hydrogencarbonate in a boiling tube, pour in 10 cm³ of distilled water and shake the tube until the solid dissolves. Add 2–3 drops of universal indicator and note the colour produced. Repeat this procedure with sodium carbonate in a second tube. Boil the sodium hydrogencarbonate solution for about a minute, with shaking, and compare the colour of the universal indicator with that in the sodium carbonate solution.
(c) Use the procedure described in Experiment 10.2.e (page 176) to measure the volume of gas given off when 0.45–0.50 g of sodium hydrogen-carbonate is heated. Test the gas with lime water.
(d) Repeat (a), (b) and (c) using potassium hydrogencarbonate. The mass of this solid needed for (c) is 0.50–0.60 g.

Questions

1 Was carbon dioxide given on heating sodium hydrogencarbonate in (a)? Did you notice any other product?
2 Was carbon dioxide given off when dilute hydrochloric acid was added to the residue in (a)?
3 Describe your observations in (b).
4 What conclusions can you draw concerning the effect of heat on sodium hydrogencarbonate? Explain your reasoning.
5 Write a possible equation for the thermal decomposition of sodium hydrogen-carbonate.
6 Do the results of (c) agree with your equation?
7 Does potassium hydrogencarbonate behave in the same way as sodium hydrogencarbonate in this experiment?

16.5.a The conversion of carbon dioxide to carbon monoxide★★

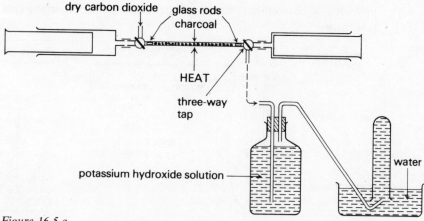

Figure 16.5.a

PROCEDURE   Dry some pieces of charcoal by covering them with sand in an evaporating basin and heating strongly for a few minutes. When cool, pack them into a silica glass tube and set up the apparatus shown in Figure 16.5.a. Flush out the apparatus several times with carbon dioxide from a Kipp's apparatus (the gas must first be bubbled through water and then through concentrated sulphuric acid). Collect 50 cm³ of the gas in one of the syringes. Heat the tube *strongly* and pass the gas from one syringe to the other until no further change in volume occurs. Allow the apparatus to cool, twist the syringe plunger and note the reading.

Expel the gas *slowly* from the syringe through the three-way tap, bubble it through concentrated potassium hydroxide solution and collect a tube-full over water as shown. Pour 1 cm³ of lime water into the tube, quickly stopper it and shake it well. Remove the stopper and apply a flame to the mouth of the tube. When combustion ceases, replace the stopper and shake the tube again. Collect a second tube-full of the gas and test it with universal indicator.

Questions

1   Why was the carbon dioxide from the Kipp's apparatus passed through (a) water, (b) concentrated sulphuric acid?
2   The equation for the reaction is

$$CO_2(g) + C(s) \rightarrow 2CO(g).$$

Did the reaction go to completion? Give a reason for your answer.
3   Why was the carbon monoxide bubbled through potassium hydroxide solution before being collected?
4   What colour was the carbon monoxide flame? Where might you see this flame in everyday life?
5   What happened to the lime water (a) before combustion of the carbon monoxide, (b) after combustion? Explain these results.
6   Is carbon monoxide an acidic, neutral or basic oxide?

16.5.b   The reaction of carbon monoxide with metal oxides ★ ★

The reaction between methanoic acid and concentrated sulphuric acid provides a convenient means of generating carbon monoxide in the laboratory.

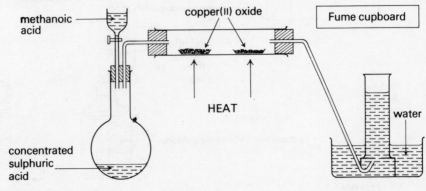

Figure 16.5.b

Because carbon monoxide is very poisonous, this experiment should be carried out in a fume cupboard or well ventilated laboratory. It must not be performed by pupils.

PROCEDURE Add methanoic acid dropwise to concentrated sulphuric acid in the apparatus shown in Figure 16.5.b so that a steady stream of bubbles is collected in the gas jar. When the jar is full, replace it with another and test the gas with a flame. Continue in this way until the gas burns quietly (why?) and then heat the copper(II) oxide until no further change is seen. Test the gas collected during this heating with lime water but, since it will contain unchanged carbon monoxide, make sure that none of it escapes into the air. Repeat the experiment using lead(II) oxide in place of the copper(II) oxide.

Questions

1 The formula for methanoic acid is HCOOH. Write an equation for its decomposition to carbon monoxide. What is the function of the concentrated sulphuric acid in this process?
2 What appeared to have happened to the copper(II) oxide?
3 What happened to the lime water? What did this indicate?
4 Write an equation for the reaction. What type of reaction is this?
5 Answer questions 2, 3 and 4 for the reaction with lead(II) oxide.

16.1 Allotropes of carbon

Diamond and graphite

Diamond and graphite do not seem at all similar and yet both consist of pure carbon. This may be proved by burning equal masses of the two substances in excess oxygen, when the same mass of carbon dioxide is obtained from each but no other product is formed.

Property	Diamond	Graphite
Appearance	Colourless crystalline solid; transparent	Grey-black crystalline solid; opaque
Density/g cm^{-3}	3.52	2.26
Hardness	One of the hardest substances known	Soft and flaky
Electrical conductivity	Low	High

Table 16.1 Some physical properties of diamond and graphite

Although they are chemically identical, diamond and graphite have widely differing physical properties, as can be seen from Table 16.1. Since both substances consist only of carbon atoms, these differences must be due to variations in the way in which the atoms are packed together.

Both substances have very high melting points, which indicates that they have giant atomic lattices (8.5). X-ray diffraction shows that a diamond consists of a three-dimensional network in which each carbon atom is surrounded tetrahedrally by four others (see Figure 16.1). The bonding in this network must be very strong because diamonds are so hard. Graphite, on the other hand, is made up of layers of hexagons of carbon atoms. The bonding in each layer is strong but the layers are only held together by weak van der Waals' forces. This explains why graphite is soft and can be used as a lubricant – the layers of carbon atoms slide easily over one another.

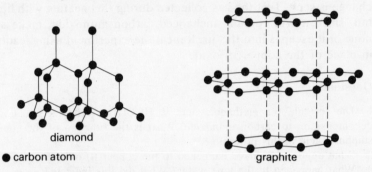

diamond

● carbon atom   graphite

Figure 16.1 The structures of diamond and graphite

In graphite each carbon atom is bonded to only three others, even though four electrons are available for bonding. The extra electrons are delocalised, i.e. they are not attached to any particular atoms but belong to the structure as a whole. Delocalised electrons are free to move from one hexagon to the next within a layer and thus conduct an electric current. You should compare this with the conduction of electricity by metals (page 143). The low density of graphite compared with that of diamond is accounted for by its more open structure.

Different forms of an element in the same state, such as diamond and graphite, are called *allotropes* (from the Greek *allos tropos*, meaning 'other form'). The element is said to exhibit *allotropy*.

Allotropy is the existence of an element in more than one form *in the same state*.

The final proviso is important in distinguishing allotropy from straightforward melting and boiling.

Other forms of carbon

All other forms of carbon were referred to in the past as *amorphous* (meaning 'without shape'). This name is misleading since it is now known that they consist of minute fragments of graphite. The different forms are made by heating various carbon-containing compounds in the absence of air and usually contain impurities. Since volatile products are given off during the heating, the process is often called *destructive distillation*.

The destructive distillation of wood gives *wood charcoal*, together with such substances as methanol and ethanoic acid (Experiment 16.1.b), and at one time this was the method of manufacturing these two important organic compounds. In a similar way *coke* is obtained from coal, and *animal charcoal* from bones; burning oil in a limited supply of air gives finely divided *lampblack*. These different forms of carbon have a variety of uses, as you will discover in the next section.

16.2 Reactions and uses of carbon

1 DIAMOND Diamonds are used as gemstones and, because they are so hard, in cutting tools, rock borers, etc.

2 GRAPHITE The slippery nature of graphite makes it a useful dry lubricant. It is also used as a moderator in atomic reactors, as an electrical conductor and, mixed with clay, to make pencil 'leads'.

3 CARBON AS A REDUCING AGENT When carbon is heated with the oxides of the less reactive metals it reduces them to the metal and is itself oxidised to carbon dioxide.

e.g. $$2CuO(s) + C(s) \rightarrow 2Cu(s) + CO_2(g) \quad \text{(see Experiment 16.2.a)}$$

In industry, coke is used to extract a number of metals such as zinc and iron from their oxide ores.

4 ANIMAL CHARCOAL Animal charcoal adsorbs coloured material on its surface (see Experiment 16.2.b). It is used to purify substances in the laboratory and to decolourise brown sugar.

5 WOOD CHARCOAL Gases are readily adsorbed by wood charcoal (see Experiment 16.2.c), particularly if it is first heated to remove any gas which is already adsorbed. It is used in gas masks.

6 LAMPBLACK This is used in printer's ink and shoe polish.

7 CARBON FIBRE A relatively new use of carbon is in strengthening plastics by means of threads of carbon fibre.

The Cambridge University crew leading the 1975 Boat Race using carbon fibre oars

16.3 Carbon hydrides

There are many carbon hydrides, the simplest one being methane, CH_4. This, along with several others, is considered in detail in Chapter 21. For the purpose of comparison with the hydrides of neighbouring elements in the periodic table, it is worth mentioning here that methane is a colourless, neutral gas which is insoluble in water. It burns in air and reacts with chlorine and bromine, but is otherwise rather inert.

16.4 Carbon dioxide

Laboratory preparation

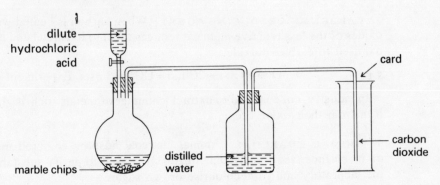

Figure 16.4.1 The laboratory preparation of carbon dioxide

Carbon dioxide may be prepared by the action of almost any acid on any carbonate but the reaction most commonly employed is that between dilute hydrochloric acid and marble chips (calcium carbonate). The gas is passed through water to remove hydrogen chloride and is collected by downward delivery, since it is denser than air (see Figure 16.4.1). A burning splint placed at the mouth of each jar is extinguised when the jar is full.

$$CaCO_3(s) + 2HCl(aq) \rightarrow CaCl_2(aq) + H_2O(l) + CO_2(g)$$

Alternatively, the reaction may be carried out in a Kipp's apparatus (page 352).

You must remember that very little carbon dioxide is produced when dilute *sulphuric* acid is added to marble chips. This is because the marble very quickly becomes coated with an insoluble layer of calcium sulphate which prevents further reaction.

Test for carbon dioxide

When carbon dioxide is passed into lime water (calcium hydroxide solution), a white precipitate of calcium carbonate is formed.

$$Ca(OH)_2(aq) + CO_2(g) \rightarrow CaCO_3(s) + H_2O(l)$$

If the passage of carbon dioxide is continued the reaction proceeds further and the precipitate disappears, owing to the formation of soluble calcium hydrogencarbonate.

$$CaCO_3(s) + H_2O(l) + CO_2(g) \rightleftharpoons Ca(HCO_3)_2(aq)$$

Physical properties

Carbon dioxide is a colourless, odourless gas. It is about $1\frac{1}{2}$ times as dense as air, and is slightly soluble in water. Under pressure the gas is easily converted to the solid form.

Chemical properties

COMBUSTION IN CARBON DIOXIDE   Carbon dioxide does not burn and will only support the combustion of substances which are hot enough to decompose it into its elements. Thus burning methylbenzene is extinguished by the gas but burning magnesium continues to burn, producing white clouds of magnesium oxide and black specks of carbon (Experiment 16.4.b). In this reaction the carbon dioxide is behaving as an oxidising agent and is itself being reduced.

$$2Mg(s) + CO_2(g) \rightarrow 2MgO(s) + C(s)$$

REACTION WITH WATER   A solution of carbon dioxide in water is weakly acidic because about 1% of the dissolved gas reacts with the water to form carbonic acid. The existence of separate molecules of this acid (H_2CO_3) is doubtful and the reaction is probably as shown:

$$CO_2(g) + water \rightleftharpoons CO_2(aq)$$
$$CO_2(aq) + H_2O(l) \rightleftharpoons 2H^+(aq) + CO_3^{2-}(aq).$$

If the solution is heated in an attempt to concentrate the acid the carbon dioxide is given off into the atmosphere. Pure carbonic acid is unobtainable.

Since carbon dioxide is an acidic oxide it is more soluble in alkalis than in water (Experiment 16.4.b). Thus sodium hydroxide solution will readily absorb the gas owing to the formation of sodium carbonate.

$$CO_2(g) + 2NaOH(aq) \rightarrow Na_2CO_3(aq) + H_2O(l)$$

If the carbon dioxide is in excess then further reaction occurs to give sodium hydrogencarbonate. (Compare this with the reaction between the gas and lime water.)

$$Na_2CO_3(aq) + H_2O(l) + CO_2(g) \rightleftharpoons 2NaHCO_3(aq)$$

Uses of carbon dioxide

1   To make 'fizzy' drinks. In these, the carbon dioxide is dissolved under pressure. Removing the top from the bottle releases this pressure and bubbles of the gas come out of solution.

2   In fire extinguishers. The compressed gas may itself be directed on to

the fire to blanket the flames, or it may be used to eject water, foam or powder from the extinguisher.

3   As a refrigerant. Solid carbon dioxide sublimes at $195 \, K$ $(-78\,°C)$, leaving no residue and is called 'Drikold' or 'Dri-ice'.

4   In the manufacture of sodium carbonate (20.2).

5   Health salts and baking powder contain sodium hydrogencarbonate and a solid acid such as tartaric acid. These compounds react together in the presence of water to produce carbon dioxide. (Can you explain why no reaction occurs until water is added?) In the case of baking powder more gas is evolved during cooking.

Manufacture of carbon dioxide

Although it is a by-product in many processes, the usual method of manufacturing carbon dioxide is by heating limestone (calcium carbonate) in a limekiln (Figure 16.4.2).

$$CaCO_3(s) \rightarrow CaO(s) + CO_2(g)$$

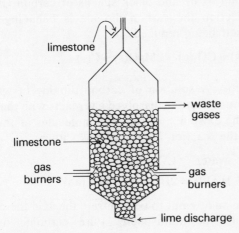

Figure 16.4.2 The limekiln

Carbonates

PREPARATION   Potassium, sodium and ammonium carbonates are soluble in water, all others being insoluble. The insoluble carbonates can be obtained as precipitates in double decomposition reactions by mixing together solutions containing the required metal ions and carbonate ions (as described in 5.4).

e.g. $ZnSO_4(aq) + Na_2CO_3(aq) \rightarrow ZnCO_3(s) + Na_2SO_4(aq)$

or, ionically, $Zn^{2+}(aq) + CO_3^{2-}(aq) \rightarrow ZnCO_3(s)$

The soluble carbonates are prepared from carbon dioxide and the corresponding alkali.

e.g. $CO_2(g) + 2KOH(aq) \rightarrow K_2CO_3(aq) + H_2O(l)$

As already mentioned, excess carbon dioxide gives the hydrogencarbonate.

THE ACTION OF ACID All carbonates react with acids to give a salt, water and carbon dioxide.

e.g. $$CuCO_3(s) + H_2SO_4(aq) \rightarrow CuSO_4(aq) + H_2O(l) + CO_2(g)$$

or, ionically, $$CO_3^{2-}(s) + 2H^+(aq) \rightarrow H_2O(l) + CO_2(g)$$

THE ACTION OF HEAT The thermal stability of a carbonate depends on the position of the metal in the reactivity series. Thus, at temperatures which are normally obtainable in the laboratory, potassium and sodium carbonates do not decompose but those of metals lower down in the series split up to give the metal oxide and carbon dioxide (Experiment 16.4.c). The lower the metal the more easily is its carbonate decomposed.

e.g. $$PbCO_3(s) \rightarrow PbO(s) + CO_2(g)$$

Hydrogencarbonates

Hydrogencarbonates contain replaceable hydrogen and therefore are acid salts (5.4). However, their solutions are *alkaline*. This apparent contradiction is explained in 5.6.

PREPARATION Very few hydrogencarbonates are known. The only common solid ones are those of potassium and sodium; calcium and magnesium hydrogencarbonates are commonly found but only exist in solution. All are prepared by passing excess carbon dioxide through a solution or suspension of the corresponding hydroxide or carbonate. (Equations have already been given.)

THE ACTION OF ACID Like carbonates, all hydrogencarbonates react with acids to give a salt, water and carbon dioxide.

e.g. $$KHCO_3(s) + HCl(aq) \rightarrow KCl(aq) + H_2O(l) + CO_2(g)$$

or, ionically, $$HCO_3^-(s) + H^+(aq) \rightarrow H_2O(l) + CO_2(g)$$

THE ACTION OF HEAT Hydrogencarbonates decompose very readily on heating to give the corresponding carbonate, water and carbon dioxide. This reaction occurs even on boiling their solutions (Experiment 16.4.d).

e.g. $$2NaHCO_3(aq) \rightarrow Na_2CO_3(aq) + H_2O(l) + CO_2(g)$$

or, ionically, $$2HCO_3^-(aq) \rightarrow CO_3^{2-}(aq) + H_2O(l) + CO_2(g)$$

Because of this, if solid potassium or sodium hydrogencarbonate is required, its solution must be allowed to evaporate at room temperature. However, if this procedure is carried out with a solution of calcium or magnesium hydrogencarbonate the solid which is formed is the *carbonate*, because the above decomposition takes place even at room temperature.

Tests for carbonates and hydrogencarbonates

On heating, all hydrogencarbonates and most carbonates give off carbon dioxide: both salts give carbon dioxide on the addition of dilute hydrochloric acid. All hydrogencarbonates are soluble in water: they may be distinguished from the few soluble carbonates by boiling their solutions with universal indicator. Carbon dioxide is evolved and the indicator changes from green to purple.

The carbon cycle

Carbon dioxide is present in the atmosphere to the extent of about 0.03% by volume, this figure remaining fairly constant owing to the delicate balance between the processes evolving the gas and those absorbing it. The ways in which carbon atoms circulate in nature are shown in the *carbon cycle* (Figure 16.4.3).

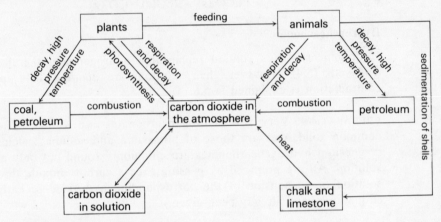

Figure 16.4.3 The carbon cycle

Virtually all of the carbon compounds on the earth have originated from carbon dioxide in the atmosphere and it may interest you to speculate on the history of the carbon atoms in your body. Perhaps they were once part of a dinosaur!

16.5   Carbon monoxide

Laboratory preparation

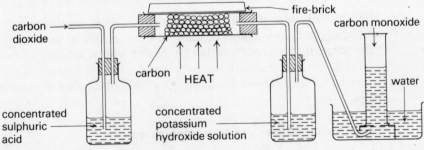

Figure 16.5.1 The laboratory preparation of carbon monoxide

When carbon dioxide is passed over strongly-heated carbon using the apparatus of Experiment 16.5.a, or that shown in Figure 16.5.1, reduction of the carbon dioxide and oxidation of the carbon occurs, giving carbon monoxide.

$$CO_2(g) + C(s) \rightarrow 2CO(g)$$

The gas is passed through concentrated potassium hydroxide solution to remove any unchanged carbon dioxide (how does it do this?) and is then collected over water. Because of the poisonous nature of the gas, care must be taken to ensure that none escapes into the atmosphere.

An alternative and more convenient method of preparation is to dehydrate methanoic acid with concentrated sulphuric acid, as in Experiment 16.5.b.

$$HCOOH(l) - H_2O(l) \rightarrow CO(g)$$

Physical properties

Carbon monoxide is a colourless, odourless gas which is slightly less dense than air. It is almost insoluble in water.

Chemical properties

COMBUSTION   The gas burns in air with a blue flame to form carbon dioxide (Experiment 16.5.a).

$$2CO(g) + O_2(g) \rightarrow 2CO_2(g)$$

REDUCING ACTION   Oxides of the less active metals are readily reduced on heating with carbon monoxide (Experiment 16.5.b).

$$\overbrace{CuO(s) + CO(g) \rightarrow Cu(s) + CO_2(g)}^{\text{oxidation}}$$

e.g. $$CuO(s) + CO(g) \rightarrow Cu(s) + CO_2(g)$$ reduction

ACIDIC NATURE   Carbon monoxide is insoluble in water and is usually considered to be a neutral oxide. However, when heated under pressure it will react with concentrated sodium hydroxide solution to give sodium methanoate and thus it does have very weak acidic properties.

$$NaOH(aq) + CO(g) \rightarrow HCOONa(aq)$$

Uses of carbon monoxide

1 In the manufacture of methanol.
2 As a reducing agent in the extraction of iron (20.6).

The poisonous nature of carbon monoxide

Carbon monoxide is an extremely poisonous gas: approximately 0.2% in the atmosphere causes loss of consciousness and 1% causes death. It is more dangerous than many other poisonous gases because it is odourless.

It combines with the haemoglobin (red corpuscles) in the blood to form cherry-red carboxyhaemoglobin. Normally oxygen is carried round the bloodstream as bright red oxyhaemoglobin, which is unstable and readily gives up its oxygen where required. Carboxyhaemoglobin is much more stable than oxyhaemoglobin and thus the blood can no longer act as an oxygen carrier. Death occurs because the body cells are starved of oxygen.

Carbon monoxide and air pollution

Carbon monoxide is formed whenever carbon or carbon-containing compounds are burned in a limited air supply. Car exhaust fumes contain the gas because of incomplete combustion of petrol in the engine (21.2). It is also present in the flue gases from solid fuel fires or boilers, as a result of the series of reactions shown in Figure 16.5.2.

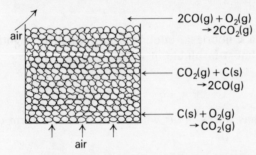

Figure 16.5.2 Reactions occurring in a solid fuel fire

At the base of the fire the coal or coke burns in a plentiful air supply to form carbon dioxide. The carbon dioxide is reduced to carbon monoxide in the middle of the fire by the hot carbon in the fuel. At the top of the fire the carbon monoxide burns to reform carbon dioxide, provided that there is plenty of air and the surface of the fuel is hot enough to ignite the gas. If one, or both of these conditions is not met then the carbon monoxide will pass into the chimney. Tragic results have occurred where old chimneys have become porous over the years and allowed fumes to enter upstairs rooms.

Air pollution can only be prevented by the design of more efficient car engines and boilers. Modern solid fuel domestic boilers are designed to burn fuel slowly, which leads to a saving in fuel but tends to produce more carbon monoxide.

16.6   Tetrachloromethane (carbon tetrachloride)

Tetrachloromethane, CCl_4, is the simplest chloride of carbon. It is a colourless, non-inflammable liquid which is insoluble in water. It is a useful solvent (e.g. in 'Thawpit') and is also used in fire extinguishers, although it can form poisonous products when heated in air. Unlike most covalent chlorides tetrachloromethane is not hydrolysed by water.

16.7   Silicon and the other group IV elements

All the elements in group IV of the periodic table have four electrons in the outermost shells of their atoms and thus would be expected to possess similar chemical properties. The metallic nature of the elements in any group increases as the group is descended (7.5): this feature is perhaps most noticeable in group IV where the non-metal carbon is followed by the metalloids silicon and germanium and finally the metals tin and lead. Germanium and tin are not discussed in this book but the chemistry of lead is dealt with in Chapter 20.

Silicon

Like carbon, silicon can be obtained by heating its oxide (sand) with magnesium. It is a black solid with a high melting point and has the same structure as diamond. It is rather inert but in the powdered form will react with halogens and with alkalis.

Silicon hydrides

The simplest silicon hydride, silane (SiH_4), is a colourless, neutral gas which is insoluble in water. Unlike methane, it is spontaneously inflammable in air.

Silicon(IV) oxide (silica)

The structure of silicon(IV) oxide is described in section 8.5. The compound occurs naturally in the form of quartz. It is insoluble in water as anyone who has been to the seaside should know (sand is impure quartz) but in spite of its insolubility, it is an acidic oxide. Thus it will dissolve in molten sodium hydroxide to form sodium silicate, which on acidification gives silicon(IV) oxide again. You should compare these reactions with those of carbon dioxide.

Uses of silicon(IV) oxide

1   To make silicon carbide (carborundum) which is used as an abrasive.
2   To make concrete.
3   In the manufacture of glass. Ordinary *soft glass* (soda glass) is made by heating a mixture of silicon(IV) oxide, sodium carbonate and calcium carbonate to a temperature of 1773 K (1500°C), when a mixture of sodium and calcium silicates is produced. *Coloured glass* is made by the addition of metal oxides to soda glass (e.g. cobalt(II) oxide will give blue glass, nickel(II) and chromium(III) oxides will give green glass). If some of the silicon is replaced by boron, *Pyrex* is obtained. *Silica glass* is made by cooling molten silicon(IV) oxide. It does not melt so readily as soda glass and has a low coefficient of expansion so that it can easily withstand variations in temperature.

Silicon tetrachloride

This is a colourless liquid. Like most covalent chlorides, but *unlike* tetrachloromethane, it is hydrolysed by water. For this reason it fumes in moist air, the products of hydrolysis being silicic(IV) acid and hydrogen chloride.

$$SiCl_4(l) + 3H_2O(l) \rightarrow H_2SiO_3(aq) + 4HCl(aq)$$

Questions on Chapter 16

1 Two of the allotropes of carbon are diamond and graphite.
(a) By considering the structures of these allotropes, explain the differences in their physical properties.
(b) How can it be shown that they are allotropes of carbon? (O & C)
2 (a) Draw a labelled diagram of the apparatus you would use to prepare and collect carbon dioxide. Name the chemicals used and give the equation for the reaction.
(b) Explain the following statements.
   (i) A blue flame is frequently seen above a red hot coal fire.
   (ii) It is dangerous to keep a motor car engine running in a closed garage.
   (iii) When carbon dioxide is passed into a concentrated solution of sodium hydroxide, a white precipitate is slowly formed. (C)
3 (a) Describe, with the aid of a diagram, how you would convert carbon dioxide to carbon monoxide. How would you test the product to see whether the reduction was complete?
(b) Explain the formation of carbon monoxide in the combustion of hydrocarbons and state why its formation is potentially dangerous.
(c) Describe one reaction in which carbon monoxide acts as a reducing agent. (O & C)
4 (a) Making use of calcium carbonate, hydrochloric acid and charcoal, describe, with the help of a diagram, the preparation and collection of carbon monoxide.
(b) A volatile liquid X is a compound of carbon and sulphur only. 0.2 mole of X contains 2.4 g of carbon and 12.8 g of sulphur. Calculate the relative molecular mass and formula of compound X.
(c) X burns in oxygen to form a mixture of two gases both of which are completely absorbed by sodium hydroxide solution.
   (i) Give the formulae of the two gases formed.
   (ii) Give the formulae of the compounds formed in solution when these gases react with sodium hydroxide. (C)

17 Nitrogen and phosphorus

Nitrogen and phosphorus are the only group V elements studied in this book. Although both are typical non-metals and their atoms have the same number of electrons in their outermost shells (how many?) they show many differences in properties. In this chapter we will be mainly concerned with the compounds of nitrogen and will refer only briefly to those of phosphorus.

Experiments

17.1.a To isolate nitrogen from the air ★ ★

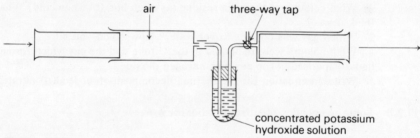

Figure 17.1.a

PROCEDURE   Pass 100 cm³ of air through concentrated potassium hydroxide solution using gas syringes as shown in Figure 17.1.a. Transfer the syringe full of air to the apparatus used in Experiment 3.1.e and pass the air over heated copper until no further change in volume occurs. Expel the residual gas through the three-way tap and collect it over water in test tubes. Note the smell of this gas and test separate samples with a burning splint, lime water and universal indicator.

Questions

1 Why was the air passed through potassium hydroxide solution?
2 Which gas was removed by the hot copper?
3 What impurities remained in the nitrogen collected in the test tubes?
4 Describe how the gas behaved in your tests.

17.2.a The laboratory preparation of nitrogen dioxide ★ ★

PROCEDURE   Heat some lead(ii) nitrate crystals in the apparatus shown in Figure 17.2.a. Allow the first few bubbles of gas to escape before placing the gas jar in position. When the reaction has finished, test the gas

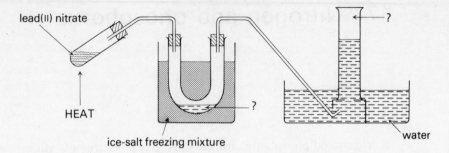

lead(II) nitrate

HEAT

ice-salt freezing mixture

?

water

Figure 17.2.a

in the jar with a glowing splint. Inspect the liquid in the U-tube and then tip a few drops of it into each of four gas jars. Quickly close the jars with cover slips and look for a colour change as the liquid vaporises. Avoid breathing the gas which forms.

Questions

1 What did you hear when the lead(II) nitrate was heated?
2 What colour was the solid residue (a) when hot, (b) when cold? What do you think it was?
3 Which gas was collected over water? How could you tell?
4 What colours were the liquid in the U-tube and the gas which formed as this liquid vaporised? (The gas was nitrogen dioxide.)
5 Write an equation for the thermal decomposition of lead(II) nitrate.

17.2.b Some properties of nitrogen dioxide ★ ★

PROCEDURE (a) Lower a burning splint into a jar of nitrogen dioxide.
(b) Lower a piece of burning magnesium ribbon into a jar of nitrogen dioxide, taking the usual precautions against looking directly at the flame.
(c) Pour a little distilled water containing universal indicator into a jar of nitrogen dioxide. Quickly replace the cover slip and shake the jar.

Questions

1 What colour was the nitrogen dioxide?
2 Did the nitrogen dioxide support combustion of (a) the splint, (b) the magnesium?
3 Was the nitrogen dioxide very soluble in water?
4 Was the solution of nitrogen dioxide acidic, neutral or alkaline?

17.2.c To prepare nitrogen oxide and investigate some of its properties ★ ★

PROCEDURE Collect four jars of nitrogen oxide as described on page 320 and use them in the following tests.
(a) Place a jar of the gas in front of a white screen and remove the cover slip. (Avoid breathing the product of this reaction.)
(b) Lower a burning splint into a jar of the gas.
(c) Lower a piece of burning magnesium ribbon into a jar of the gas,

taking the usual safety precautions against looking directly at the flame. (d) Add a little iron(II) sulphate solution to a jar of the gas, replace the cover slip and shake the jar.

Questions

1 Which gas did you see in the reaction flask? Why was this not collected in the gas jar? What colour was the nitrogen oxide?
2 What colour was the solution which remained in the flask at the end of the reaction? What do you think this was?
3 Did the gas support the combustion of the splint and/or the magnesium? What can you say about the thermal stability of nitrogen oxide compared with that of nitrogen dioxide?
4 What did you see in (d)?

17.2.d To prepare dinitrogen oxide and investigate some of its properties ★ ★

PROCEDURE Collect a jar of dinitrogen oxide over water as described on page 321. Lower a glowing splint into the gas and note the result. **N.B.** If heating is continued until only a little ammonium nitrate remains there is a serious risk of explosion.

Questions

1 What colour was the dinitrogen oxide?
2 What was the effect of the gas on a glowing splint?
3 How does the thermal stability of the gas compare with that of nitrogen oxide and nitrogen dioxide? Explain your answer.
4 Write an equation for the reaction you would expect to occur if burning magnesium were lowered into a jar of dinitrogen oxide.

17.3.a The laboratory preparation of ammonia ★ ★

PROCEDURE Collect several gas jars of ammonia by warming a mixture of equal masses of ammonium chloride and calcium hydroxide as shown in Figure 17.3.1. Hold a piece of moist universal indicator paper at the mouth of each jar so that you can tell when the jar is full.

Questions

1 Where have you encountered the smell of ammonia before?
2 What was the purpose of the calcium oxide?
3 How does the density of ammonia compare with that of air? Give a reason for your answer.
4 What was the effect of the gas on the universal indicator paper? What can you conclude from this?
5 Why can ammonia not be dried by bubbling it through concentrated sulphuric acid?
6 Write an equation for the preparation of ammonia, given that the other products are calcium chloride and water.

17.3.b To investigate the solubility of ammonia in water ★ ★

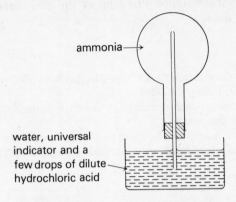

ammonia→

water, universal
indicator and a
few drops of dilute—
hydrochloric acid

Figure 17.3.b

PROCEDURE Replace the gas jar shown in Figure 17.3.1 with a dry 500 cm³ round-bottomed flask and fill it *completely* with ammonia. Insert a bung and glass tube in the mouth of the flask and set up the apparatus shown in Figure 17.3.b. Pour a little propanone (**caution:** inflammable) on to the base of the flask and repeat if necessary until a few drops of universal indicator solution enter the flask.

Questions

1 What did you observe? How soluble is ammonia in water?
2 Why might it be dangerous to prepare ammonia solution by dipping a delivery tube from an ammonia generator into a beaker of water?
3 What did the colour of the universal indicator tell you about the nature of aqueous ammonia solution?
4 Place a drop of propanone on your hand and note what happens. Now explain the purpose of the propanone in the above experiment.

17.3.c Two characteristic reactions of ammonia ★

PROCEDURE (a) Dip a glass rod into concentrated hydrochloric acid and then hold the rod near the mouth of a jar of ammonia. Repeat, using concentrated nitric acid in place of the hydrochloric acid.
(b) Add 1 drop of aqueous ammonia solution to 5 cm³ of copper(II) sulphate solution in a test tube and stir with a glass rod. Continue adding small volumes of ammonia solution, with stirring, until no further change occurs.

Questions

1 What did you see in (a)? Name the product in each case and write equations to represent the reactions.
2 Do you think that the reaction between ammonia and hydrogen chloride would be useful in identifying (a) ammonia, (b) hydrogen chloride, (c) both, (d) neither? Explain your answer.
3 What did you see after adding one drop of ammonia solution in (b)? Name the products and write an equation for the reaction. (Remember that ammonia solution behaves as if it were a solution of ammonium hydroxide.)

4 What happened when you added excess ammonia solution in (b)? Suggest a use for this reaction.

17.3.d   Does ammonia burn? ★ ★

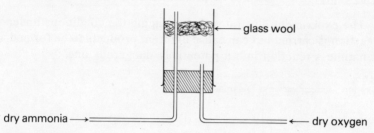

Figure 17.3.d

PROCEDURE Pass dry ammonia and oxygen through the apparatus shown in Figure 17.3.d and ignite the ammonia at the end of the tube. Hold a flask of cold water in the flame for a few seconds and then inspect it closely. Finally, turn off the oxygen supply and note whether the ammonia continues burning.

Questions

1 Why was the glass wool used?
2 What did you see on the surface of the flask? Suggest the most likely products of the combustion of ammonia in oxygen and write an equation for the reaction.
3 Did the ammonia burn in air?

17.3.e   The catalytic oxidation of ammonia by the oxygen of the air ★ ★

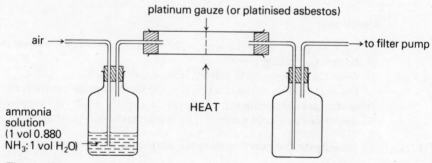

Figure 17.3.e

PROCEDURE Draw a mixture of air and ammonia through the apparatus shown in Figure 17.3.e by means of a filter pump. Heat the catalyst and look for a product in the wash-bottle.

Questions

1 The product of the catalytic oxidation of ammonia is actually nitrogen oxide. Which compound *appeared* to be the product of the reaction? Explain how you came to this conclusion and also how the product which you saw was formed from the nitrogen oxide.

2 Write an equation for the catalytic oxidation of ammonia.

17.3.f   The catalytic oxidation of ammonia by pure oxygen ★ ★

This experiment must not be performed by pupils.

The experiment illustrates that changing the conditions under which a reaction is carried out can cause different products to be formed and turn a harmless reaction into a potentially dangerous one.

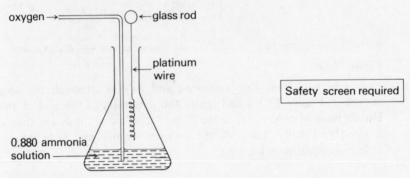

Figure 17.3.f

PROCEDURE Bubble oxygen fairly slowly into about 1 cm depth of 0.880 ammonia solution in a conical flask. Suspend a spiral of platinum wire from a glass rod, heat it in a bunsen flame and lower it into the flask (see Figure 17.3.f). (**Caution:** keep your hand away from the mouth of the flask.) Observe all that happens and then investigate the effect of speeding up the flow of oxygen slightly.

Questions

1 What happened to the platinum wire? What does this indicate about the nature of the reaction taking place?
2 Were the final products stable? How could you tell?
3 The first product of the reaction was the same as in the previous experiment. Suggest a possible sequence of events to explain what you observed. (Hint: remember that ammonium nitrate is thermally unstable, as is ammonium nitrite.)

17.3.g   To investigate the reaction between ammonia and copper(II) oxide ★ ★

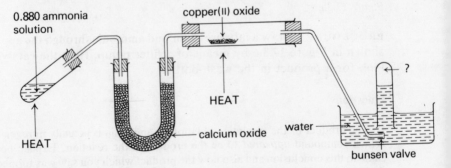

Figure 17.3.g

PROCEDURE Set up the apparatus shown in Figure 17.3.g. Warm the 0.880 ammonia solution and sweep the air from the apparatus with the ammonia which is evolved, then place the test tube in position. Heat the copper(II) oxide and collect several tubes of gas. Look for a product on the cooler parts of the combustion tube and test the gas in the test tubes as in Experiment 17.1.a

Questions

1 What was the purpose of the bunsen valve?
2 What appeared to happen to the copper(II) oxide? What sort of reaction was this?
3 What did you see on the cooler parts of the combustion tube?
4 Name the gas collected in the test tubes. How did you identify it?
5 Write an equation for the reaction.

17.3.h Some reactions of ammonium salts

PROCEDURE (a) Heat a spatula measure of ammonium chloride in a test tube, smell cautiously any gas which is evolved and test it with universal indicator paper. Repeat the experiment using in turn, ammonium sulphate and ammonium carbonate but in the latter case also test for gas *before* heating.
(b) *Just* fill the rounded part of a test tube with ammonium nitrate and heat it until all of the salt has gone. (**Caution:** keep your hands clear of the mouth of the tube and make sure that the tube is not pointing at anyone.)
(c) Warm a spatula measure of each of the ammonium salts in (a) and (b) with 5 cm³ of sodium hydroxide solution in a test tube. Test for the evolution of gas with universal indicator paper and by smell as in (a).

Questions

1 Which gas was given in (a)? How could you tell?
2 Describe what happened in (b).
3 Which gas was given in (c)? Did all of the ammonium salts give this gas?

17.4.a The action of heat on concentrated nitric acid ★ ★

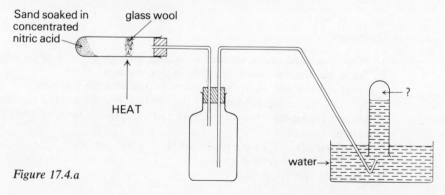

Sand soaked in concentrated nitric acid

glass wool

HEAT

water→

← ?

Figure 17.4.a

PROCEDURE   Pour 2 cm³ of concentrated nitric acid or fuming nitric acid on to 2 cm depth of sand in a boiling tube. Place a tuft of glass wool in the middle of the tube and assemble the rest of the apparatus as shown in Figure 17.4.a. Heat the glass wool and observe the wash-bottle closely. Allow the first bubbles which emerge from the delivery tube to escape into the atmosphere and then collect a test tube full of gas. Test this gas with a glowing splint.

Questions

1   What do you think was the purpose of the glass wool? (Hint: was it recovered unchanged at the end of the reaction?)
2   What did you see in the wash-bottle? Why was the wash-bottle reversed?
3   Explain why the colour of the gas in the wash-bottle was different from that collected in the test tube.
4   What was the gas in the test tube? How could you tell?
5   Write an equation for the thermal decomposition of nitric acid.

17.4.b   The oxidising action of concentrated nitric acid ★ ★

In reactions (a) and (b), fuming nitric acid will give better results than concentrated nitric acid.

PROCEDURE   (a) Warm some sawdust in an evaporating basin and carefully add a few drops of concentrated nitric acid. When the reaction ceases, add a few more drops of the acid.
(b) Add some powdered sulphur to about 10 cm³ of concentrated nitric acid in an evaporating basin in a fume cupboard. Heat the dish gently until a reaction starts and then turn off the bunsen. When the reaction has finished pour off a little of the liquid into a test tube and add some barium chloride solution.
(c) Add a few drops of concentrated nitric acid to a copper turning in a test tube and observe the result. Investigate the reaction between concentrated nitric acid and magnesium, aluminium and iron in a similar way.
(d) Add a little concentrated nitric acid to some freshly prepared iron(II) sulphate solution.

Questions

1   What happened to the sawdust on adding the concentrated nitric acid?
2   What was the purpose of the acid in this reaction?
3   What did you see when the acid was heated with the sulphur?
4   What was the sulphur changed to? Explain how you reached this conclusion.
5   Discuss the reaction between sulphur and nitric acid in terms of oxidation and reduction.
6   Which metals in (c) were oxidised? How could you tell?
7   Which metals in (c) did not react?
8   What did you see in (d)?

17.4.c The acidic properties of dilute nitric acid

PROCEDURE   Carry out Experiments 5.2.a, 5.2.b and 5.2.c using dilute nitric acid as the only acid.

Questions

1   Is dilute nitric acid a strong or a weak acid? Give reasons for your answer.
2   Sum up the acidic properties of dilute nitric acid, giving a chemical equation and an ionic equation to illustrate each property. Are there any reactions where this acid does not behave in the expected manner?

17.4.d To investigate the effect of heat on nitrates

PROCEDURE   Heat a little sodium nitrate in an ignition tube and test for the evolution of oxygen by means of a glowing splint. (Make sure that fragments of charred wood do not drop into the tube.) Look also for the formation of nitrogen dioxide. Repeat the experiment using samples of calcium nitrate, copper(II) nitrate, zinc nitrate and potassium nitrate, noting the relative ease with which the compounds decompose. In some cases the salts are hydrated; do not test for oxygen until the water of crystallisation has been driven off.

Questions

1   Which of the nitrates, if any, (a) did not give oxygen, (b) gave oxygen only, (c) gave oxygen and nitrogen dioxide?
2   What were the solid residues? (The colours may give you a clue in some cases.)
3   What connection is there between the thermal stability of a metal nitrate and the position of the metal in the reactivity series?
4   Write equations for the thermal decomposition of potassium nitrate and copper(II) nitrate.

17.4.e Tests for nitrates

If a solid gives off oxygen or a mixture of oxygen and nitrogen dioxide on heating it is probably a nitrate. The following tests will confirm its identity.

PROCEDURE   (a) Add a spatula measure of zinc nitrate and a few copper turnings, or a little copper powder, to 1 cm^3 of concentrated sulphuric acid in a test tube. Warm *gently* (**care**) and note all that happens. Carry out a similar experiment with the other nitrates mentioned in the previous experiment.
N.B. To dispose of each mixture, first allow it to cool and then pour it into plenty of water in the sink. *Never* pour the *hot* mixture into water.
(b) Dissolve a spatula measure of zinc nitrate and about half as much iron(II) sulphate in 5 cm^3 of distilled water in a test tube. Tilt the tube and *carefully* add concentrated sulphuric acid dropwise until it fills the rounded part (see Figure 17.4.e). Look closely at the junction between the

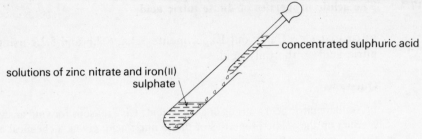

concentrated sulphuric acid

solutions of zinc nitrate and iron(II) sulphate

Figure 17.4.e

two layers of liquid. Repeat the procedure using sodium nitrate in place of the zinc nitrate.

Questions

1   What appeared to be evolved in each case in (a)? Can you explain this? (Hint: what is formed when concentrated sulphuric acid is warmed with a nitrate?)
2   What did you see at the junction of the two liquid layers in (b)?
3   Why would the procedure in (b) be less satisfactory for, say, lead(II) nitrate? (Hint: what do you know about the solubility of lead(II) sulphate?)

17.1   Nitrogen

About 78% by volume of the air is nitrogen and it was from this source that the Swedish chemist, Carl Scheele, first isolated the gas in 1772. Nitrogen is also found in ammonium salts and nitrates: its name comes from the Greek *nitron*, meaning 'nitre' (i.e. potassium nitrate).

Preparation

In the laboratory, nitrogen is prepared by removing carbon dioxide and oxygen from the air (Experiment 17.1.a) or by heating an aqueous solution containing ammonium chloride and sodium nitrite. Atmospheric nitrogen differs in density from 'chemical' nitrogen by about 0.5% owing to the presence of the noble gases in the former. Indeed, it was this difference in density which led to the discovery of the noble gases by Lord Rayleigh and Sir William Ramsay in the last few years of the nineteenth century.

Physical properties

Nitrogen is a colourless, odourless gas, a little less dense than air and slightly soluble in water.

Chemical properties

The two atoms in a nitrogen molecule are held together by a triple covalent bond. This bond is very strong, which explains why the gas is so unreactive. However, nitrogen does undergo a few reactions: for example, it forms nitrogen oxide when sparked with oxygen; it combines with hydrogen under special conditions to form ammonia (see page 324); and it forms nitrides on heating with a few metals, such as magnesium and calcium,

e.g. $$3Mg(s) + N_2(g) \rightarrow Mg_3N_2(s).$$

A nitrogen atom contains five electrons in its outermost shell and it usually completes its octet by forming three covalent bonds. However, in the metal nitrides it accepts three electrons and forms the N^{3-} ion. Nitrogen atoms never give up five electrons to form N^{5+} ions (why not?).

Uses of nitrogen

1 The main use of nitrogen is in the manufacture of ammonia by the Haber process.
2 Liquid nitrogen is used as a refrigerant.

Manufacture of nitrogen

Industrially, nitrogen is obtained by the fractional distillation of liquid air (18.1).

17.2 The oxides of nitrogen

Nitrogen dioxide

Nitrogen dioxide is a dark brown gas with a pungent, choking smell and is about $1\frac{1}{2}$ times as dense as air. It is extremely poisonous and is one of the air pollutants emitted in car exhaust fumes. (Initially nitrogen oxide is produced by the combination of nitrogen and oxygen at the high temperature which exists inside the engine. The nitrogen oxide then combines with more oxygen to form nitrogen dioxide.) In the laboratory the gas may be prepared by the action of *concentrated* nitric acid on copper (Figure 17.2.1.)

$$Cu(s) + 4HNO_3(aq) \rightarrow Cu(NO_3)_2(aq) + 2H_2O(l) + 2NO_2(g)$$

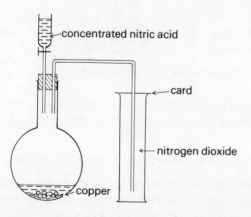

concentrated nitric acid

card

nitrogen dioxide

copper

Figure 17.2.1 The laboratory preparation of nitrogen dioxide

Nitrogen dioxide can also be made by heating lead(II) nitrate: the crystals *decrepitate* (i.e. decompose with small explosions) forming lead(II) oxide and giving off a mixture of nitrogen dioxide and oxygen (Experiment 17.2.a).

$$2Pb(NO_3)_2(s) \rightarrow 2PbO(s) + 4NO_2(g) + O_2(g)$$

At room temperature, 'nitrogen dioxide' really consists of a mixture of nitrogen dioxide and dinitrogen tetraoxide (N_2O_4) in equilibrium with one another (14.3). Cooling this mixture in an ice/salt bath produces liquid dinitrogen tetraoxide, which should be pale yellow but is usually green owing to the presence of impurities. As the temperature rises, the dinitrogen tetraoxide dissociates into dark brown nitrogen dioxide, the process being complete at 150°C.

$$N_2O_4(g) \rightleftharpoons 2NO_2(g)$$

Nitrogen dioxide itself dissociates into nitrogen oxide and oxygen fairly readily on heating and it supports combustion better than nitrogen oxide does. Both a splint and magnesium continue to burn in the gas (Experiment 17.2.b).

$$4Mg(s) + 2NO_2(g) \rightarrow 4MgO(s) + N_2(g)$$

The gas is very soluble in water, the resultant solution being a mixture of dilute nitric and nitrous acids.

$$2NO_2(g) + H_2O(l) \rightarrow HNO_3(aq) + HNO_2(aq)$$

Nitrogen oxide

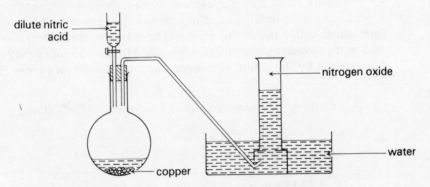

Figure 17.2.2 The laboratory preparation of nitrogen oxide

Nitrogen oxide is prepared by the action of dilute nitric acid (1 volume of concentrated acid:1 volume of water) on copper and is collected over water.

$$3Cu(s) + 8HNO_3(aq) \rightarrow 3Cu(NO_3)_2(aq) + 4H_2O(l) + 2NO(g)$$

The colourless gas is immediately oxidised to brown fumes of nitrogen dioxide on contact with oxygen at room temperature and hence its smell is unknown. The reaction reverses at higher temperatures.

$$2NO(g) + O_2(g) \rightleftharpoons 2NO_2(g)$$

Nitrogen oxide only supports combustion if the flame of the burning substance is hot enough to decompose it into its elements, the oxygen so produced permitting combustion to continue. Thus a burning splint is extinguished but magnesium continues to burn, forming magnesium oxide and nitrogen (Experiment 17.2.c).

$$2Mg(s) + 2NO(g) \rightarrow 2MgO(s) + N_2(g)$$

With an aqueous solution of iron(II) sulphate, nitrogen oxide forms a brown/black solution of the 'brown ring complex'. The production of this substance is made use of in the 'brown ring test' for nitrates (Experiment 17.4.e).

Dinitrogen oxide

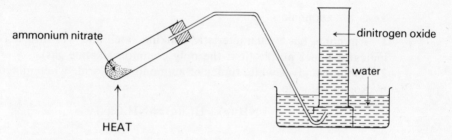

Figure 17.2.3 The laboratory preparation of dinitrogen oxide

Dinitrogen oxide is a colourless gas with a faint, sweetish smell. It is prepared by heating ammonium nitrate and is collected over water. Care must be taken to stop heating while plenty of ammonium nitrate still remains, to avoid the risk of an explosion.

$$NH_4NO_3(l) \rightarrow N_2O(g) + 2H_2O(g)$$

This gas is the least thermally stable of the three common oxides of nitrogen and will even relight a glowing splint (Experiment 17.2.d), owing to its decomposition into nitrogen and oxygen. It is used as an anaesthetic ('laughing gas').

17.3 Ammonia

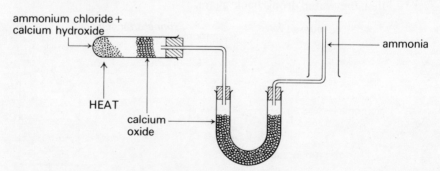

Figure 17.3.1 The laboratory preparation of ammonia

Preparation

Ammonia may be prepared by warming any ammonium salt with any base, but usually ammonium chloride and calcium hydroxide are used. The other products of the reaction are then calcium chloride and water.

$$2NH_4Cl(s) + Ca(OH)_2(s) \rightarrow CaCl_2(s) + 2H_2O(g) + 2NH_3(g)$$

The gas is dried by passing it over calcium oxide because it reacts with the more common drying agents, anhydrous calcium chloride and concentrated sulphuric acid. It is collected by upward delivery, a moist piece of universal indicator paper placed at the mouth of each jar turning purple when the jar is full (Experiment 17.3.a). Alternatively, ammonia may be obtained by warming its concentrated aqueous solution (0.880 ammonia solution).

Tests for ammonia

1 Ammonia has a characteristic choking smell and will turn universal indicator paper purple (it is the only common alkaline gas).
2 Ammonia gives white fumes of ammonium chloride when mixed with hydrogen chloride.

$$NH_3(g) + HCl(g) \rightleftharpoons NH_4Cl(s)$$

Physical properties

Ammonia is a colourless gas with a pungent choking smell and is about half as dense as air. It is very soluble in water, as can be demonstrated by the fountain experiment (Experiment 17.3.b). Because of its great solubility in water (over 700 cm^3 dissolve in 1 cm^3 of water at room temperature) special precautions have to be taken to prevent sucking back if an aqueous solution of the gas has to be made. Either a bunsen valve (see Experiment 17.3.g) or the funnel arrangement must be used (Figure 17.3.2). The bunsen valve consists of a piece of rubber tubing with a slit cut in it and closed at one end with a short length of glass rod. As the pressure inside the apparatus builds up gases escape through the slit but no water can enter the tube. If the funnel arrangement is used, water rises in the funnel but the level in the beaker drops. Air enters under the rim of the funnel and equalises the pressure inside and outside the apparatus so that the water drops back again.

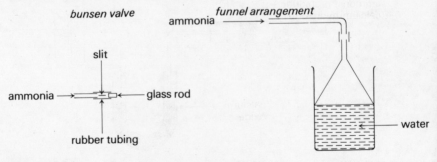

Figure 17.3.2

Chemical properties

AMMONIA SOLUTION An aqueous solution of ammonia is alkaline because a little of the dissolved gas reacts with the water to form ammonium ions and hydroxide ions. The existence of ammonium hydroxide (NH_4OH) molecules in the solution is doubtful.

$$NH_3(g) + water \rightleftharpoons NH_3(aq)$$
$$NH_3(aq) + H_2O(l) \rightleftharpoons NH_4^+(aq) + OH^-(aq)$$

Since it contains hydroxide ions, ammonia solution neutralises acids (giving ammonium salts). It also precipitates insoluble metallic hydroxides from solutions containing the metal ions (double decomposition). For example, if ammonia solution is added to copper(II) sulphate solution a pale blue precipitate of copper(II) hydroxide is formed (Experiment 17.3.c).

$$Cu^{2+}(aq) + 2OH^-(aq) \rightarrow Cu(OH)_2(s)$$

If excess ammonia solution is added, a further reaction takes place giving a royal blue solution containing the complex tetraamminecopper(II) ion. This reaction may be used to test for the presence of copper(II) ions.

$$Cu(OH)_2(s) + 4NH_3(aq) \rightarrow [Cu(NH_3)_4]^{2+}(aq) + 2OH^-(aq)$$

COMBUSTION AND CATALYTIC OXIDATION Ammonia does not burn in air: it does, however, burn in oxygen, forming nitrogen and steam (Experiment 17.3.d).

$$4NH_3(g) + 3O_2(g) \rightarrow 2N_2(g) + 6H_2O(g)$$

Air *will* oxidise ammonia if a platinum catalyst is present, the reaction being exothermic. The initial products are nitrogen oxide and steam but the nitrogen oxide is immediately oxidised to nitrogen dioxide by more oxygen. The catalytic oxidation of ammonia by the oxygen of the air is the first stage in the manufacture of nitric acid by the Ostwald process (17.4).

$$4NH_3(g) + 5O_2(g) \rightarrow 4NO(g) + 6H_2O(g)$$
$$2NO(g) + O_2(g) \rightleftharpoons 2NO_2(g)$$

If the reaction is carried out using pure oxygen and excess ammonia, as in Experiment 17.3.f, the resulting nitrogen dioxide reacts with the excess ammonia in the presence of water to form ammonium nitrate and ammonium nitrite. These thermally unstable compounds decompose explosively on contact with the hot catalyst.

REDUCING ACTION The oxides of metals low down in the reactivity series are reduced to the metals by heating in a stream of ammonia. The other products of the reaction are nitrogen and steam.

e.g. $$3CuO(s) + 2NH_3(g) \rightarrow 3Cu(s) + N_2(g) + 3H_2O(g)$$

Ammonia also reduces chlorine (19.2).

Uses of ammonia

1   To make nitrogenous fertilisers, such as ammonium sulphate (see 17.4, *The nitrogen cycle*).
2   To make nitric acid by the Ostwald process (17.4).
3   To make nylon and plastic foams for cavity wall insulation.
4   Ammonia solution is used in cleaning – it softens the water and neutralises acidic stains left by perspiration.

Manufacture of ammonia by the Haber process

Because of the world food shortage, vast quantities of ammonia are needed for the manufacture of fertilisers. The gas is also employed in the manufacture of explosives. Thus, it may be used both to preserve life and to destroy it.

In 1918 the German, Fritz Haber, was awarded the Nobel Prize in Chemistry for his earlier development of a process for the manufacture of ammonia from its elements. The process, in which nitrogen and hydrogen are heated under pressure in the presence of an iron catalyst, was improved by another German Nobel Prize winner, Karl Bosch, and was employed to produce the explosives used by their country in the First World War.

The Haber process may be divided into three stages: (1) obtaining the nitrogen and hydrogen, (2) removing the impurities and (3) producing the ammonia.

(1) Nowadays the nitrogen and hydrogen are obtained by heating methane (North Sea Gas) under pressure with steam and a limited air supply in the presence of a nickel catalyst. The reaction also produces carbon monoxide and carbon dioxide.

$$2CH_4(g) + \underbrace{O_2(g) + 4N_2(g)}_{\text{air}} \rightarrow 2CO(g) + 4H_2(g) + 4N_2(g)$$

$$CH_4(g) + H_2O(g) \rightarrow CO(g) + 3H_2(g)$$
$$CH_4(g) + 2H_2O(g) \rightarrow CO_2(g) + 4H_2(g)$$

The reaction with air is exothermic while those with steam are endothermic. By balancing them, the need for external heating is eliminated and the products produced more cheaply.

(2) The mixture of gases plus more steam is passed over an iron oxide catalyst at 773 K (500°C) to oxidise the carbon monoxide to carbon dioxide and produce more hydrogen.

$$H_2O(g) + CO(g) \rightleftharpoons H_2(g) + CO_2(g) \qquad \text{(Bosch process)}$$

The carbon dioxide is dissolved in water under pressure and any residual carbon monoxide is absorbed in ammoniacal copper(I) chloride solution. The gases are then dried.

(3) The nitrogen and hydrogen (1 volume : 3 volumes) are compressed to about 2.5×10^7 Pa (250 atm) and heated to 773 K (500°C) in a heat exchanger. The heat is supplied by the gases leaving the catalyst chamber and this makes the process more economical. A yield of about 10% is

obtained on passing the gases over an iron catalyst and the product is liquefied or dissolved in water. The unchanged gases are recirculated.

$$N_2(g) + 3H_2(g) \rightleftharpoons 2NH_3(g) \qquad \text{(exothermic)}$$

The effect of changes in temperature and pressure on the position of equilibrium is discussed in 14.4.

Ammonia manufacture

Ammonium salts

All ammonium salts are soluble in water and are usually prepared by the titration method (5.4). They decompose on heating and some, such as ammonium chloride, sublime. All of them give off ammonia when warmed with sodium hydroxide solution (Experiment 17.3.h) and this reaction is used to test for their presence.

$$NH_4^+(aq) + OH^-(aq) \rightarrow NH_3(g) + H_2O(l)$$

Ammonium salts find a variety of uses. For example, ammonium sulphate and ammonium nitrate are used as fertilisers, ammonium chloride in dry batteries and ammonium carbonate in smelling salts.

17.4   Nitric Acid

Preparation

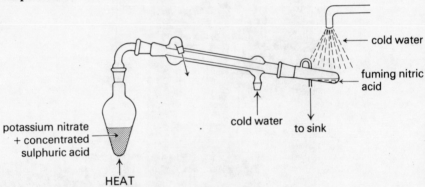

Figure 17.4.1 The laboratory preparation of nitric acid

Nitric acid is usually prepared in the laboratory by heating potassium nitrate with concentrated sulphuric acid, in the apparatus shown in Figure 17.4.1. The use of all-glass apparatus is necessary because cork and rubber are rapidly attacked by nitric acid vapour.

$$KNO_3(s) + H_2SO_4(l) \rightarrow KHSO_4(s) + HNO_3(l)$$

Heating must be as gentle as possible to minimise the thermal decomposition of the product. The acid produced by this method is called *fuming nitric acid* and contains no water; ordinary concentrated nitric acid contains about one-third water by mass.

Physical properties

The *pure* acid is a colourless liquid which boils at 359 K (86°C). Often it is yellow for reasons discussed below.

Chemical properties

THERMAL DECOMPOSITION   When heated, or even on standing, concentrated nitric acid decomposes into water, nitrogen dioxide and oxygen (Experiment 17.4.a). The reaction is catalysed by glass.

$$4HNO_3(aq) \rightarrow 2H_2O(l) + 4NO_2(g) + O_2(g)$$

OXIDISING PROPERTIES Because it so easily gives up oxygen, concentrated nitric acid is a powerful oxidising agent. It converts non-metals such as carbon and sulphur to their oxides and metals to the corresponding nitrates. The acid is usually itself reduced to nitrogen dioxide (Experiment 17.4.b).

e.g. $Cu(s) + 4HNO_3(aq) \rightarrow Cu(NO_3)_2(aq) + 2H_2O(l) + 2NO_2(g)$

$C(s) + 4HNO_3(aq) \rightarrow \quad CO_2(g) \quad + 2H_2O(l) + 4NO_2(g)$

Aluminium and iron are rendered 'passive' (i.e. unreactive) by concentrated nitric acid owing to the formation of a protective layer of oxide.

ACIDIC PROPERTIES Dilute nitric acid shows the usual properties of a strong acid (5.2) *but* it is reduced by metals to one or more of the oxides of nitrogen and does not generally give off hydrogen (Experiment 17.4.c). If *very* dilute nitric acid is used, magnesium *will* liberate hydrogen from it.

Uses of nitric acid

1 To make fertilisers, such as ammonium nitrate.
2 To make explosives, such as nitroglycerine.
3 To make dyestuffs.

Manufacture of nitric acid by the Ostwald process

The basis for this process is the catalytic oxidation of ammonia (page 323). Ammonia is mixed with excess air and the gases are cleaned and heated in a heat exchanger. (The heat is provided by the gases leaving the catalyst chamber.) The mixture is then passed over a hot platinum alloy catalyst, where the ammonia and oxygen combine exothermically to form nitrogen oxide and water. The rate of flow of gases is carefully controlled to keep the temperature of the catalyst at 1173 K (900°C – bright red heat) without external heating.

$$4NH_3(g) + 5O_2(g) \rightarrow 4NO(g) + 6H_2O(g) \qquad \text{(exothermic)}$$

The gases are cooled in the heat exchanger and more air is added so that the nitrogen oxide is oxidised to nitrogen dioxide.

$$2NO(g) + O_2(g) \rightarrow 2NO_2(g)$$

The nitrogen dioxide is mixed with excess air and passed up towers down which hot water (to decompose any nitrous acid) flows. The overall reaction may be represented by the equation

$$4NO_2(g) + O_2(g) + 2H_2O(l) \rightarrow 4HNO_3(aq)$$

Finally, the acid is concentrated by evaporation. The various stages of the process are shown in Figure 17.4.2.

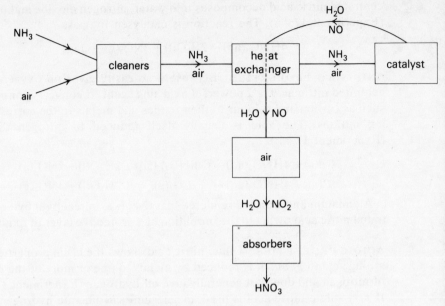

Figure 17.4.2 Flow diagram for the manufacture of nitric acid

Nitrates

PREPARATION All nitrates are soluble in water. Those of potassium, sodium and ammonium are prepared by the titration method and the remainder by the insoluble base method (5.4).

e.g. $\qquad KOH(aq) + HNO_3(aq) \rightarrow KNO_3(aq) + H_2O(l)$

$\qquad PbO(s) + 2HNO_3(aq) \rightarrow Pb(NO_3)_2(aq) + H_2O(l)$

ACTION OF HEAT ON NITRATES The thermal stability of a metal nitrate depends on the position of the metal in the reactivity series. Potassium and sodium nitrates melt and evolve oxygen on heating, the residue being the corresponding *nitrite*.

e.g. $\qquad 2NaNO_3(l) \rightarrow 2NaNO_2(l) + O_2(g)$

All other nitrates give off both oxygen and nitrogen dioxide when they are heated and leave a residue of the metal oxide (Experiment 17.4.d). The lower the metal in the reactivity series, the more easily its nitrate is decomposed.

e.g. $\qquad 2Cu(NO_3)_2(s) \rightarrow 2CuO(s) + 4NO_2(g) + O_2(g)$

Ammonium nitrate is an exception to this rule: on heating it decomposes to produce dinitrogen oxide and steam (17.2). If heating is continued too long the last few drops of molten ammonium nitrate explode.

Tests for nitrates

1 All nitrates form nitric acid when warmed with concentrated sulphuric

acid. If copper powder is also present the nitric acid will react with this to evolve brown fumes of nitrogen dioxide.

2 The brown ring test may be used to detect nitrates in solution (see Experiment 17.4.e).

The nitrogen cycle and the importance of artificial fertilisers

Nitrogen is essential to plants and animals for the production of protein. In nature the element is recirculated and the main ways in which this is done are shown in Figure 17.4.3.

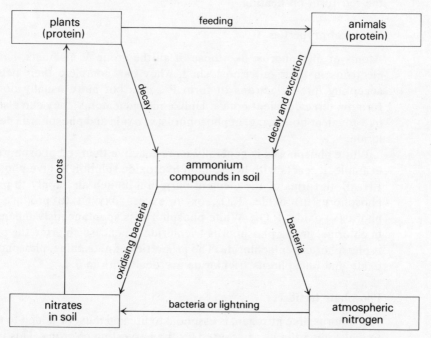

Figure 17.4.3 The nitrogen cycle

The energy in a lightning flash can cause atmospheric nitrogen to combine with oxygen to produce nitrogen oxide. This eventually enters the soil as nitric acid (how?). Bacteria in the roots of leguminous plants, such as peas, beans and lupins, convert atmospheric nitrogen to nitrates as shown at the bottom of the cycle. However, man's removal of plants and animals from the land for use as food, together with modern methods of sewage disposal, have resulted in the top part of the cycle no longer returning sufficient quantities of nitrogen to the soil. Thus it is essential to restore the balance by using artificial fertilisers. Most plants can only take in nitrogen in the form of nitrates but, as ammonium compounds are oxidised to nitrates by bacteria, either nitrates or ammonium compounds can be used as fertilisers. In addition to simply keeping the balance, the use of fertilisers improves the yield of crops on poor land. This is of tremendous importance in today's world, where many thousands of people are short of food.

17.5   Phosphorus

Phosphorus is widely found in nature in the form of phosphates. Two main allotropes of the element are known: *white phosphorus*, which melts at 317 K (44°C), is poisonous and spontaneously inflammable in air; and *red phosphorus*, which sublimes at 690 K (417°C) and is non-poisonous. The differences in properties of the two forms are due to differences in structure. White phosphorus consists of tetrahedral P_4 molecules while red phosphorus is thought to have a fairly weak giant atomic structure in which these tetrahedra are linked together. White phosphorus changes to the red form on heating.

Chemical properties

Atoms of phosphorus, like those of all the group V elements, have five electrons in their outermost shell. They can complete their octets by accepting three electrons to form P^{3-} ions but more usually do so by forming three covalent bonds. Unlike nitrogen atoms, they can also form five covalent bonds, e.g. in phosphorus(v) oxide and phosphorus pentachloride.

White phosphorus is generally more reactive than red phosphorus: for example, it reacts with hot sodium hydroxide solution to give phosphine, PH_3. Both forms of the element burn in a limited air supply to produce phosphorus(III) oxide, P_4O_6, or in excess oxygen to produce phosphorus(v) oxide, P_4O_{10}. White phosphorus is spontaneously inflammable in chlorine, giving phosphorus trichloride; in excess chlorine the product is phosphorus pentachloride. The properties of phosphine, phosphorus(v) oxide and phosphorus trichloride are dealt with in 7.4.

Phosphate fertilisers

Phosphorus, like nitrogen, is essential to life and must be added to the soil to make good the losses caused by the harvesting of crops. This is often done by adding calcium phosphate in the form of bone meal or 'basic slag' from steelworks. The more soluble calcium dihydrogenphosphate is made by heating phosphate rock with fairly concentrated sulphuric acid. The product is known as 'superphosphate' and has the composition $(Ca(H_2PO_4)_2 + CaSO_4)$.

Questions on Chapter 17

1   State what would happen if the following tests were carried out on nitrogen gas.
(a) Litmus solution was added to the gas.
(b) A lighted taper was placed in the gas.
(c) Lime water was added to the gas.
(d) Oxygen was added to the gas and a spark passed through the mixture for some time.

One substance which is reported to release nitrogen compounds into soil is dust from a leather factory. Describe how you would try to find out if a sample of such dust actually did contain compounds of nitrogen.

Give a brief description of the industrial preparation of ammonia (Haber process). (SCE)

2   The following equation represents the thermal decomposition of lead nitrate.

$$2Pb(NO_3)_2(s) \rightarrow 2PbO(s) + 4NO_2(g) + O_2(g)$$

(a) Nitrogen dioxide boils at $22°C$; oxygen boils at $-183°C$. Sketch and label the apparatus you would use to separate the nitrogen dioxide and oxygen obtained in this reaction.

(b) Briefly state how you would obtain a sample of lead from lead(II) oxide.

(c) Calculate the volume of oxygen that would be obtained from the decomposition of 0.1 mole of lead nitrate. (Assume that 1 mole of gas occupies $24 \, dm^3$ at the conditions of the experiment.)

(d) Name the products formed when potassium nitrate is decomposed by heat. Suggest why the nitrates of lead and potassium behave differently on heating.

Give a careful explanation of the following changes.

A little concentrated sulphuric acid was added to some potassium nitrate crystals in a test tube and the mixture warmed. Colourless, oily drops were seen to condense on the wall of the tube. In addition, brown fumes were observed as the temperature was raised. (AEB)

3   (a) Give an outline of the manufacture of nitric acid from ammonia.

(b) If you were provided with concentrated nitric acid, copper and water but *no other* chemical substances, describe briefly reactions by which nitrogen oxide (NO), nitrogen dioxide (NO_2) and oxygen could be produced. (Diagrams and descriptions of apparatus are *not* required.)

(c) State briefly *three* differences in properties between the oxides NO and NO_2.

(C)

4   (a) Give the *names* of two compounds which will react together to give ammonia gas. Write down the equation for the reaction.

(b) (i) By means of a diagram, show how you could make an aqueous solution of ammonia from the gas. (ii) Write down the formulae of the ions produced when an aqueous solution of the gas is made.

(c) Describe and explain what you would see when aqueous ammonia is added to aqueous solutions of (i) iron(III) chloride, (ii) copper(II) sulphate. In (i) give the ionic equation for the reaction. (O & C)

5   (a) Describe in outline the manufacture of ammonia from nitrogen and hydrogen.

(b) Calcium nitrate and sodium nitrate can both be used as nitrogenous fertilisers. Compare their nitrogen contents by calculating for *each* the mass of the anhydrous compound that contains 14 g of nitrogen.

(c) What products are formed when dry ammonia is passed over heated copper(II) oxide? Draw a diagram of the apparatus you would use to carry out this reaction and to obtain samples of *each* of the products. (No further description is required and you are *not* required to show the production of the ammonia.) (C)

6   (a) Draw and label a diagram of the apparatus suitable for the laboratory preparation of a sample of nitric acid.

(b) Give a *brief* account of the chemistry of the manufacture of nitric acid from atmospheric nitrogen.

(c) Describe and explain the *one* reaction in each case which you would select to demonstrate (i) nitric acid reacting simply as an acid, (ii) nitric acid reacting as an oxidising agent.

Explain why you classify the latter reaction as an oxidation and indicate how you would recognise the oxidation product.

(d) Pure nitric acid is colourless. Explain why a fresh sample of the concentrated acid in the laboratory commonly develops a deepening yellow colour with time.

(W)

18 Oxygen and sulphur

Oxygen is the most abundant element on earth, making up almost 50% of the combined mass of the earth's crust, the seas and the atmosphere. Sulphur, on the other hand, makes up only 0.05% of this mass but it is a far more important element than this small percentage suggests.

Experiments

18.1.a   The decomposition of hydrogen peroxide

PROCEDURE   Add half a spatula measure of powdered manganese(IV) oxide to 5 cm³ of 20 volumes hydrogen peroxide solution in a test tube. Fit the tube with a bung and delivery tube and collect the gas given off in the same way as described for hydrogen in Experiment 4.4.a (page 52). Hold a glowing splint in the mouth of a tube of the gas and note the result.

Questions

1   Hydrogen peroxide solution decomposes very slowly by itself at room temperature. Did the manganese(IV) oxide speed up this decomposition? Did the manganese(IV) oxide appear to be used up during the reaction?
2   What happened to the glowing splint? Which gas did this indicate?

18.1.b   The reaction of oxygen with metals ★

The part of this experiment which involves sodium must not be performed by pupils.

PROCEDURE   Collect a boiling tube full of oxygen, either from a cylinder or by means of the apparatus illustrated in Figure 18.1 (page 343), and stopper it firmly. Wrap a 2.5 cm length of magnesium ribbon round the bowl of a combustion spoon so that about 2 cm of it hangs below. Ignite the ribbon and lower it into the oxygen. (**Caution:** do not look directly at the flame; use tinted glass if available.) When the reaction has finished, remove the spoon, quickly add 2–3 cm³ of distilled water and re-stopper the tube. Shake the tube and test the liquid with universal indicator paper. Carry out the same procedure with a tuft of iron wool and then with a piece of sodium. (**Caution:** do not handle the sodium.)

Questions

1   In each case, describe what you saw when the metal was placed in the oxygen.
2   Did these elements react more vigorously or less vigorously with oxygen than they did with air? Can you think of a reason for this?

3   When an element combines with oxygen it forms an *oxide*. What happened to the universal indicator paper in each case and what did this tell you about the nature of the oxides of these metals?

4   Did any of the oxides not dissolve in water?

18.1.c   The reaction of oxygen with non-metals ★

The part of this experiment which involves phosphorus must not be performed by pupils.

PROCEDURE   Ignite a small quantity of powdered sulphur in a combustion spoon and lower it into a tube-full of oxygen. Treat the product with water and universal indicator paper as in the previous experiment. Carry out the same procedure with a small piece of white phosphorus and then with a piece of carbon in place of the sulphur. (**Caution:** do not handle the phosphorus.) The carbon must be glowing red-hot before it is placed in the oxygen.

Questions

1 to 3   As in the previous experiment.

18.1.d   Which metallic oxides are amphoteric?

Some metallic oxides behave as both acids and bases: they are said to be *amphoteric*.

PROCEDURE   (a) Add half a spatula measure of magnesium oxide to 10 cm³ of dilute nitric acid in a boiling tube. Boil the liquid, with shaking, for a short time and note whether the solid dissolves. Repeat the procedure using *freshly prepared* aluminium oxide, zinc oxide, iron(III) oxide, lead(II) oxide and copper(II) oxide in place of the magnesium oxide. (b) Repeat (a) using 10 cm³ portions of sodium hydroxide solution in place of the dilute nitric acid.

Questions

1   Which oxides, if any, acted as (a) acids only, (b) bases only, (c) both acids and bases? How did you reach your conclusions?

2   Is there any connection between the behaviour of the metals and their positions in (a) the reactivity series, (b) the periodic table?

18.2.a   To investigate the nature of some metallic hydroxides

PROCEDURE   Pour magnesium sulphate solution into a test tube to a depth of 1 cm, add a few drops of sodium hydroxide solution, shake the tube and observe the result. Fill the tube with sodium hydroxide solution and tip the contents into another tube to ensure thorough mixing. Repeat the experiment but this time, after adding the first few drops of sodium hydroxide solution, fill the tube with dilute nitric acid. Carry out similar investigations using solutions of aluminium chloride, zinc sulphate, iron(III) chloride, lead(II) nitrate and copper(II) sulphate in place of the magnesium sulphate solution.

Questions

1   Describe what happened on adding the first few drops of sodium hydroxide solution to each of the metallic salt solutions. Name the solid product and give both a chemical equation and an ionic equation for each reaction.

2   What type of reactions are these? Which metallic hydroxides cannot be prepared in this way? Give a reason for your answer.

3   Which metallic hydroxides, if any, were (a) acidic only, (b) basic only, (c) amphoteric? What led you to these conclusions?

18.2.b   The effect of heat on metallic hydroxides

PROCEDURE   (a) One-quarter-fill a *dry* test tube with calcium hydroxide and set up the apparatus shown in Figure 4.2.a (page 49). Heat the tube for about 2 minutes, gently at first and then more strongly, and if any liquid is given off collect it in the cooled tube. Do not allow the end of the delivery tube to dip into the liquid in case sucking back occurs. Identify the liquid by means of a suitable test.

(b) Precipitate some copper(II) hydroxide from copper(II) sulphate solution as in the previous experiment and then heat the contents of the test tube to boiling point.

Questions

1   Name the liquid collected in (a). How did you identify it?

2   What were the solid residues in parts (a) and (b)? (The colour should give you a clue in the case of the copper compound.)

3   Write an equation for each reaction.

4   Which hydroxide decomposed the more easily? Is there a relationship between the thermal stability of metallic hydroxides and the position of the metals in the reactivity series?

18.3.a   Some properties of hydrogen peroxide solution

PROCEDURE   (a) Place 5 cm³ of 20 volumes hydrogen peroxide solution in a boiling tube. Heat the tube gently and collect the gas evolved over water. Test it with a glowing splint.

(b) Place 1 cm depth of potassium iodide solution and an equal volume of dilute sulphuric acid in a test tube. Add hydrogen peroxide solution dropwise, with shaking, until no further change occurs.

(c) Repeat (b) using 1 cm depth of potassium manganate(VII) solution in place of the potassium iodide solution.

Questions

1   What happened to the glowing splint in (a)? Can you explain this result?

2   Write an equation for the reaction that occurred in (a).

3   Suggest one other way of bringing about the decomposition of hydrogen peroxide solution into the same products.

4   What did you see in (b)? What must have happened to the iodide ions? Is this oxidation or reduction?

5   In what way did the colour of the potassium manganate(VII) change? What type of reagent does this indicate?
6   Does hydrogen peroxide behave as (a) an oxidising agent, (b) a reducing agent, (c) both?

18.4.a   The different forms of sulphur★

PROCEDURE   (a) Half-fill a 250 cm³ beaker with tap water and heat it to a temperature of 80°C. Turn off the bunsen and stand a boiling tube containing 20 cm³ of methylbenzene (**caution:** highly inflammable) in the hot water. Add a spatula measure of powdered roll sulphur to the methylbenzene and stir for five minutes. Filter into a second boiling tube and allow the filtrate to cool. Examine the resulting crystals of sulphur with a hand lens.

20 cm length of glass tubing

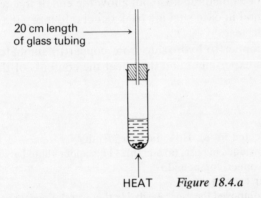

HEAT   *Figure 18.4.a*

(b) Great care is needed with this part of the experiment, because of the risk of fire. Add a spatula measure of powdered roll sulphur to the mixture obtained in (a) and set up the apparatus shown in Figure 18.4.a. Warm the mixture gently over a *small* flame until the liquid boils and then *immediately* turn off the bunsen and wrap the tube in a cloth so that it does not cool down too quickly. Examine the crystals which form and keep them for a few days to see if their appearance changes.

(c) Place 3 cm depth of powdered roll sulphur in a boiling tube. Heat this gently, with shaking, and note the changes in colour and viscosity (thickness) which take place. When the sulphur is almost boiling, pour it into a 250 cm³ beaker half-filled with cold water (if the sulphur ignites, cover the beaker immediately with a damp cloth). Look for a product on the surface of the water and after a few seconds remove the sulphur from the water and test its flexibility. Knead it for a minute or two and observe any changes in its properties. Examine it again after a few days.

Questions

1   Describe the shape of the crystals obtained in (a) and draw a diagram of one of them.
2   Answer question 1 for the crystals obtained in (b). What happened to these crystals after some days?

3 Describe carefully all of the changes which you observed while the sulphur was being heated in (c).
4 What was the appearance of the sulphur after it had been cooled in water? How did the properties of the sample change on kneading and on being left for a few days?
5 Which of the three forms of sulphur was the most stable at room temperature?
6 What is the term used to describe different forms of the same element in the same state?

18.4.b The reaction of sulphur with metals

PROCEDURE Place a spatula measure of a mixture of equal volumes of flowers of sulphur and powdered iron in an ignition tube. Heat the mixture and look for signs of a reaction. Examine the residue when it has cooled down. Repeat the experiment using separate mixtures of flowers of sulphur with copper powder and zinc powder but *in the case of the zinc use only enough mixture to just fill the rounded part of the tube.* Keep the residues for the next experiment.

Questions

1 Describe what you saw in each case.
2 The product formed when an element combines with oxygen is called an oxide. What do you think the products of these reactions are called?
3 Is the violence of the reaction related to the position of the metals in the reactivity series? How would you expect magnesium to react with flowers of sulphur? (Do not try this reaction.)

18.5.a To investigate the effect of dilute acids on metal sulphides

This experiment should be carried out in a well-ventilated laboratory.

PROCEDURE Add about 1 cm^3 of dilute hydrochloric acid to one of the tubes from the previous experiment. Cautiously smell the gas which is evolved (see Experiment 5.3.a) and then hold a piece of filter paper which has been dipped into lead(II) ethanoate solution in the mouth of the tube until it changes colour. As soon as you have completed your tests place the tube in the fume cupboard. Finally, dip the filter paper into a little 20 volumes hydrogen peroxide solution in a test tube. Repeat the whole procedure with the other products of Experiment 18.4.b; in one case use dilute sulphuric acid in place of the hydrochloric acid.

Questions

1 Which gas was produced? Write an equation for each reaction.
2 What did you see when the lead(II) ethanoate reacted with the gas? Which compound of lead was formed?
3 What happened when the filter paper was placed in the hydrogen peroxide solution? Explain the reaction which took place.

18.5.b Some properties of hydrogen sulphide ★

These reactions must be carried out in the fume cupboard.

PROCEDURE   (a) Collect a gas jar full of hydrogen sulphide as described on page 351 and hold a lighted splint near its mouth.

(b) Bubble hydrogen sulphide from a Kipp's apparatus into the following liquids in test tubes. In each case the gas should be passed for about 10 seconds.

(i) $2 cm^3$ of distilled water. Test the solution with universal indicator paper. Keep the solution for a few days and note any changes which occur.

(ii) $1 cm^3$ of potassium manganate(VII) solution acidified with an equal volume of dilute sulphuric acid.

(iii) $1 cm^3$ of potassium dichromate(VI) solution acidified with an equal volume of dilute sulphuric acid.

(iv) $2 cm^3$ of bromine water.

(v) $2 cm^3$ of 20 volumes hydrogen peroxide solution.

(vi) $2 cm^3$ of concentrated nitric acid (**care**).

(vii) $2 cm^3$ of concentrated sulphuric acid (**care**).

(viii) $2 cm^3$ of sulphur dioxide solution.

Questions

1   Describe what you saw in (a). What products do you think were formed?

2   What did the universal indicator paper tell you about the nature of hydrogen sulphide solution? What happened to the solution on standing?

3   Describe what you saw in the reactions in (b). Which substance was formed in every case?

4   In the reactions mentioned in question 3, did hydrogen sulphide act as a reducing agent, an oxidising agent, or neither? Explain your answer.

5   Write equations for the reactions between hydrogen sulphide and (i) bromine water, (ii) hydrogen peroxide solution, (iii) sulphur dioxide solution.

18.5.c To investigate the action of hydrogen sulphide on solutions of metallic salts ★

PROCEDURE   Place $5 cm^3$ of solutions of copper(II) sulphate, zinc sulphate and lead(II) nitrate (**caution:** poisonous) in separate test tubes and bubble hydrogen sulphide through each of them for about 10 seconds in a fume cupboard.

Questions

1   What type of reaction are these?

2   Describe what you saw in each case and name the solid product. Write a chemical equation and an ionic equation for each reaction.

3   Can you think of a use for these reactions in the laboratory?

4   Explain why this method could not be used to prepare sodium sulphide. How *could* this compound be made?

18.6.a **To investigate the reaction of barium chloride with sulphites and sulphates in aqueous solution.**

PROCEDURE   (a) Add a few drops of barium chloride solution to 5 cm^3 portions of solutions of sodium sulphite and sodium sulphate in separate test tubes.

(b) Repeat (a) but this time add 2 cm^3 of dilute hydrochloric acid before adding the barium chloride solution.

Questions

1   What did you see in (a)? What type of reaction took place?
2   Name the solid products and write equations for the reactions in (a).
3   What difference did the addition of the dilute hydrochloric acid make? Which procedure, (a) or (b), must be used to test for a sulphate in aqueous solution?

18.6.b **To prepare sulphur dioxide solution and investigate its properties**

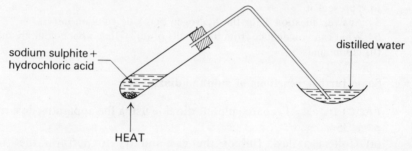

sodium sulphite + hydrochloric acid

distilled water

HEAT

Figure 18.6.b

PROCEDURE   (a) *Warm* a spatula measure of hydrated sodium sulphite with 10 cm^3 of dilute hydrochloric acid in a 150×19 mm test tube fitted with a bung and delivery tube and pass the resultant sulphur dioxide into 50 cm^3 of distilled water in an evaporating basin (see Figure 18.6.b). Hold the basin in your hand so that you can lower it away from the delivery tube at the first sign of sucking back. Test the sulphur dioxide solution with universal indicator paper.

(b) To 5 cm^3 portions of sulphur dioxide solution in separate test tubes, add the following liquids *in the order stated* and note what happens.

(i) 2 cm^3 of barium chloride solution.

(ii) 2 cm^3 of dilute hydrochloric acid followed by a few drops of barium chloride solution. Keep another 5 cm^3 portion of the sulphur dioxide solution for several days and then treat it in a similar manner.

(iii) 2 cm^3 of dilute sulphuric acid and an equal volume of potassium manganate(VII) solution.

(iv) 2 cm^3 of dilute sulphuric acid and an equal volume of potassium dichromate(VI) solution.

(v) 2 cm^3 of bromine water followed by an equal volume of dilute hydrochloric acid and a few drops of barium chloride solution.

(vi) 5 cm^3 of 20 volumes hydrogen peroxide solution followed by 2 cm^3 of dilute hydrochloric acid and a few drops of barium chloride solution.

(c) Heat 5 cm³ of iron(III) chloride solution to boiling in a boiling tube and then bubble sulphur dioxide into the hot liquid for 30 seconds. (Carry out this operation in a fume cupboard.) Note the colour of the resultant liquid and then add 2 cm³ of dilute hydrochloric acid and a few drops of barium chloride solution.

Questions

1   Was the sulphur dioxide solution acidic, alkaline or neutral? Which ions were present in the solution? Write an equation for the reaction between sulphur dioxide and water and name the product formed.
2   What happened to the sulphur dioxide solution on standing for several days? How could you tell?
3   Describe what you saw in parts (iii) and (iv) of (b). What did these results tell you about the sulphur dioxide solution?
4   What happened on adding the bromine water in (v)? What did the barium chloride test tell you? Explain the reaction between the sulphur dioxide solution and the bromine water in terms of oxidation and reduction and write an equation to represent it.
5   Answer question 4 for the hydrogen peroxide solution in (vi).
6   What can you deduce from the results of (c)? (Hint: what colour are most iron (II) compounds?)

18.6.c   Some further reactions of sulphur dioxide ★ ★

PROCEDURE   Prepare sulphur dioxide using the apparatus described on page 354.
(a) Collect a flask full of the gas and try to perform the fountain experiment (Experiment 17.3.b) with it.
(b) Add 5 cm³ of concentrated nitric acid to a gas jar full of sulphur dioxide and replace the cover slip. Shake the jar gently and note the result. Finally, add a little barium chloride solution to the jar.
(c) Place a red or blue flower in a jar of the gas and observe it over a period of about 20 minutes.
(d) Place a jar of dry sulphur dioxide and one of hydrogen sulphide together, remove the cover slips and invert the jars several times to mix the gases. Add a few drops of water and repeat the inversion.
(e) Plunge (i) a burning splint, (ii) a 10 cm length of burning magnesium ribbon, into separate gas jars of sulphur dioxide. (Do not look directly at the flame of the burning magnesium.)

Questions

1   Is sulphur dioxide very soluble in water? How does its solubility compare with that of ammonia and hydrogen chloride?
2   What did you see when the concentrated nitric acid reacted with the sulphur dioxide in (b)? What did the barium chloride test tell you?
3   What happened to the flower in (c)?
4   Did the dry sulphur dioxide and hydrogen sulphide react in (d)? What did you see on adding the water? Explain the reaction in terms of oxidation and reduction and write an equation to represent it.
5   What happened to (i) the burning splint, (ii) the burning magnesium? Explain

the differences observed. (Hint: compare these reactions with similar ones involving carbon dioxide.)

6 Explain the reaction between sulphur dioxide and magnesium in terms of oxidation and reduction and write an equation to represent it.

7 Does sulphur dioxide behave as (a) an oxidising agent, (b) a reducing agent, (c) both?

18.7.a To prepare sulphur trioxide ★ ★

PROCEDURE Set up the apparatus (which must be *absolutely dry*) as shown in Figure 18.7 (page 357), heat the platinised asbestos and sweep the moist air from the apparatus with a stream of dry oxygen. Pass a mixture of sulphur dioxide and oxygen over the heated platinised asbestos and adjust the rate of flow of the gases so that there are two bubbles of sulphur dioxide to every one bubble of oxygen, both rates being slow. When a reasonable amount of product has been obtained in the cooled receiver, turn off the gas supplies and remove the receiver from the apparatus. Note the appearance of the product and then add distilled water to it, a few drops at a time, until it has all dissolved (**caution:** vigorous reaction). Test the resulting solution with (i) universal indicator paper, (ii) dilute hydrochloric acid followed by barium chloride solution.

Questions

1 What was the purpose of the platinum and why was it finely deposited on the asbestos instead of being used in lump form?

2 What did you observe when the sulphur dioxide and oxygen reacted together? Describe the appearance of the product in the cooled receiver.

3 What was formed when the sulphur trioxide reacted with the water? How did you reach this conclusion?

4 Write equations for all the reactions which occurred during the experiment.

18.8.a The reaction of concentrated sulphuric acid with water ★

PROCEDURE (a) Half-fill a 100 cm³ beaker with water and measure its temperature. Add 5 cm³ of concentrated sulphuric acid, stir gently with the thermometer and note any change in temperature.

(b) Half-fill a 100 cm³ beaker with concentrated sulphuric acid and mark the level of the liquid. Leave the beaker for a few days and then inspect the level again.

Questions

1 Was the dissolving of concentrated sulphuric acid in water an exothermic change or an endothermic change?

2 Why would it be dangerous to add water to concentrated sulphuric acid rather than to add the acid to water? (Hint: the boiling points of water and concentrated sulphuric acid are 100 °C and 330 °C respectively.)

3 In what way did the volume of the concentrated sulphuric acid in (b) change over a period of time? Can you explain this? What use is made of this property of concentrated sulphuric acid?

18.8.b The reaction of concentrated sulphuric acid with hydrated copper(II) sulphate and with sugar ★ ★

PROCEDURE   (a) Place a spatula measure of hydrated copper(II) sulphate ($CuSO_4 \cdot 5H_2O$) in a test tube. Add 5 cm³ of concentrated sulphuric acid, warm the tube *gently* and observe the result.
(b) Add 2 cm³ of distilled water to a 2 cm depth of sugar in a 150 cm³ beaker. Stir the mixture well, stand the beaker on an asbestos tile and pour in 10 cm³ of concentrated sulphuric acid (**care**).

Questions

1   How did the appearance of the copper(II) sulphate crystals change in (a)? What do you think was formed?
2   What was the function of the concentrated sulphuric acid in (a)?
3   What appeared to be formed in (b)? Sugar has the formula $C_{12}H_{22}O_{11}$ but contains *no* water. What must the concentrated sulphuric acid have done to the sugar molecules?

18.8.c To investigate the action of concentrated sulphuric acid on metals and non-metals ★ ★

PROCEDURE   (a) Add a few copper turnings to 5 cm³ of concentrated sulphuric acid in a boiling tube and heat gently until a gas is evolved. Smell the gas *cautiously* (see Experiment 5.3.a) and test it with a moist potassium dichromate(VI) paper. *Allow the mixture to cool* and then pour it into a 100 cm³ beaker half-filled with water.
(b) Repeat (a) using, in turn, a spatula measure of powdered sulphur and one of powdered carbon in place of the copper turnings, but do not pour the products into water.

Questions

1   Which gas was evolved in all three reactions? How could you tell?
2   Consider the oxidation number of sulphur in sulphuric acid and in this gas and decide whether the acid was oxidised or reduced during the reactions. Is hot concentrated sulphuric acid an oxidising agent or a reducing agent?
3   Name one other product formed in the reaction between the acid and copper. On what evidence did you base this conclusion?
4   Explain the reaction between concentrated sulphuric acid and (i) copper, (ii) carbon, in terms of oxidation and reduction. Give an equation for each reaction.

18.8.d The acidic properties of dilute sulphuric acid

PROCEDURE   Carry out Experiments 5.2.a, 5.2.b and 5.2.c using dilute sulphuric acid as the only acid.

Questions

1   Is dilute sulphuric acid a strong or a weak acid? Give reasons for your answer.
2   Sum up the acidic properties of dilute sulphuric acid, giving a chemical equation and an ionic equation to illustrate each property. Are there any reactions where this acid does not behave in the expected manner?

18.1 Oxygen and oxides

Oxygen makes up about 21% by volume of the air, 86% by mass of the sea and is widely distributed in rocks, clays and minerals. It was discovered independently by the Swedish chemist Carl Scheele, in 1772, and the Englishman Joseph Priestley, in 1774. Antoine Lavoisier recognised it to be an element and named it oxygen from the Greek *oxys genon*, meaning 'acid former', because he believed that combustion in oxygen always produced acids. (Was he correct?)

Preparation

Many oxygen-rich compounds decompose on heating to yield oxygen as one of the products. Examples of substances which behave in this way include metal nitrates, potassium chlorate(v), potassium manganate(vII), mercury(II) oxide and red lead oxide. The gas is also produced at the anode during the electrolysis of many aqueous solutions (11.2).

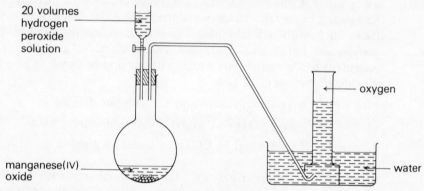

Figure 18.1 The laboratory preparation of oxygen

The most convenient method of preparing oxygen in the laboratory is by adding 20 volumes hydrogen peroxide solution to manganese(IV) oxide in the apparatus shown in Figure 18.1 and collecting the gas over water. The decomposition of the hydrogen peroxide is normally very slow at room temperature but is speeded up on contact with the manganese(IV) oxide. The latter is recovered chemically unchanged at the end of the reaction and therefore acts as a *catalyst* (13.5).

$$2H_2O_2(aq) \rightarrow 2H_2O(l) + O_2(g)$$

Test for oxygen

Oxygen relights a glowing splint.

Physical properties

Oxygen is a colourless, odourless gas, a little more dense than air and slightly soluble in water.

Chemical properties

Oxygen is found in group VI of the periodic table and therefore its atoms have six electrons in their outermost shells. They can complete their octets by gaining two electrons to give the O^{2-} ion, by forming two covalent bonds, or by forming one covalent bond and gaining one electron as in the OH^- ion.

REACTION WITH METALS   Most metals react to give their oxides when heated in oxygen. Thus sodium burns with a yellow flame, magnesium with a brilliant white flame and iron with a shower of bright sparks (Experiment 18.1.b).

$$4Na(s) + O_2(g) \rightarrow 2Na_2O(s) \qquad \text{sodium oxide}$$
$$2Mg(s) + O_2(g) \rightarrow 2MgO(s) \qquad \text{magnesium oxide}$$
$$3Fe(s) + 2O_2(g) \rightarrow Fe_3O_4(s) \qquad \text{'magnetic iron oxide'}$$
$$\text{(iron(II) diiron(III) oxide)}$$

REACTION WITH NON-METALS   Oxygen combines directly with all non-metals except the halogens and the noble gases. For example, sulphur burns in it with a bright blue flame to give mainly sulphur dioxide, phosphorus burns with a brilliant yellow flame to give mainly phosphorus(v) oxide and carbon burns with an orange flame to give carbon dioxide (Experiment 18.1.c).

$$S(s) + O_2(g) \rightarrow SO_2(g) \qquad \text{sulphur dioxide}$$
$$P_4(s) + 5O_2(g) \rightarrow P_4O_{10}(s) \qquad \text{phosphorus(v) oxide}$$
$$C(s) + O_2(g) \rightarrow CO_2(g) \qquad \text{carbon dioxide}$$

REACTION WITH HYDROGEN   Oxygen combines with hydrogen to form water as the only product. This reaction is discussed in 4.4.

$$2H_2(g) + O_2(g) \rightarrow 2H_2O(g)$$

Types of oxide

ACIDIC OXIDES   These react with bases to form salts and with alkalis to form salts and water only. Many of them combine with water to form acids, in which case they are referred to as *acid anhydrides* (e.g. sulphur dioxide is the acid anhydride of sulphurous acid).

$$SO_2(aq) + H_2O(l) \rightleftharpoons H_2SO_3(aq)$$

Acidic oxides are usually the oxides of non-metals, but not all non-metallic oxides are acidic.

BASIC OXIDES   These react with an acid to form a salt and water only. Generally they do not combine with water, but those which do, form alkalis.

e.g. $$Na_2O(s) + H_2O(l) \rightarrow 2NaOH(aq)$$

Basic oxides are the oxides of metals, but not all metallic oxides are basic.

AMPHOTERIC OXIDES Some metals, particularly those in the middle groups of the periodic table, form amphoteric oxides. Such oxides have the properties of both acidic and basic oxides, i.e. they react with both alkalis and acids to form salts and water only. The ones which are most commonly encountered are those of aluminium, zinc and lead(II).

e.g.
$$ZnO(s) + 2HCl(aq) \rightarrow ZnCl_2(aq) + H_2O(l)$$
(base)

$$ZnO(s) + 2NaOH(aq) + H_2O(l) \rightarrow Na_2[Zn(OH)_4](aq)$$
(acid)   sodium zincate

There is a *general* pattern connecting acidity of the oxides with the position of the elements in the periodic table. Elements to the left of the table form basic oxides and those to the right form acidic oxides (see Table 18.1). In between these come elements which form amphoteric oxides.

Form alkalis with water	Insoluble bases	Amphoteric	Acidic, but insoluble in water	Form acids with water
Na_2O, K_2O, CaO	Fe_2O_3, CuO	ZnO, Al_2O_3, PbO	SiO_2	CO_2, NO_2, SO_2, SO_3

Table 18.1 The nature of some commonly encountered oxides

NEUTRAL OXIDES A few non-metals form neutral oxides, i.e. oxides which react with neither acids nor bases. These non-metals are chiefly in groups IV and V. Examples of this type of oxide include carbon monoxide, nitrogen oxide and dinitrogen oxide.

PEROXIDES Peroxides are only formed by the very electropositive metals of groups I and II. They give hydrogen peroxide when treated with dilute acids.

e.g.
$$Na_2O_2(s) + H_2SO_4(aq) \rightarrow Na_2SO_4(aq) + H_2O_2(aq)$$

Uses of oxygen

1 Vast quantities are used in steel-making to remove impurities from cast iron (20.6).
2 In oxyacetylene blowpipes which are used for cutting and welding steel.
3 As an aid to breathing, e.g. in hospitals and in climbing.
4 In rocket fuels.

Manufacture of oxygen

Oxygen is manufactured by the fractional distillation of liquid air. Before liquefaction, water vapour and carbon dioxide are removed from the air

The launch of Explorer 47 from Cape Kennedy on 22 September 1972

because they would solidify later in the process and block the pipes. The remaining air is compressed, cooled and then allowed to expand again. When the air expands, its temperature falls, and it is passed back to cool fresh gases leaving the compressor. It is then itself returned to the compressor and the cycle is repeated, the temperature falling lower each time until eventually liquefaction occurs. Helium and neon remain as gases and may be separated at this stage. The pale blue liquid air is fractionally distilled and gives firstly nitrogen, b.p. 77 K ($-196\,°C$), and then oxygen, b.p. 90 K ($-183\,°C$). If desired the remaining noble gases may be removed from the oxygen by further fractional distillation.

18.2 Metal hydroxides

Preparation

Most metal hydroxides are insoluble in water and may be prepared by double decomposition. Solutions containing the required metal ions and hydroxide ions are mixed together and the resulting precipitate of the insoluble hydroxide filtered off.

e.g. $\qquad FeCl_3(aq) + 3NaOH(aq) \rightarrow Fe(OH)_3(s) + 3NaCl(aq)$

or $\qquad Fe^{3+}(aq) + 3OH^-(aq) \rightarrow Fe(OH)_3(s)$

Soluble hydroxides can be formed by the addition of the metal oxide to water (20.2) or by the reaction of a solution of the metal carbonate with the correct amount of calcium hydroxide.

e.g. $\qquad Na_2CO_3(aq) + Ca(OH)_2(s) \rightarrow CaCO_3(s) + 2NaOH(aq)$

Filtration, followed by evaporation, produces the solid hydroxide.

Properties

All metal hydroxides dissolve in acids to give a salt and water only but some dissolve in alkalis as well. Such hydroxides are amphoteric, the commonest examples being those of aluminium, zinc and lead(II).

e.g. $\qquad Zn(OH)_2(s) + 2HCl(aq) \rightarrow ZnCl_2(aq) + 2H_2O(l)$
      base
$\qquad Zn(OH)_2(s) + 2NaOH(aq) \rightarrow Na_2[Zn(OH)_4](aq)$
      acid

The hydroxides of potassium and sodium are stable to heat but those of metals below sodium in the reactivity series decompose into the corresponding oxide and water on heating. The lower the metal in the series, the more easily its hydroxide is decomposed.

e.g. $\qquad Cu(OH)_2(s) \rightarrow CuO(s) + H_2O(l)$

18.3 Hydrogen peroxide

Hydrogen peroxide is one of the two hydrides of oxygen, the other being water.

Physical properties

Pure hydrogen peroxide is a colourless liquid. It is very unstable and for this reason is generally used in aqueous solution.

Chemical properties

DECOMPOSITION Hydrogen peroxide solution decomposes slowly into water and oxygen at room temperature but the reaction can be speeded up by the action of heat or a catalyst (Experiments 18.1.a and 18.3.a).

$$2H_2O_2(aq) \rightarrow 2H_2O(l) + O_2(g)$$

Commercial solutions of hydrogen peroxide usually contain an inhibitor (negative catalyst) to make them more stable.

ACIDIC PROPERTIES Hydrogen peroxide is a weak acid. Its salts are peroxides.

OXIDISING ACTION When acidified hydrogen peroxide solution is mixed with potassium iodide solution the iodide ions are oxidised to iodine and the hydrogen peroxide is reduced to water (Experiment 18.3.a).

$$H_2O_2(aq) + 2H^+(aq) + 2e^- \longrightarrow 2H_2O(l) \quad \text{(reduction)}$$
$$2I^-(aq) - 2e^- \longrightarrow I_2(s) \quad \text{(oxidation)}$$

Adding $\quad \overline{H_2O_2(aq) + 2H^+(aq) + 2I^-(aq) \rightarrow 2H_2O(l) + I_2(s)}$ (overall reaction)

Other examples of the oxidising action of hydrogen peroxide solution include its reaction with hydrogen sulphide (18.5) and with sulphur dioxide (18.6). It will also oxidise dyes, which makes it a useful bleaching agent. Another process which involves the oxidising properties of hydrogen peroxide solution is the restoration of old paintings. The lead paints have usually darkened over the years because of their conversion to black lead(II) sulphide by the hydrogen sulphide in the atmosphere. Careful treatment with hydrogen peroxide solution oxidises the lead(II) sulphide to white lead(II) sulphate (Experiment 18.5.a).

$$PbS(s) + 4H_2O_2(aq) \rightarrow PbSO_4(s) + 4H_2O(l)$$

REDUCING ACTION When mixed with oxidising agents more powerful than itself, hydrogen peroxide solution can act as a reducing agent. For example, it will decolourise an acidified solution of potassium manganate(VII).

Volume strength

The concentration of hydrogen peroxide solution is usually quoted in terms of *volume strength*.

> The **volume strength** of a solution of hydrogen peroxide is the number of volumes of oxygen, measured at s.t.p., which can be obtained by the decomposition of one volume of the solution.

Thus $1\,dm^3$ of 20 volumes hydrogen peroxide solution will decompose to give $20\,dm^3$ of oxygen at s.t.p.

Uses of hydrogen peroxide

1 As a rocket fuel.
2 To bleach hair and clothing.
3 In toothpaste and antiseptics.

18.4 Sulphur

Sulphur occurs free in nature and has been known since ancient times. It is also found in sulphide ores and in sulphates.

Allotropes of sulphur

Sulphur, like carbon, exhibits allotropy (see 16.1). Octahedral crystals of *rhombic sulphur (α-sulphur)* crystallise from a solution of sulphur in methylbenzene if the temperature is kept below 369 K (96 °C). Above this temperature, needle-shaped crystals of *monoclinic sulphur (β-sulphur)* separate out (Experiment 18.4.a).

Figure 18.4.1 Crystals of (a) rhombic and (b) monoclinic sulphur

If monoclinic sulphur is kept at room temperature for a few days it gradually changes to the rhombic form. Similarly, rhombic sulphur slowly changes to the monoclinic form at temperatures above 96 °C.

$$\text{rhombic sulphur} \underset{\text{below 96 °C}}{\overset{\text{above 96 °C}}{\rightleftharpoons}} \text{monoclinic sulphur}$$

The temperature at which one allotrope of an element changes to another is called the **transition temperature**.

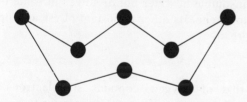

Figure 18.4.2 A sulphur molecule

Both allotropes of sulphur consist of S_8 molecules in which the atoms are arranged in puckered rings (Figure 18.4.2) but these molecules are packed together in different ways in the two types of crystal. The differences in physical properties which arise from this variation in molecular packing are shown in Table 18.4.

Property	Rhombic (α-) sulphur	Monoclinic (β-) sulphur
Temperature at which stable/K (°C)	Up to 369 (96)	369–392 (96–119)
Melting point/K (°C)	386 (113)	392 (119)
Density/g cm^{-3}	2.07	1.96

Table 18.4 Some physical properties of rhombic and monoclinic sulphur

In addition to rhombic and monoclinic sulphur there are two other stable solid forms of the element. These are *flowers of sulphur* which is produced as a fine powder when sulphur vapour is rapidly cooled, and *colloidal sulphur*, which is produced in a number of ways (e.g. when sodium thiosulphate(VI) solution is mixed with a dilute acid in Experiment 13.3.b). Both of these types of sulphur are amorphous.

The action of heat on sulphur

Sulphur melts at 386 K (113 °C) to give a mobile amber liquid in which the S_8 rings are moving with random motion. As the temperature rises the liquid darkens and at about 453 K (180 °C) it becomes very viscous (thick). This is because the increased energy of the collisions between the S_8 rings causes them to break open and join together to form long chains containing up to 10^6 atoms. These chains become entangled with one another, thus preventing the liquid from flowing. If the temperature is increased still further the liquid once more becomes mobile because the vibration of the atoms within the chains causes them to break up into shorter lengths. Slow cooling reverses these changes but rapid cooling gives rise to a black pliable solid called *plastic sulphur*. This contains long chains of atoms which have not had time to reform S_8 rings. On being kneaded, or left for a few days, the chains of atoms break up into S_8 rings and the plastic sulphur gradually hardens into the rhombic form.

Physical properties

Rhombic sulphur, the stable form at room temperature, is a yellow non-metallic solid. It is insoluble in water but dissolves in some organic solvents, such as methylbenzene. Its other important physical properties have been considered already.

Chemical properties

A sulphur atom, like that of oxygen, contains six electrons in its outermost shell and it usually completes its octet by forming two covalent bonds or by gaining two electrons to give the S^{2-} ion. However, unlike oxygen, it can expand its octet and form up to six covalent bonds, as in molecules of sulphur trioxide or sulphuric acid.

Sulphur combines directly when heated with non-metals such as oxygen, chlorine and hydrogen. It also combines with most metals. If the metal is finely powdered and is fairly high in the reactivity series the reaction can be violent (Experiment 18.4.b). In each case the metal sulphide is formed.

e.g. $$Zn(s) + S(s) \rightarrow ZnS(s)$$

Uses of sulphur

1 To make sulphuric acid.
2 To vulcanise (i.e. harden) rubber.
3 To make calcium hydrogensulphite for use in the paper industry.
4 As a fungicide.

5 In the manufacture of matches and gunpowder.
6 To make ointments and drugs.

Extraction of sulphur

Most of the world's supply of sulphur comes from Texas and Louisiana, where it is extracted from underground deposits by the *Frasch process*. Three concentric pipes are sunk down to the deposits as shown in Figure 18.4.3. Superheated water at 443 K (170 °C) is pumped down the outer pipe and hot compressed air down the centre one; a froth of air, water and molten sulphur is forced up the middle pipe and is led off to settling tanks where the sulphur separates out.

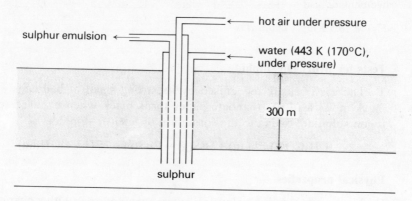

Figure 18.4.3 The Frasch process for the extraction of sulphur

18.5   Hydrogen sulphide

Preparation

Hydrogen sulphide is prepared in the laboratory by the action of a dilute acid on a metal sulphide: generally hydrochloric acid and iron(II) sulphide are used.

$$FeS(s) + 2HCl(aq) \rightarrow FeCl_2(aq) + H_2S(g)$$

The apparatus used in the preparation of carbon dioxide (see Figure 16.4.1) can be employed here also, but it is more usual to generate hydrogen sulphide in a Kipp's apparatus (Figure 18.5). When the tap on this apparatus is opened, the gas escapes and the acid flows down from the reservoir and rises into the middle chamber to react with the iron(II) sulphide. On closing the tap the pressure inside the apparatus builds up and forces the acid out of the middle chamber and back into the reservoir. This stops the reaction until the tap is opened once more. The hydrogen sulphide is bubbled through distilled water to remove hydrogen chloride and then collected by downward delivery. A moist lead(II) ethanoate paper placed at the mouth of each jar turns black when the jar is full.

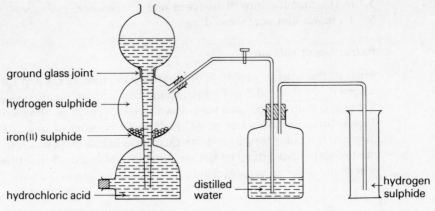

ground glass joint
hydrogen sulphide
iron(II) sulphide
hydrochloric acid
distilled water
hydrogen sulphide

Figure 18.5 Kipp's apparatus

Tests for hydrogen sulphide

1 The gas is easily recognised by its strong smell of bad eggs.
2 A moist lead(II) ethanoate paper turns black when exposed to hydrogen sulphide, owing to the formation of lead(II) sulphide.

$$(CH_3COO)_2Pb(aq) + H_2S(g) \rightarrow PbS(s) + 2CH_3COOH(aq)$$

Physical properties

Hydrogen sulphide is a colourless, very poisonous gas with a characteristic smell (see above). It is a little more dense than air and slightly soluble in water.

Chemical properties

COMBUSTION   Hydrogen sulphide burns with a blue flame. In a plentiful air supply the products of the reaction are sulphur dioxide and steam, but if the air supply is limited sulphur is formed.

$$2H_2S(g) + 3O_2(g) \rightarrow 2SO_2(g) + 2H_2O(g)$$

$$2H_2S(g) + O_2(g) \rightarrow 2S(s) + 2H_2O(g)$$

When a jar of the gas is burned (Experiment 18.5.b) both of these reactions take place.

REDUCING ACTION   Hydrogen sulphide is very readily oxidised to sulphur and therefore acts as a powerful reducing agent. For example, an aqueous solution of the gas is oxidised by dissolved oxygen from the air to give a white precipitate of amorphous sulphur.

$$2H_2S(aq) + O_2(aq) \rightarrow 2H_2O(l) + 2S(s)$$

Acidified solutions of purple potassium manganate(VII) and orange potassium dichromate(VI) turn colourless and green respectively when treated with the gas. These colour changes (which are tests for the presence of reducing agents) are masked to some extent by the precipitate of amorphous sulphur which is formed by the oxidation of the hydrogen sulphide.

Other examples of the reducing action of hydrogen sulphide are given below. Notice that in every case sulphur is precipitated.

1   Reddish-brown bromine water is reduced to a colourless solution of hydrogen bromide; the hydrogen sulphide is oxidised to sulphur.

$$Br_2(aq) + H_2S(g) \rightarrow 2HBr(aq) + S(s)$$

2   Hydrogen peroxide solution is reduced to water; a yellow precipitate of sulphur is seen.

$$H_2O_2(aq) + H_2S(g) \rightarrow 2H_2O(l) + S(s)$$

3   Concentrated nitric acid is reduced to brown fumes of nitrogen dioxide; sulphur is precipitated.

$$2HNO_3(aq) + H_2S(g) \rightarrow 2H_2O(l) + 2NO_2(g) + S(s)$$

4   Concentrated sulphuric acid is reduced to sulphur and the hydrogen sulphide is oxidised to the same product. Thus concentrated sulphuric acid cannot be used to dry the gas.

$$H_2SO_4(l) + 3H_2S(g) \rightarrow 4H_2O(l) + 4S(s)$$

5   Sulphur dioxide solution oxidises hydrogen sulphide to sulphur and is itself reduced to the same product. This is one of the few reactions in which sulphur dioxide acts as an oxidising agent (see page 356).

$$SO_2(aq) + 2H_2S(g) \rightarrow 2H_2O(l) + 3S(s)$$

REACTION WITH WATER   Hydrogen sulphide is slightly soluble in water, producing a weakly acidic solution.

$$H_2S(g) + water \rightleftharpoons H_2S(aq)$$
$$H_2S(aq) \rightleftharpoons H^+(aq) + HS^-(aq) \rightleftharpoons 2H^+(aq) + S^{2-}(aq)$$

As can be seen from the equations, aqueous hydrogen sulphide is a dibasic acid: it gives rise to two series of salts, hydrogensulphides and sulphides.

Metallic sulphides

Sodium, potassium and ammonium sulphides are soluble in water and must be prepared from hydrogen sulphide and the corresponding alkali. However, most sulphides are insoluble in water. They can be prepared by direct combination of the metal and sulphur (Experiment 18.4.b) or by bubbling hydrogen sulphide through a solution of the appropriate metal salt (double decomposition).

e.g. $$CuSO_4(aq) + H_2S(g) \rightarrow CuS(s) + H_2SO_4(aq)$$

Some sulphides are soluble in the acid produced by the reaction and in these cases ammonia solution must be added to neutralise the acid and increase the yield. Zinc sulphide and iron(II) sulphide fall into this category.

Many metallic sulphides have characteristic colours and their production in analysis is useful in identifying the metal. For example, zinc sulphide is white and cadmium sulphide is bright yellow.

Test for sulphides

Metal sulphides react with dilute acids to give hydrogen sulphide.

18.6 Sulphur dioxide

Preparation

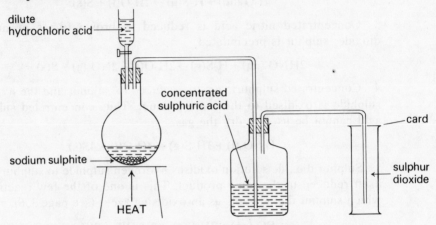

Figure 18.6 The laboratory preparation of sulphur dioxide

Sulphur dioxide is prepared by warming a dilute acid with a sulphite in the apparatus shown in Figure 18.6. Usually dilute hydrochloric acid and sodium sulphite are chosen. The gas is dried by passing it through concentrated sulphuric acid and collected by downward delivery, a moist potassium dichromate(VI) paper placed at the mouth of each jar turning from orange to green when the jar is full.

$$Na_2SO_3(s) + 2HCl(aq) \rightarrow 2NaCl(aq) + H_2O(l) + SO_2(g)$$

Alternatively, concentrated sulphuric acid can be heated with copper turnings to generate sulphur dioxide. However, this procedure is potentially more hazardous than that using a sulphite and dilute acid.

$$Cu(s) + 2H_2SO_4(l) \rightarrow CuSO_4(s) + 2H_2O(g) + SO_2(g)$$

Tests for sulphur dioxide

1 Sulphur dioxide has a characteristic choking smell.
2 A moist potassium dichromate(VI) paper turns green in sulphur dioxide owing to reduction.

Physical properties

Sulphur dioxide is a colourless, poisonous gas with a characteristic smell (see above). It is about twice as dense as air and is very soluble in water.

The gas is readily liquefied under slight pressure (b.p. 263 K, $-10°C$) and small cylinders of the liquid are often used in laboratories.

Chemical properties

REACTION WITH WATER   Sulphur dioxide is very soluble in water, giving a solution of sulphurous acid, H_2SO_3. It is the acid anhydride of sulphurous acid (page 344).

$$SO_2(g) + water \rightleftharpoons SO_2(aq)$$
$$SO_2(aq) + H_2O(l) \rightleftharpoons H^+(aq) + HSO_3^-(aq) \rightleftharpoons 2H^+(aq) + SO_3^{2-}(aq)$$

Sulphurous acid is a weak dibasic acid and gives two types of salts, hydrogensulphites and sulphites. Many of the reactions involving sulphur dioxide will not take place in the absence of water, showing that they are really reactions of sulphurous acid.

REDUCING ACTION   (Experiments 18.6.b and 18.6.c) An aqueous solution of sulphur dioxide is readily oxidised to sulphuric acid and therefore acts as a powerful reducing agent. The change can be brought about by dissolved oxygen from the air.

$$2H_2SO_3(aq) + O_2(aq) \rightarrow 2H_2SO_4(aq)$$

The sulphuric acid may be detected by the addition of dilute hydrochloric acid followed by barium chloride solution. A white precipitate of barium sulphate is obtained by double decomposition (see page 361).

Sulphur dioxide answers the usual tests for reducing agents, turning acidified potassium manganate(VII) solution colourless, and acidified potassium dichromate(VI) solution from orange to green. In contrast to hydrogen sulphide it does not give a precipitate of sulphur in these reactions.

Further examples of the reducing action of sulphur dioxide are described below. In each case sulphuric acid is one of the products.

1   Reddish-brown bromine water is decolourised, owing to the formation of hydrobromic acid.

$$Br_2(aq) + H_2SO_3(aq) + H_2O(l) \rightarrow 2HBr(aq) + H_2SO_4(aq)$$

2   Hydrogen peroxide solution is reduced to water.

$$H_2O_2(aq) + H_2SO_3(aq) \rightarrow H_2SO_4(aq) + H_2O(l)$$

3   Sulphur dioxide reduces iron(III) ions to pale green iron(II) ions in aqueous solution.

$$2Fe^{3+}(aq) + H_2SO_3(aq) + H_2O(l) \rightarrow 2Fe^{2+}(aq) + H_2SO_4(aq) + 2H^+(aq)$$

4   Brown fumes of nitrogen dioxide are given when sulphur dioxide is bubbled into concentrated nitric acid.

$$2HNO_3(aq) + SO_2(g) \rightarrow H_2SO_4(aq) + 2NO_2(g)$$

BLEACHING ACTION   Sulphur dioxide bleaches by reduction. Firstly it combines with any moisture present (the dry gas will not bleach) and then it reduces the dye to form a colourless product (Experiment 18.6.c).

$$\underset{\text{coloured}}{\text{dye}} + H_2SO_3(aq) \rightarrow \underset{\text{colourless}}{(\text{dye} - \text{oxygen})} + H_2SO_4(aq)$$

Oxygen from the air, especially in the presence of sunlight, may replace the oxygen removed during bleaching and thus restore the original colour. This is why old newspapers turn brown.

OXIDISING ACTION Sulphur dioxide can react as an oxidising agent with reducing agents more powerful than itself. For example, magnesium burns in it to form white magnesium oxide and yellow specks of sulphur. (Compare this with the reaction between magnesium and carbon dioxide, described on page 301.)

$$2Mg(s) + SO_2(g) \rightarrow 2MgO(s) + S(s)$$

Sulphur dioxide also oxidises hydrogen sulphide in the presence of moisture. The dry gases do not react.

$$SO_2(aq) + 2H_2S(g) \rightarrow 2H_2O(l) + 3S(s)$$

Uses of sulphur dioxide

1   As an intermediate in the production of sulphuric acid.
2   As a bleaching agent.
3   To preserve food, e.g. jam and fruit juice.

Air pollution

Both hydrogen sulphide and sulphur dioxide pollute the atmosphere, their concentrations being higher in industrial areas than in the country. Because hydrogen sulphide so readily forms metal sulphides, it darkens lead paints and tarnishes clean metal surfaces. Sulphur dioxide is acidic and corrodes metal and stonework: in addition it is particularly harmful to people with respiratory complaints. The air in smokeless zones does not necessarily contain less of these gases than the air in other areas because both are produced by smokeless fuels such as anthracite.

Sulphites

The most commonly encountered sulphite is that of sodium. It is prepared by saturating a solution of sodium hydroxide with sulphur dioxide to produce sodium hydrogensulphite solution and then adding to this an equal volume of sodium hydroxide solution.

$$NaOH(aq) + SO_2(aq) \rightarrow NaHSO_3(aq)$$
$$NaOH(aq) + NaHSO_3(aq) \rightarrow Na_2SO_3(aq) + H_2O(l)$$

The solution is evaporated and crystallised in the usual way. The carbonates and sulphides of potassium and sodium can be made in a similar manner from carbon dioxide or hydrogen sulphide and the corresponding alkali.

Air pollution damage to St Paul's Cathedral

Test for sulphites

All sulphites yield sulphur dioxide when warmed with dilute hydrochloric acid.

18.7 Sulphur trioxide

Preparation

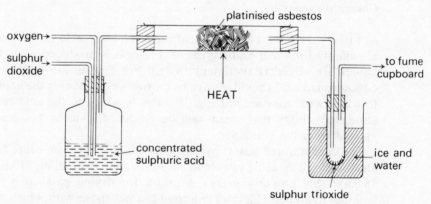

Figure 18.7 The laboratory preparation of sulphur trioxide

Sulphur trioxide is prepared in the laboratory by the catalytic oxidation of sulphur dioxide using the apparatus shown in Figure 18.7. The platinum catalyst is spread out on asbestos to give a large surface area on which gases can combine. White needle-shaped crystals of the product form in the cooled receiver, provided that no moisture is present in the apparatus.

$$2SO_2(g) + O_2(g) \rightleftharpoons 2SO_3(g)$$

Properties

The solid fumes strongly in moist air and reacts violently with water to form sulphuric acid. Thus sulphur trioxide is the acid anhydride of sulphuric acid.

$$SO_3(s) + H_2O(l) \rightarrow H_2SO_4(aq)$$

Uses of sulphur trioxide

Almost all of the sulphur trioxide produced industrially is converted to sulphuric acid.

18.8 Sulphuric acid

Preparation

Sulphuric acid can be prepared in the laboratory by *cautiously* adding water to sulphur trioxide but it is better to dissolve the sulphur trioxide in concentrated sulphuric acid and then pour the liquid into water (see *Manufacture of sulphuric acid*).

Physical properties

The concentrated sulphuric acid which is used in the laboratory contains about 2% water. It is a colourless oily liquid which has a density of $1.84\,g\,cm^{-3}$ and boils at 611 K (338°C).

Chemical properties

AFFINITY FOR WATER *Concentrated* sulphuric acid has a considerable affinity for water and a great deal of heat is produced when the two liquids are mixed (Experiment 18.8.a). For this reason dilution of the concentrated acid should always be carried out by adding the acid slowly to water with constant stirring. If water is added to the acid there is a strong possibility that steam will be produced, causing hot acid to be ejected from the container.

The concentrated acid is hygroscopic, i.e. it absorbs moisture from the air, thereby increasing in volume and becoming more dilute (Experiment 18.8.a). This property makes it useful for drying gases (e.g. sulphur dioxide, hydrogen chloride, chlorine) but not those with which it reacts (e.g. ammonia, hydrogen sulphide).

Concentrated sulphuric acid is capable of removing chemically combined water from hydrated salts. Thus it converts blue crystals of copper(II) sulphate-5-water to the white anhydrous form of the salt. It also removes the *elements of water* (i.e. hydrogen and oxygen atoms in the ratio of $2:1$) from compounds which do not themselves contain water molecules. For example, sugar, $C_{12}H_{22}O_{11}$, is changed into a black mass of carbon when mixed with the concentrated acid (Experiment 18.8.b).

$$CuSO_4 \cdot 5H_2O(s) \rightarrow CuSO_4(s) + 5H_2O(l)$$
$$C_{12}H_{22}O_{11}(s) \rightarrow 12C(s) + 11H_2O(l)$$

In these reactions concentrated sulphuric acid is acting as a *dehydrating agent*.

A **dehydrating agent** is a substance which is capable of removing chemically combined water or the elements of water from other compounds.

Do not confuse *dehydration* with *drying*. A drying agent only removes water that is *not* chemically combined.
N.B. Concentrated sulphuric acid will dehydrate flesh. If any is spilt on the skin it must be washed off immediately with *plenty* of water.

OXIDISING ACTION *Concentrated* sulphuric acid is a powerful oxidising agent, particularly when hot (Experiment 18.8.c).

e.g.
$$Cu(s) + 2H_2SO_4(l) \rightarrow CuSO_4(s) + 2H_2O(l) + SO_2(g)$$
$$S(s) + 2H_2SO_4(l) \rightarrow 2H_2O(l) + 3SO_2(g)$$
$$C(s) + 2H_2SO_4(l) \rightarrow CO_2(g) + 2H_2O(l) + 2SO_2(g)$$

Notice that in each case the concentrated sulphuric acid is itself reduced to sulphur dioxide and water.

DISPLACEMENT OF ACIDS *Concentrated* sulphuric acid is involatile (i.e. has a high boiling point) and can displace more volatile acids when heated with their salts.

e.g.
$$KNO_3(s) + H_2SO_4(l) \rightarrow KHSO_4(s) + HNO_3(g)$$

The nitric acid is distilled off, any excess sulphuric acid remaining behind in the flask (see page 326).

ACIDIC PROPERTIES *Dilute* sulphuric acid behaves as a typical strong acid (see Chapter 5). It is a dibasic acid and gives two types of salts, sulphates and hydrogensulphates.

Uses of sulphuric acid

1 To manufacture fertilisers, such as ammonium sulphate and 'superphosphate'.
2 In the production of synthetic fibres and plastics.
3 To 'pickle' steel (i.e. clean the surface prior to plating with another metal).
4 In petroleum refining.
5 To make soapless detergents.
6 In car batteries.

7   To make dyes, drugs, explosives and many other compounds.

Manufacture of sulphuric acid by the contact process

This process may conveniently be divided into three stages:
1   the preparation and purification of sulphur dioxide,
2   the catalytic oxidation of sulphur dioxide to sulphur trioxide, and
3   the conversion of the sulphur trioxide into sulphuric acid.

(1) The simplest way of producing sulphur dioxide is to burn sulphur in excess air.

$$S(s) + O_2(g) \rightarrow SO_2(g)$$

The gases are cleaned (e.g. by electrostatic precipitation) to remove catalyst poisons and dried by passing them through concentrated sulphuric acid. They are then heated in a heat exchanger by hot gases leaving the catalyst chamber.

(2) The sulphur dioxide combines with oxygen from the air in the presence of a complex vanadium catalyst made from vanadium(v) oxide and alkali metal sulphates. The catalyst is spread out on silica gel to give a greater surface area. Although platinum is used as the catalyst in the laboratory preparation of sulphur trioxide (Experiment 18.7.a) it is too expensive and too easily poisoned to be used on an industrial scale.

$$2SO_2(g) + O_2(g) \rightleftharpoons 2SO_3(g)$$

The forward reaction is exothermic and the rate of flow of gases is adjusted to keep the temperature at 723 K (450°C) without external heating. The various factors influencing the position of equilibrium are discussed in 14.4.

(3) The hot sulphur trioxide is passed back to the heat exchanger to heat the incoming gases and then dissolved in concentrated sulphuric acid. It is not dissolved in water because the reaction is too violent and inefficient. The resulting product, fuming sulphuric acid (oleum), is then diluted with water to give the ordinary concentrated acid.

$$H_2SO_4(l) + SO_3(g) \rightarrow H_2S_2O_7(l) \qquad \text{(oleum)}$$
$$H_2S_2O_7(l) + H_2O(l) \rightarrow 2H_2SO_4(l)$$

The whole process is summed up in Figure 18.8. Notice that the initial letters of the first five boxes are in alphabetical order.

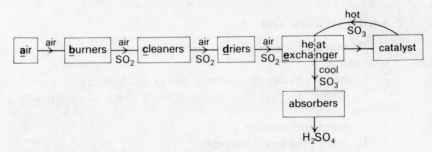

Figure 18.8 Flow diagram for the manufacture of sulphuric acid

Sulphates and hydrogensulphates

These may be prepared by the standard methods mentioned in 5.4. All of the common sulphates are soluble in water except for those of calcium, barium and lead.

The sulphates of metals which are fairly low in the reactivity series decompose on strong heating. Generally the metal oxide and sulphur trioxide are obtained but sometimes the reaction is more complicated. For example, pale green crystals of iron(II) sulphate-7-water first lose their water of crystallisation to form the white anhydrous salt and then decompose to give sulphur dioxide, sulphur trioxide and reddish-brown iron(III) oxide.

$$FeSO_4 \cdot 7H_2O(s) \rightarrow FeSO_4(s) + 7H_2O(g)$$

$$2FeSO_4(s) \rightarrow Fe_2O_3(s) + SO_2(g) + SO_3(g)$$

N.B. Sulphates are *normal salts* and hydrogensulphates are *acid salts* (see page 79).

Test for sulphates

If dilute hydrochloric acid followed by a few drops of barium chloride solution is added to a solution of a sulphate, a white precipitate of barium sulphate is obtained by double decomposition.

e.g. $\qquad Na_2SO_4(aq) + BaCl_2(aq) \rightarrow BaSO_4(s) + 2NaCl(aq)$

or $\qquad SO_4^{2-}(aq) + Ba^{2+}(aq) \rightarrow BaSO_4(s)$

There are a number of other substances, such as sulphites and carbonates, which give white precipitates of the corresponding barium salts by double decomposition with barium chloride solution. However, in the presence of dilute hydrochloric acid these precipitates do not form.

Questions on Chapter 18

1 (a) Describe in outline the commercial preparation of oxygen from the air. Mention *three* large scale uses of oxygen.
(b) How do the processes of respiration and the burning of fuels (i) resemble one another, (ii) differ from one another?
(c) 'When we burn coal, we are making use of energy that came originally from the sun.' Explain this statement. (C)
2 Explain the terms (i) basic hydroxide, (ii) amphoteric hydroxide, (iii) alkaline hydroxide. Give the name and formula of *one* hydroxide of *each* type, the three examples being different.

For *each* of the hydroxides you have chosen (a) give a brief account of the reactions by which it can be prepared from its metal, and (b) describe its behaviour when heated. (C)
3 Hydrogen peroxide is a weak dibasic acid, miscible with water.
(a) Explain the terms 'weak' and 'dibasic'.
(b) Write down the empirical formula of this compound.
(c) Suggest the formula of calcium peroxide, and give the equation for its reaction with hydrochloric acid.

(d) Explain why hydrogen peroxide bleaches hair.

(e) Describe an experiment for obtaining, and collecting, a sample of oxygen from hydrogen peroxide. (O & C)

4 (a) (i) You are provided with a finely powdered mixture of potassium nitrate and sulphur. Describe how you would obtain pure crystals of potassium nitrate from this mixture.

(ii) Starting from powdered sulphur, describe how you would prepare samples of *two* crystalline allotropes of this element. In each case give a sketch showing the crystalline form.

(b) Describe the changes which occur when powdered sulphur is gently heated in a test tube up to its boiling point. (O & C)

5 Give a reaction by which hydrogen sulphide may conveniently be prepared in the laboratory. (*Neither* diagrams *nor* method of collection are required.)

How could you show that both sulphur and sulphur dioxide may be formed when hydrogen sulphide burns in air?

Describe *two* reactions other than burning in air in which hydrogen sulphide behaves as a reducing agent.

Explain how you could use hydrogen sulphide to distinguish between solutions of magnesium nitrate, zinc nitrate and lead(II) nitrate. (C)

6 A pupil added concentrated sulphuric acid to a substance and noted that a colourless gas was given off which decolourised bromine water. He concluded that (i) the gas was sulphur dioxide, and (ii) that the sulphur dioxide had been produced by the reduction of the sulphuric acid.

(a) He may have been wrong in his first conclusion. Why?

(b) Name *one* substance which would produce sulphur dioxide when acted upon by concentrated sulphuric acid *without involving the reduction of the sulphuric acid.*

Write a balanced equation for the reaction, and use it to show that reduction of the acid has not occurred.

(c) (i) A solution of barium chloride was added to a solution of sodium sulphite and a white precipitate was obtained. What is this precipitate likely to be?

(ii) When dilute hydrochloric acid was added to the precipitate it reacted and a pungent smelling gas was given off. What is this gas?

(iii) A solution of sulphur dioxide in water was warmed with a little nitric acid. Barium chloride solution was added and a white precipitate was formed which did *not* dissolve when dilute hydrochloric acid was added. What is the precipitate?

What happened to the solution of sulphur dioxide when it was acted upon by the nitric acid?

(d) Name *two* substances which are used for the manufacture of sulphur dioxide in industry. (SCE)

7 Describe how you would prepare sulphur dioxide from sodium sulphite. No diagram is required, but you should say how you would collect a sample of the gas.

Sulphur dioxide is said to be an *acid anhydride*. Explain this statement.

Describe the chemical reaction which takes place when the gas is passed into (a) chlorine water, (b) potassium dichromate(VI) solution.

Explain why these reactions are classified as *redox reactions*. (O & C)

8 In the 'Contact process', pure sulphur dioxide is converted to gaseous sulphur trioxide which, on further treatment, gives concentrated sulphuric acid.

(a) State the conditions used in the conversion to sulphur trioxide, and the equation for the reaction that takes place.

(b) Why must the sulphur dioxide be purified before use?

(c) The conversion is exothermic. What would be the effect of increasing the temperature at which the reaction is carried out?

(d) How is concentrated sulphuric acid obtained from sulphur trioxide?

Explain why gaseous sulphur trioxide is not dissolved in water directly. Give reactions (*one* in each case) in which sulphuric acid reacts as (i) an acid, (ii) an oxidising agent, (iii) a dehydrating agent. (AEB)

19 The halogens

Fluorine, chlorine, bromine and iodine are found in group VII of the periodic table and are collectively known as the halogens ('salt formers'). The following experiments are mainly concerned with chlorine and its compounds but you will find a comparison of chlorine with fluorine, bromine and iodine later in the chapter.

Experiments

19.1.a   To investigate the action of concentrated sulphuric acid on rock salt

PROCEDURE   Fill the rounded part of a test tube with rock salt and add a few drops of concentrated sulphuric acid (**care**). Cautiously smell the product and then blow gently across the mouth of the tube. Dip a glass rod into ammonia solution and place it near the mouth of the tube. Repeat this using silver nitrate solution instead of ammonia solution. Finally, test the gas with moist universal indicator paper.

Questions

1   What is rock salt?
2   What did you see when the acid was added to the rock salt? Name the gaseous product.
3   What happened when ammonia was mixed with the gas? Name the new substance obtained and write an equation for its formation.
4   What did you see when the silver nitrate solution reacted with the original gas? Name the products obtained and write an equation for the reaction.
5   Was the gas acidic, alkaline or neutral? How could you tell?

19.1.b   To investigate the solubility of hydrogen chloride in water★★

PROCEDURE   Replace the gas jar in Figure 19.1 by a 500 cm³ round-bottomed flask and collect the gas until the flask is *completely filled*. Carry out the same procedure as described for Experiment 17.3.b.

Questions

1   What did you observe? How soluble is hydrogen chloride in water?
2   Why might it be dangerous to prepare hydrogen chloride solution by dipping a delivery tube from a hydrogen chloride generator into a beaker of water?
3   What did the colour of the universal indicator tell you about the nature of aqueous hydrogen chloride solution?
4   What is the common name for hydrogen chloride solution?

5 Place a drop of propanone on your hand and note what happens. Now explain the purpose of the propanone in the above experiment.

19.1.c Some reactions of dilute hydrochloric acid

PROCEDURE Carry out Experiments 5.2.a, 5.2.b and 5.2.c, using dilute hydrochloric acid as the only acid.

Questions

1 Is dilute hydrochloric acid a strong or a weak acid? Give reasons for your answer.
2 Sum up the acidic properties of dilute hydrochloric acid giving a chemical equation and an ionic equation to illustrate each property. Are there any reactions where this acid does not behave in the expected manner?

19.1.d The reaction between iron(III) chloride and water

PROCEDURE Fill the rounded part of a test tube with anhydrous iron(III) chloride and add 1 cm³ of distilled water. Boil the solution, test the vapour with universal indicator paper and then with a drop of ammonia solution on a glass rod.

Questions

1 What did you observe in the solution as boiling proceeded?
2 Which iron compound do you think has been formed?
3 Which gas was evolved?

19.1.e Tests for chlorides

PROCEDURE (a) Add a spatula measure of hydrated magnesium chloride to 1 cm³ of concentrated sulphuric acid in a test tube. Warm *gently* (**care**). Blow across the mouth of the tube and test for the presence of hydrogen chloride as in Experiment 19.1.a. Repeat the experiment with other metallic chlorides.
N.B. To dispose of each mixture, first allow it to cool and then pour it into plenty of water in the sink. *Never* pour the *hot* mixture into water.
(b) Dissolve a spatula measure of hydrated magnesium chloride in about 5 cm³ of distilled water in a test tube. Add 1 cm³ of dilute nitric acid followed by a few drops of silver nitrate solution and shake the tube. Finally, fill the tube with ammonia solution and tip the liquid into another tube to mix it thoroughly. Repeat the whole procedure with other chlorides.

Questions

1 Was hydrogen chloride evolved for each chloride used in (a)? How did you detect it?
2 Describe and explain what you saw on adding the silver nitrate solution in (b). What happened when the ammonia solution was added?
3 What was the purpose of the dilute nitric acid?

19.2.a   The reaction of chlorine with metals ★ ★

PROCEDURE   Prepare four gas jars of chlorine as described on page 374.

(a) Heat some iron wool to redness in a combustion spoon and then lower it into a jar of chlorine.

(b) Take a small piece of sodium (a cube of side 3 mm), dry it between filter papers and place it in a combustion spoon. Heat the sodium and immediately it starts to burn, plunge the spoon into a jar of chlorine.

(c) Using tongs, place a piece of Dutch metal in a jar of chlorine.

(d) Lower a strip of burning magnesium ribbon into a jar of chlorine (do not look directly at the flame).

Questions

1   Look at some iron(II) and iron(III) salts. Which iron compound do you think has been formed in (a)? Write an equation for the reaction.

2   What would you expect to be formed in the reaction between sodium and chlorine? Write an equation for the reaction.

3   Can you explain the colour of the product in (b)?

4   What did you observe when Dutch metal reacted with chlorine?

5   Dutch metal is made from copper and zinc. Would you expect these metals to react as violently as they did? Why do you think Dutch metal reacts so readily?

6   Name the products formed in (c).

7   Describe what you saw in (d). Name the product formed. Write an equation for the reaction.

8   In each of these experiments, atoms of one element lose electrons while those of the other gain them. Explain the reactions in terms of oxidation and reduction.

19.2.b   To prepare an anhydrous metal chloride ★ ★

Experiment 19.1.d shows that iron(III) chloride reacts with water. For this reason, special conditions are required to prepare the anhydrous salt; it cannot be made simply by heating the hydrated salt to drive off the water of crystallisation.

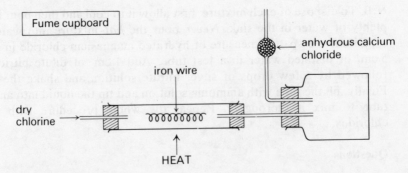

Figure 19.2.b

PROCEDURE   Pass a fast stream of chlorine through the apparatus shown in Figure 19.2.b in a fume cupboard until all the air has been displaced and then heat the iron until the reaction begins.

Questions

1 What did you observe? Name the product formed.
2 By what process did the product get from the combustion tube into the flask?
3 Why was anhydrous calcium chloride used in the guard tube?
4 How could you convert anhydrous iron(II) chloride to anhydrous iron(III) chloride?

19.2.c The reaction of chlorine with a non-metal ★ ★

This experiment must not be performed by pupils.

PROCEDURE Take a small piece of white phosphorus (a cube of side 3 mm) and dry it between filter papers. Using tongs, place the phosphorus in a combustion spoon and lower it into a gas jar of chlorine. If no reaction is apparent, heat the phosphorus *gently* over a bunsen flame and then lower it into the jar once more.

Questions

1 What did you observe?
2 There are two products, phosphorus trichloride and phosphorus pentachloride. Which would you expect to predominate if an excess of chlorine were used?

19.2.d The reaction of chlorine with hydrogen ★ ★

These experiments must not be performed by pupils.

These experiments show how the violence of a reaction depends upon the conditions under which it is carried out.

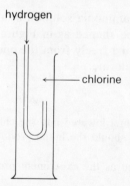

Figure 19.2.d

PROCEDURE The chlorine for the next three experiments should be collected over brine.
(a) Ignite a jet of hydrogen at the end of a delivery tube and lower it into the gas jar of chlorine as shown in Figure 19.2.d. (**Caution:** first check that no air is present in the hydrogen as described in Experiment 4.3.b.) Test the product by blowing across the mouth of the jar.
(b) Fill a test tube with equal volumes of chlorine and hydrogen over brine. Apply a flame to the mouth of the tube.
(c) Fill a second tube in the same way, loosely stopper the tube and hold a

strip of burning magnesium ribbon near it. (The tube should be behind a safety screen and it is advisable to wear a glove. Take the usual precautions with respect to looking at the magnesium flame.)

Questions

1 What was the product in each case? How could you tell?
2 Which reaction was the quietest and which the most violent?
3 What provided the energy to start the reaction in (c)?

19.2.e The reaction of chlorine with hydrocarbons ★ ★

The previous experiment shows that chlorine has a great affinity for hydrogen. This experiment is designed to see if chlorine can remove hydrogen from compounds containing it. Wax and turpentine are hydrocarbons, i.e. they contain hydrogen and carbon only.

PROCEDURE   Observe what happens when a burning taper is plunged into a gas jar of chlorine. Soak some glass wool in warm turpentine (**caution:** inflammable) and, using tongs, drop this into a second jar of chlorine. Blow gently across the mouth of each jar.

Questions

1 What do you think has been formed in each experiment? What led you to this conclusion?
2 Explain the reaction in terms of oxidation and reduction.

19.2.f The reaction of chlorine with ammonia ★ ★

PROCEDURE   Place 1 cm depth of 0.880 ammonia solution in a boiling tube fitted with a bung and a delivery tube shaped as in Figure 19.2.d. Heat gently and when the ammonia is issuing freely from the end of the delivery tube, lower it into a gas jar of chlorine.

Questions

1 What did you observe when the tube was first lowered into the chlorine?
2 Since chlorine is an oxidising agent, what should the initial products have been? Write an equation for this reaction.
3 What other product appeared to be formed as the experiment proceeded? Write an equation for the second reaction.
4 Add the two equations together to obtain the overall equation for the reaction.
5 How would you expect chlorine to react with hydrogen sulphide? Check your answer by experiment.

19.2.g The effect of chlorine water on coloured substances★

When chlorine is passed into water, about two-thirds of it simply dissolves: the rest reacts with the water to form hydrochloric acid and chloric(I) acid.

$$H_2O(l) + Cl_2(aq) \rightleftharpoons HCl(aq) + HOCl(aq)$$

PROCEDURE   Prepare a solution of chlorine in water by bubbling the gas into $100 \, cm^3$ of water contained in a beaker. Immerse various coloured substances (e.g. litmus paper, flowers, pieces of dyed cloth) in the chlorine water and leave the mixture for a few days.

Questions

1   What happened to the coloured substances?
2   Which chemical in the chlorine water do you think is responsible for the above change?
3   From your knowledge of the chemistry of chlorine, what type of reaction do you think this is?
4   How does this reaction differ from the action of sulphur dioxide solution on coloured substances?
5   Can you think of any disadvantage of using chlorine water to bring about the above change?

19.3.a   To investigate the reactions of the halogens with water ★

The purpose of this and the next two experiments is to find out if there is a connection between the behaviour of the three halogens, chlorine, bromine and iodine.

PROCEDURE   (a) Test a little chlorine water and bromine water with universal indicator paper. Add a few crystals of iodine (caution: corrosive, avoid skin contact) to $5 \, cm^3$ of distilled water in a test tube, test the liquid with universal indicator paper, heat to boiling and test again.
(b) Add $1 \, cm^3$ of sodium hydroxide solution to $5 \, cm^3$ of bromine water in a test tube and shake the tube. Place a few iodine crystals in a second test tube, add $5 \, cm^3$ of sodium hydroxide solution and heat the mixture carefully.

Questions

1   Describe what you saw in each case in (a). What conclusions can you draw about the reaction of the halogens with water?
2   Did the bromine and iodine react with the sodium hydroxide solution in (b)? Give a reason for your answer.
3   Do the halogens seem to undergo similar reactions with water, or are the reactions unconnected?
4   Which of the halogens seems the least reactive towards water?

19.3.b   The reaction of bromine and iodine with iron ★ ★

This experiment should not be performed by pupils.

PROCEDURE   (a) Set up the apparatus shown in Figure 19.3.b in a fume cupboard and heat the iron wool strongly. If the bromine does not vaporise, flick the bunsen flame on to the bottom of the tube a few times until it does.
(b) Repeat the experiment using iodine crystals in place of the rocksil and bromine.

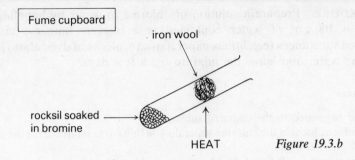

Fume cupboard

iron wool

rocksil soaked
in bromine

HEAT

Figure 19.3.b

Questions

1 What did you observe (a) with bromine and (b) with iodine?

2 Do these reactions show any similarity to that between iron and chlorine? Suggest what is formed in each case.

3 Which of the halogens reacted the most vigorously and which the least vigorously? Do these results agree with those of the previous experiment?

19.3.c Displacement reactions

PROCEDURE Proceed as in Experiment 11.1.e.

Questions

1 As in Experiment 11.1.e.

2 As in Experiment 11.1.e.

3 Which halogen is the most powerful oxidising agent and which the weakest?

19.4.a The action of concentrated sulphuric acid on bromides and iodides

This experiment must be carried out in a well-ventilated laboratory.

PROCEDURE Proceed as in Experiment 19.1.a but use sodium bromide and then sodium iodide in place of the rock salt. Omit the test with the silver nitrate solution and wash the mixtures down the sink as quickly as possible.

Questions

1 What products would you expect from the reactions between concentrated sulphuric acid and (a) sodium bromide, (b) sodium iodide? (Hint: what happened with the sodium chloride in Experiment 19.1.a?)

2 In each case describe what you saw and name the coloured products. By what process are these coloured substances obtained from the products you named in question 1?

3 From the results of this experiment and of Experiment 19.1.a, which of the hydrogen halides do you think is (a) the most stable, (b) the least stable and (c) the best reducing agent?

4 What did you observe when the ammonia solution was brought near the mouth of each tube? Name the products formed.

19.4.b Testing halides with silver nitrate solution

PROCEDURE Proceed as in part (b) of Experiment 19.1.e using in turn sodium chloride, sodium bromide and sodium iodide in place of the magnesium chloride.

Questions

1 Give the colour and name of each of the solid products obtained when silver nitrate solution was added to the halide solutions.
2 What differences were observed when the ammonia solution was added to each mixture?
3 Would you expect to obtain a precipitate if silver nitrate solution were added to tetrachloromethane? Explain your answer.

19.1 Hydrogen chloride

Preparation

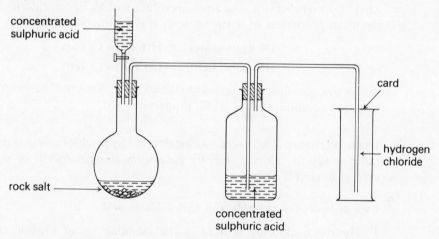

concentrated
sulphuric acid

card

hydrogen
chloride

rock salt

concentrated
sulphuric acid

Figure 19.1 The laboratory preparation of hydrogen chloride

Hydrogen chloride is usually made in the laboratory by the action of concentrated sulphuric acid on rock salt (impure sodium chloride) in the apparatus shown in Figure 19.1. Pure sodium chloride may be used but the small crystals tend to cause excessive frothing of the mixture.

$$H_2SO_4(l) + NaCl(s) \rightarrow NaHSO_4(s) + HCl(g)$$

The gas is dried by passing it through concentrated sulphuric acid and collected by downward delivery, a moist universal indicator paper placed at the mouth of each jar turning red when the jar is full.

Tests for hydrogen chloride

Hydrogen chloride has a pungent choking smell and forms steamy fumes in moist air. It will turn universal indicator paper red and gives a thick white precipitate of silver chloride with silver nitrate solution (Experiment 19.1.a).

$$HCl(aq) + AgNO_3(aq) \rightarrow AgCl(s) + HNO_3(aq)$$

Physical properties

It is a colourless gas which, as mentioned (p. 371), has a pungent choking smell and gives steamy fumes in moist air. It is slightly more dense than air and is very soluble in water (Experiment 19.1.b). Thus, if a solution of the gas is required the funnel arrangement must be used (Figure 17.3.2).

Chemical properties

REACTION WITH AMMONIA Thick white fumes of ammonium chloride are formed when ammonia and hydrogen chloride mix (Experiment 19.1.a).

$$NH_3(g) + HCl(g) \rightleftharpoons NH_4Cl(s)$$

ACIDIC PROPERTIES An aqueous solution of hydrogen chloride shows the usual properties of a strong acid: it is hydrochloric acid.

$$HCl(g) + water \rightleftharpoons H^+(aq) + Cl^-(aq)$$
or $$HCl(g) + H_2O(l) \rightleftharpoons H_3O^+(aq) + Cl^-(aq)$$

When the gas dissolves in methylbenzene it does *not* ionise and the resultant solution has no acidic properties (5.5).

WITH OXIDISING AGENTS Concentrated hydrochloric acid is oxidised fairly easily to chlorine, e.g. by potassium manganate(VII) or manganese(IV) oxide (19.2).

Uses of hydrogen chloride and hydrochloric acid

1   Hydrogen chloride is used in the manufacture of organic chlorine compounds such as chloroethene (vinyl chloride), the starting material for polyvinyl chloride (PVC).
2   Hydrochloric acid is used to 'pickle' metals, i.e. clean the surface prior to electroplating or galvanising.
3   Hydrochloric acid is used in the manufacture of dyes, drugs and photographic materials.

Manufacture of hydrogen chloride

The gas is made industrially by burning hydrogen in chlorine, both materials being obtained from the electrolysis of brine in the mercury cathode cell (20.2).

Chlorides

PREPARATION All of the common metallic chlorides are soluble in water except those of silver and lead and they can be made by the standard methods of salt preparation described in 5.4. Anhydrous chlorides which react with water are best prepared by direct combination (Experiment 19.2.b).

HYDROLYSIS Most non-metallic chlorides and many metallic chlorides are hydrolysed by water (see 5.6 and 7.4).

e.g. $FeCl_3(aq) + 3H_2O(l) \rightarrow Fe(OH)_3(s) + 3HCl(aq)$ (Experiment 19.1.d).

Tests for chlorides

1 Chlorides react when warmed with concentrated sulphuric acid to give steamy fumes of hydrogen chloride.

e.g. $NaCl(s) + H_2SO_4(l) \rightarrow NaHSO_4(s) + HCl(g)$

2 Chlorides react with silver nitrate in aqueous solution to give a white precipitate of silver chloride. Dilute nitric acid must be present to prevent the precipitation of other insoluble silver compounds.

e.g. $NaCl(aq) + AgNO_3(aq) \rightarrow AgCl(s) + NaNO_3(aq)$

The precipitate is soluble in excess ammonia solution.

19.2   Chlorine

Chlorine is found in huge quantities in sea water and underground salt deposits in the form of sodium chloride and, to a lesser extent, potassium chloride. It was first prepared by Carl Scheele in Sweden in 1774 and was shown to be an element by Humphry Davy in 1810. He named it chlorine from the Greek *chloros*, meaning 'pale green'.

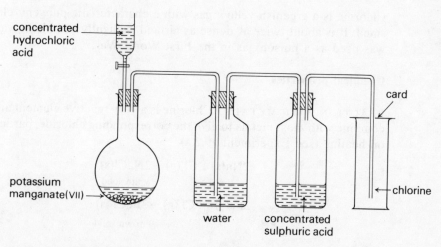

concentrated hydrochloric acid

card

potassium manganate(VII)

water

concentrated sulphuric acid

chlorine

Figure 19.2 The laboratory preparation of chlorine

Preparation

Chlorine is prepared in the laboratory by the oxidation of concentrated hydrochloric acid. The reaction can be brought about by strong oxidising agents, such as potassium manganate(VII) or manganese(IV) oxide.

$$2KMnO_4(s) + 16HCl(aq) \rightarrow 2KCl(aq) + 2MnCl_2(aq) + 8H_2O(l) + 5Cl_2(g)$$

A suitable apparatus for this preparation is shown in Figure 19.2. The chlorine is bubbled through water to remove unchanged hydrogen chloride, dried by means of concentrated sulphuric acid and collected by downward delivery. A moist universal indicator paper placed at the mouth of each jar turns red and is then bleached when the jar is full.

If manganese(IV) oxide is used as the oxidising agent, the mixture must be gently heated.

$$MnO_2(s) + 4HCl(aq) \rightarrow MnCl_2(aq) + Cl_2(g) + 2H_2O(l)$$

Note that the oxidation number of the manganese has changed from $+4$ to $+2$ (reduction).

As an alternative, chlorine can be prepared directly from sodium chloride. If a mixture of sodium chloride, concentrated sulphuric acid and manganese(IV) oxide is warmed, the first two compounds react to form hydrogen chloride which is then oxidised to chlorine by the manganese(IV) oxide. The chlorine may be collected over saturated sodium chloride solution rather than by downward delivery if desired. (It is less soluble in sodium chloride solution than in water.)

Tests for chlorine

Apart from its colour and smell (see *Physical properties*), chlorine can be identified by the fact that it bleaches moist universal indicator paper and turns moist starch–iodide paper blue–black.

Physical properties

Chlorine is a greenish-yellow gas with a characteristic pungent, choking smell. It is about twice as dense as air and is slightly soluble in water. It was used as a poison gas in the First World War.

Chemical properties

REACTION WITH METALS   Chlorine is a very reactive element and will combine with most metals to give the corresponding chloride, particularly on heating (see Experiment 19.2.a).

e.g.
$$2Na(s) + Cl_2(g) \rightarrow 2NaCl(s)$$
white solid

$$2Fe(s) + 3Cl_2(g) \rightarrow 2FeCl_3(s)$$
brown solid

With metals having more than one oxidation number, the chloride formed is the higher one. Thus iron(III) chloride and not iron(II) chloride is obtained when iron is heated in chlorine.

REACTION WITH NON-METALS Many non-metals combine with chlorine on heating. For example, phosphorus burns in the gas to give white fumes of phosphorus trichloride and phosphorus pentachloride (Experiment 19.2.c).

$$P_4(s) + 6Cl_2(g) \rightarrow 4PCl_3(l)$$
$$P_4(s) + 10Cl_2(g) \rightarrow 4PCl_5(s)$$

Carbon, nitrogen and oxygen do not combine directly with chlorine.

REACTION WITH HYDROGEN When a jet of burning hydrogen is lowered into a gas jar full of chlorine, combustion continues and steamy fumes of hydrogen chloride are seen at the mouth of the jar. A mixture of equal volumes of hydrogen and chlorine explodes when exposed to a flame or to ultra-violet light produced, for example, by burning magnesium (Experiment 19.2.d).

$$H_2(g) + Cl_2(g) \rightarrow 2HCl(g)$$

REACTION WITH HYDROCARBONS Chlorine is an extremely powerful oxidising agent and will remove hydrogen from a number of compounds. For example, a taper burns with a red flame in the gas, producing black particles of carbon and white fumes of hydrogen chloride. Warm turpentine ignites spontaneously in chlorine giving the same products (Experiment 19.2.e).

$$C_{10}H_{16}(l) + 8Cl_2(g) \rightarrow 10C(s) + 16HCl(g)$$
turpentine

REACTION WITH AMMONIA A jet of ammonia is spontaneously inflammable in chlorine, producing nitrogen and hydrogen chloride. When the chlorine has been used up the flame goes out and the hydrogen chloride reacts with more ammonia to form white fumes of ammonium chloride (Experiment 19.2.f).

$$2NH_3(g) + 3Cl_2(g) \rightarrow N_2(g) + 6HCl(g)$$
$$6NH_3(g) + 6HCl(g) \rightarrow 6NH_4Cl(s)$$

Adding:
$$\overline{8NH_3(g) + 3Cl_2(g) \rightarrow N_2(g) + 6NH_4Cl(s)}$$
$$\left(\begin{matrix}\text{overall}\\\text{reaction}\end{matrix}\right)$$

CHLORINE WATER (AQUEOUS CHLORINE SOLUTION) When chlorine is passed into water, a yellow–green solution consisting of dissolved chlorine, hydrochloric acid and chloric(I) acid (hypochlorous acid) is formed.

$$Cl_2(g) + water \rightleftharpoons Cl_2(aq)$$
$$Cl_2(aq) + H_2O(l) \rightleftharpoons HCl(aq) + HOCl(aq)$$

The chloric(I) acid is unstable and readily gives up its oxygen. For example, it oxidises dyes, thereby converting them to colourless compounds (Experiment 19.2.g).

$$\text{dye} + \text{HOCl(aq)} \rightarrow (\text{dye} + \text{oxygen}) + \text{HCl(aq)}$$

coloured   colourless

If chlorine is passed into sodium hydroxide solution, the sodium salts of hydrochloric acid and chloric(I) acid are formed.

$$\text{Cl}_2(\text{g}) + 2\text{NaOH(aq)} \rightarrow \text{NaCl(aq)} + \text{NaOCl(aq)} + \text{H}_2\text{O(l)}$$

sodium
chlorate(I)

Uses of chlorine

1   In the manufacture of plastics (e.g. PVC), anaesthetics, insecticides (e.g. DDT), solvents and aerosol propellants, all of which are chlorinated organic compounds.
2   As a bleach in the pulp and textile industries.
3   In the treatment of sewage and in the purification of water.
4   In the manufacture of hydrogen chloride and hydrochloric acid.

Manufacture of chlorine

Chlorine is prepared industrially by the electrolysis of sodium chloride solution (20.2), using graphite anodes and a flowing mercury cathode.

19.3   A comparison of fluorine, bromine and iodine with chlorine

Atoms of the halogens all have seven electrons in their outermost shells and, since the properties of atoms depend upon their electronic configurations (7.2), we would expect them to behave similarly. They can each complete their octets by taking an electron from another atom to form a negative ion (oxidation number −1) or by sharing a pair of electrons to make a single covalent bond.

Preparation

Chlorine, bromine and iodine can all be obtained by the oxidation of the corresponding hydrogen halide; fluorine must be obtained by the electrolysis of hydrogen fluoride to which some potassium fluoride has been added.

Physical properties

The physical properties of the halogens are summarised towards the end of Table 19.3. What trends can you see?

Property	Fluorine	Chlorine	Bromine	Iodine
Atomic number	9	17	35	53
Relative atomic mass	19	35.5	80	127
Electronic configuration	2.7	2.8.7	2.8.18.7	2.8.18.18.7
Atomic radius/nm	0.072	0.099	0.114	0.133
Oxidising ability	← decreases —————————————→			
Nature of reaction with other elements	← decreasing vigour ——————→			
Ionic radius/nm	0.136	0.181	0.195	0.216
Resistance of ions to oxidation	← decreases —————————————→			
Appearance	pale yellow gas	yellow–green gas	red–brown liquid	black solid
m.p./K (°C)	50 (−223)	171 (−102)	266 (−7)	387 (114)
b.p./K (°C)	86 (−187)	238 (−35)	332 (59)	456 (183)
Solubility in water	reacts with water to give hydrogen fluoride and oxygen	moderately soluble in water, hydrolysis occurs	moderately soluble in water, slightly hydrolysed	sparingly soluble in water

Table 19.3 Some properties of the halogens

Chemical properties

The halogens combine with a large number of metals and non-metals and with hydrogen. In general, the reactions of fluorine are the most vigorous and those of iodine are the least vigorous. For example, if iron is heated in a stream of chlorine, the iron inflames and a brown solid, iron(III) chloride, is formed (Experiment 19.2.b). A similar reaction occurs with bromine vapour, giving red iron(III) bromide (Experiment 19.3.b).

$$2Fe(s) + 3Br_2(g) \rightarrow 2FeBr_3(s)$$

However, iron reacts much less vigorously with iodine and gives a mixture of products, probably consisting mainly of iron(II) iodide.

Another example of this reactivity trend is seen in the reactions between the halogens and water (Experiment 19.3.a). Although bromine water, like chlorine water, is acidic, only about 1% of the dissolved bromine actually reacts with the water. The products are hydrobromic acid and bromic(I) acid.

$$Br_2(aq) + H_2O(l) \rightleftharpoons HBr(aq) + HOBr(aq)$$

Iodine is virtually insoluble in water but does show slight acidic properties by dissolving in sodium hydroxide solution.

An explanation of the reactivity trend

The reactivity of a halogen must be related to the ease with which its atoms gain electrons to complete their octets. This in turn is connected with the size of the atoms. With increase in atomic radius, the nucleus has less hold over the electrons in the outermost shell and is less able to attract an extra electron into this shell. Thus fluorine, which has the smallest atoms, is the best 'electron attractor' (i.e. oxidising agent) of these four elements and iodine the feeblest (see Table 19.3). Once the extra electron has been gained it will be less firmly held in a large ion than in a small one. For this reason iodide ions give up their extra electrons (i.e. are oxidised) the most readily and the fluoride ions the least readily of the four. Both of these trends are illustrated in Experiment 19.3.c where chlorine atoms remove electrons from iodide ions. The reverse change is not possible.

e.g. $$Cl_2(aq) + 2I^-(aq) \rightarrow 2Cl^-(aq) + I_2(aq)$$

You will have noticed from Table 19.3 that the halide ions are larger than the corresponding atoms. This is quite simple to explain: when a further electron is added to an atom the attraction of the nucleus for the electrons remains the same but the mutual repulsion between the electrons increases. Hence the radius becomes larger.

19.4   A comparison of the hydrogen halides

The main properties of the hydrogen halides are summarised in Table 19.4.

Tests for halides

These tests are simply extensions of the ones given for chlorides in 19.1.

1   On warming with concentrated sulphuric acid, chlorides give hyd-

Property	Hydrogen fluoride	Hydrogen chloride	Hydrogen bromide	Hydrogen iodide
Appearance	colourless liquid	colourless gas	colourless gas	colourless gas
m.p./K (°C)	190 (−83)	158 (−115)	186 (−87)	222 (−51)
b.p./K (°C)	292.5 (19.5)	188 (−85)	206 (−67)	238 (−35)
Stability	decreases ⟶			
Reducing power	increases ⟶			
Solubility in water	very soluble	very soluble	very soluble	very soluble
Nature of aqueous solution	weak acid	strong acid	strong acid	strong acid
Silver salt	white solid, soluble in water	white solid, insoluble in water, soluble in excess ammonia solution	cream solid, insoluble in water, sparingly soluble in excess ammonia solution	yellow solid, insoluble in water, insoluble in excess ammonia solution

Table 19.4 Some properties of the hydrogen halides

rogen chloride, bromides give hydrogen bromide and bromine, and iodides give mainly iodine (Experiment 19.4.a). The differences are due to the relative ease of oxidation of the three hydrogen halides (see Table 19.4). Hydrogen chloride is not oxidised by the concentrated sulphuric acid. However, hydrogen bromide is fairly easily oxidised to bromine while hydrogen iodide, being a powerful reducing agent, is almost completely oxidised to iodine.

2   An aqueous solution of a halide precipitates the corresponding silver salt from silver nitrate solution acidified with dilute nitric acid.

e.g.

$$NaBr(aq) + AgNO_3(aq) \rightarrow AgBr(s) + NaNO_3(aq)$$
$$NaI(aq) + AgNO_3(aq) \rightarrow AgI(s) + NaNO_3(aq)$$

If excess ammonia solution is added, silver chloride dissolves, silver bromide dissolves slightly, but silver iodide is insoluble (Experiment 19.4.b).

N.B. Halide *ions* are producing the precipitates in these tests; *covalent* halogen compounds, such as tetrachloromethane, will not give precipitates.

3   An alternative method of identifying bromides and iodides is to shake a solution of the suspected halide with chlorine water and tetrachloromethane. Chlorine will displace bromine from bromides, imparting an orange colour to the tetrachloromethane, and will liberate iodine from iodides, imparting a purple colour to the tetrachloromethane (Experiment 11.1.e).

e.g. $$2NaBr(aq) + Cl_2(aq) \rightarrow 2NaCl(aq) + Br_2(aq)$$

or, ionically $$2Br^-(aq) + Cl_2(aq) \rightarrow 2Cl^-(aq) + Br_2(aq)$$

Questions on Chapter 19

1   (a) Describe how you would obtain a sample of hydrogen chloride gas from sodium chloride. You should give the *name* of the other chemical required and the equation for the reaction. No diagram is required.
(b) With the aid of a diagram, explain how you would obtain an aqueous solution of this gas.
(c) Describe, giving an equation in each case, the reaction of hydrochloric acid with (i) silver nitrate solution, (ii) granulated zinc.   (O & C)

2   When concentrated brine is electrolysed using inert electrodes two gases are produced. One is chlorine; the other may be designated X.
(a) (i) What is the name given to the family of elements to which chlorine belongs?
   (ii) What colour is chlorine?
   (iii) At which of the two electrodes is chlorine liberated?
   (iv) Name a substance from which both the inert electrodes may be made.
(b) When chlorine is bubbled through aqueous sodium bromide it is converted to chloride ions.
   (i) What colour change in the solution would be seen during this reaction?
   (ii) Which of the reactants is *oxidised* in the reaction?
   (iii) How would you test for the presence of chloride ions in an aqueous mixture of sodium chloride and sodium carbonate? State: test reagent(s) and result.
(c) The second product of the electrolysis, X, reacts vigorously with chlorine to form the gas Y.
   (i) Name X.
   (ii) Name Y.
   (iii) Name two other reagents from which Y can be produced readily in the laboratory.
   (iv) Give the equation for any reaction of chlorine (apart from that with X) in which Y is one of the products.
(d) (i) Give the equation for a laboratory preparation of chlorine which does not involve electrolysis.
   (ii) Using this reaction describe the apparatus you would use to prepare and collect a damp sample of chlorine. Your description may be *either* a labelled diagram *or* a verbal one.   (NI)

3   (a) Describe the reactions which occur between chlorine and the following substances: hydrogen; phosphorus; methane; water; iron; ammonia.
   In each case write equations for the reactions and give a concise statement of the necessary experimental conditions. Diagrams are *not* required.
(b) Explain the part played by water in the bleaching action of chlorine.
(c) Describe what is observed when solid specimens of sodium chloride and

sodium bromide are treated separately with concentrated sulphuric acid. How do you account for the differences between these two reactions? (W)

4 Fluorine (symbol F), chlorine, and iodine are members of the same chemical family.

(a) What name is given to this group of elements? Name *one* other element which should be included and describe its physical appearance at room temperature.

(b) The relative atomic mass of a fluorine atom is 19; its atomic number is 9. Deduce the composition of the nucleus and the electronic structure of the fluorine atom.

(c) Like the other elements in this family, fluorine forms diatomic molecules. Explain the nature of the bonding between the atoms in a fluorine molecule.

(d) Why do these elements show similarities in their chemical reactions?

(e) Small crystals of graphite (carbon) and iodine are similar in appearance. What difference in behaviour would you expect to observe when these substances are heated in separate test tubes? Briefly state how this difference is related to the structures of the two solids.

(f) A few drops of chlorine water are added to aqueous potassium iodide. The mixture is then shaken with trichloromethane. (Trichloromethane is an organic liquid which does not mix with water.) What would you expect to see (i) after the addition of the chlorine water; (ii) after shaking the mixture with trichloromethane? Explain these observations. (AEB)

5 (a) Describe one series of experiments which you would perform with chlorine, bromine and iodine, or their compounds in order to compare the ease with which the elements undergo the type of change summarised by

$$X_2 + 2e \rightarrow 2X^-.$$

(b) How do you account for the differences in the melting points and boiling points of the chlorides of magnesium and silicon, Si, shown in the table below? (The atomic numbers of magnesium and silicon are 12 and 14 respectively.)

	Magnesium chloride	Silicon chloride
Melting point/(°C)	710	−70
Boiling point/(°C)	1420	60

(c) What reactions would you expect to occur between chlorine and aqueous solutions of (i) iron(II) chloride, (ii) sulphur dioxide? In each case describe and explain one test you would apply in order to confirm that the reaction you have described has occurred. (W)

20 The metals

Some metals have been known for thousands of years while others were not isolated until the advent of electrolysis. As you will discover, the reason for this can be traced to the relative reactivities of the metals; our first task, then, is to arrange the metals in a *reactivity series*.

Experiments

20.1.a The reactivity of metals with air ★

Part (a) of this experiment must not be performed by pupils.

PROCEDURE (a)★★ Using tongs, remove a piece of sodium from the oil under which it is stored and place it on a filter paper. Cut off a 3 mm cube of the metal and return the remainder to the bottle. (Notice how quickly the surface tarnishes.) Blot the cube to remove the oil and place it on a crucible lid resting on a pipe-clay triangle on a tripod. Heat it in a *closed* fume cupboard by means of a bunsen burner, keeping well clear until the reaction has finished. Carry out the same procedure with potassium.
(b) Heat half a spatula measure of powdered copper, aluminium, zinc, iron and magnesium on separate crucible lids. (Do not look directly at the burning magnesium: use tinted glass if available.) Examine the residue at the end of each experiment.

Questions

1 Why do you think that sodium and potassium are stored under oil?
2 Did any of the metals not react? How could you tell?
3 Make a list of the metals in order of decreasing vigour of reaction with air.
4 Is it easy to decide on the correct order?

20.1.b Competition for oxygen

You probably found it rather difficult to decide on the correct order for some of the metals in the last experiment. A more reliable method is to allow pairs of metals to compete with one another for oxygen. This may be done by heating one metal with the oxide of another. If a reaction occurs, the first metal must be taking oxygen from the second, i.e. the first metal must be more reactive towards oxygen than the second. If no reaction occurs, the reverse must be true.

Parts (b) and (c) of this experiment must not be performed by pupils.

PROCEDURE (a) Heat half a spatula measure of each of the following mixtures on a crucible lid: iron filings and freshly dried copper(II) oxide, iron filings and zinc oxide, iron filings and lead(II) oxide, zinc powder and iron(III) oxide. Keep well clear during the heating and look for evidence of a reaction (e.g. evolution of heat). Allow the residue to cool and examine it carefully for signs of a change.

(b)★★ Carry out the same procedure using a mixture of magnesium powder and freshly dried copper(II) oxide and then a mixture of magnesium powder and lead(II) oxide. Use a safety screen for these reactions.

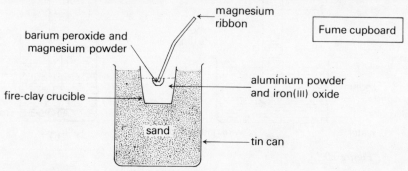

Figure 20.1.b

(c)★★ Set up the apparatus as shown in Figure 20.1.b in a fume cupboard, light the end of the magnesium ribbon and stand well back. When the reaction has finished, examine the contents of the crucible (**care**).

Record the results of all the experiments as shown in Table 20.1.b.

Mixture	Reaction	Residue
	✓ or ✗	

Table 20.1.b

Questions

1 Make a list of the metals in order of decreasing reactivity towards oxygen. If you are unable to differentiate between a particular pair, place them on the same line in the list.

2 Could aluminium metal be used to extract iron from an ore containing iron(III) oxide? Why is this method not used on a large scale?

20.1.c Can metals remove oxygen from water? ★★

You probably know that water is the oxide of hydrogen. Thus by using the same arguments as in the previous experiment you should be able to fit hydrogen into your reactivity series.

Part (a) of this experiment must not be performed by pupils.

PROCEDURE (a) Place about 200 cm³ of distilled water and a few drops of universal indicator in a 1 dm³ beaker. Using tongs, drop a 3 mm cube of sodium into the water at arm's length and immediately cover the beaker with a large watch glass. Note all that happens and then repeat the experiment using potassium in place of the sodium.

Use a sodium spoon as shown in Figure 20.1.c, part (a), to enable you to collect the gas given off from the reaction between sodium and water. Test the gas with a flame.

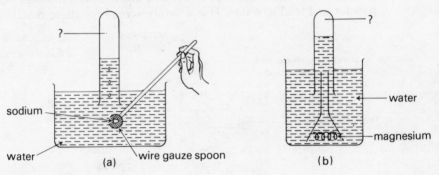

Figure 20.1.c

(b) Place a few calcium turnings (**caution:** do not touch the metal) in a test tube containing 5 cm³ of distilled water coloured with universal indicator. Collect any gas given off over water in the usual way and test it with a flame.

(c) Investigate the reaction between clean magnesium ribbon and distilled water as shown in Figure 20.1.c, part (b). Look closely at the metal for signs of a reaction and then leave the apparatus set up for several days. Using a similar arrangement, try to react magnesium with hot water. In both cases test any gas evolved with a flame.

Questions

1 What did you observe in each case in (a)? Which metal reacted the more violently?

2 Which gas was given off in (a)? How could you tell?

3 What do you think was left in solution? (The universal indicator should give you a clue.)

4 Write equations for the reactions in (a).

5 Which gas was given off in (b) and what do you think was left in the tube? Write an equation for the reaction.

6 What happened in (c)?

7 Are these metals above or below hydrogen in the reactivity series? Explain your answer.

20.1.d To investigate the reactions of some metals with steam ★ ★

PROCEDURE   Heat a coil of clean magnesium ribbon in the apparatus shown in Figure 20.1.d until the reaction appears to have finished. (The water on the rocksil will be vaporised by conducted heat.) Remove the delivery tube from the trough before you stop heating (why?). Test any gas evolved with a flame and then repeat the experiment using in turn a

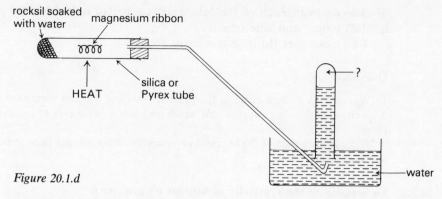

rocksil soaked with water
magnesium ribbon
silica or Pyrex tube
HEAT
?
water

Figure 20.1.d

spiral of aluminium foil, zinc powder, iron wool, lead powder and copper powder.

Questions

1 What did you observe with the magnesium? Which gas was evolved?
2 What do you think remained in the tube? (Bear in mind the temperature involved.) Write an equation for the reaction.
3 Which other metals reacted with the steam? Answer question 2 for these metals.
4 Which metals did not react? Are any of these out of position compared with the order you found in Experiment 20.1.b?
5 Which metals seemed to be below hydrogen in the reactivity series? Explain your answer.

20.1.e Can hydrogen remove oxygen from metal oxides? ★ ★

PROCEDURE   Carry out the same procedure as in Experiment 4.4.c using, in turn, iron(II) diiron(III) oxide (magnetic iron oxide), aluminium oxide, lead(II) oxide and copper(II) oxide. Add dilute sulphuric acid to the residue from the experiment with the iron oxide and test for the evolution of hydrogen.

Questions

1 Which oxides reacted with hydrogen? Write equations for the reactions.
2 What type of reactions are these?
3 Why was the residue from the experiment with the iron oxide treated with dilute sulphuric acid?
4 Did the aluminium oxide behave as you expected? Explain your answer.
5 Which reaction is reversible, i.e. capable of going both ways?
6 Use the results of the last three experiments to make a list of the metals and hydrogen in order of decreasing reactivity towards oxygen.

20.1.f Where does carbon come in the reactivity series?

A cheap reducing agent is required for extracting metals from their oxide ores on a large scale. Carbon is cheap, but which oxides will it reduce?

PROCEDURE   (a) Carry out Experiment 16.2.a using separate mixtures

of charcoal with each of the following: copper(II) oxide, lead(II) oxide, iron(III) oxide, and zinc oxide.
(b) Carry out part (b) of Experiment 16.4.b.

Questions

1 Where does carbon come in the reactivity series? Explain your answer.
2 Write equations for the reactions which took place. What type of reactions are these?
3 Explain the result of (b) in terms of competition for oxygen between carbon and magnesium.

20.2.a To investigate the reactivity of lithium with water ★

PROCEDURE Carry out part (a) of Experiment 20.1.c, using lithium in place of the sodium.

Questions

1 to 4 As in Experiment 20.1.c.
5 Does lithium show any similarity to potassium and sodium in this reaction?
6 Where does lithium come in the reactivity series?
7 Write down the names of the three elements lithium, potassium and sodium in order of decreasing reactivity.

20.3.a Some compounds of calcium

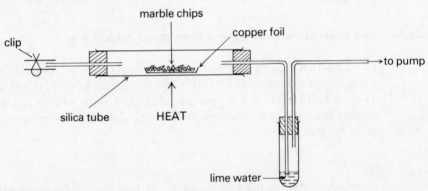

Figure 20.3.a

PROCEDURE (a) With the clip closed and the filter pump turned off, heat a few marble chips as strongly as possible in the apparatus shown in Figure 20.3.a. (The copper foil prevents the marble from reacting with the tube.) After two or three minutes open the clip and turn on the filter pump so that a *slow* stream of bubbles passes through the lime water.
(b) Using tongs, hold a marble chip just above the blue cone of a roaring bunsen flame. After two or three minutes place the chip in a crucible, allow it to cool and inspect it closely. Next, hold the crucible in the palm of your hand, add two drops of distilled water to the chip and note the result. (Be prepared to put the crucible down quickly on an asbestos board.) Inspect the chip again and then fill the crucible with distilled

water. Finally, filter the mixture into a boiling tube and blow *gently* through a drinking straw into the filtrate.

Questions

1 What is the chemical name for marble?
2 Which gas was given off in (a)? How could you tell?
3 In what way did the appearance of the marble chip change during heating? What do you think was formed?
4 Describe what happened on adding water to the cooled chip. What do you think was produced by this reaction? Was this new product very soluble in water?
5 Did any unchanged marble appear to be left at the end of the experiment? How could you have tested this residue to confirm that it was marble?
6 What was the filtrate obtained at the end of the experiment? Explain how you reached this conclusion.
7 Write equations for all the reactions which occurred.

20.3.b Which ions cause hardness of water?

PROCEDURE Measure out 25 cm^3 of distilled water into a 100 cm^3 conical flask and add 0.5 cm^3 of soap solution from a burette. Stopper the flask, shake it vigorously and stand it on the bench. If the resulting lather does not cover the surface of the water *completely* for two minutes add further 0.5 cm^3 portions of soap solution, shaking after each addition, until it does. Wash out the flask with plenty of tap water and then with a little distilled water and repeat the experiment using 25 cm^3 portions of 0.005 M solutions of calcium chloride, sodium chloride, magnesium chloride and potassium chloride in place of the distilled water.

Questions

1 Which substances made the water hard (i.e. caused it not to lather very well)?
2 Which ions in these substances were responsible for the hardness? Explain your answer.
3 Do you think that these ions reacted with the soap? What led you to this conclusion?

20.3.c To compare the hardness of various samples of water

PROCEDURE Proceed as in the previous experiment but use 50 cm^3 portions of the water samples listed in Table 20.3.c. To prepare the boiled water, half-fill a 250 cm^3 beaker with tap water, mark the level and cover the beaker with a watch glass. Boil the water gently for about 15 minutes to remove temporary hardness and then make it up to the original volume by adding distilled water. For the fourth water sample, dissolve a spatula measure of anhydrous sodium carbonate in about 100 cm^3 of tap water. Record your results as shown in the table overleaf.

Questions

1 Why was the boiled water made up to the original volume with distilled water?
2 Why was v subtracted from x, y and z in the third column of the table?
3 Did the tap water contain both temporary and permanent hardness?

Water sample	Volume of soap solution/cm³	Volume of soap solution to overcome hardness/cm³
Distilled water	v	0
Tap water	x	(x − v)
Boiled water	y	(y − v)
Tap water + sodium carbonate	z	(z − v)

Table 20.3.c

4   Which type of hardness predominated? How could you tell?
5   What was the purpose of the sodium carbonate and how effective was it?

20.4.a   Why does aluminium not react with air? ★

Aluminium reacts violently with sulphur (Experiment 2.1.e) and with iron(III) oxide (Experiment 20.1.b) and yet it does not react with water or with cold air. Why is this?

PROCEDURE   Immerse an aluminium milk top in mercury(II) chloride solution for about 30 seconds so that the oxide coating on the surface of the aluminium is replaced by aluminium amalgam (i.e. a solution of aluminium in mercury). Rinse the milk top and quickly wrap it round the bulb of a 0–100 °C thermometer. Clamp the thermometer horizontally so that the milk top does not fall off and observe it closely. At the end of the experiment wash your hands thoroughly – mercury(II) chloride is poisonous.

Questions

1   Did the aluminium react with the air after amalgamation? What evidence was there for this?
2   Did the oxide layer which was initially present on the aluminium act as a protective coating?

20.6.a   Some reactions of iron compounds

PROCEDURE   (a) Add about 1 cm³ of sodium hydroxide solution to 5 cm³ of iron(II) sulphate solution in a test tube. Stir well and note the result. Repeat this procedure using iron(III) sulphate solution in place of the iron(II) sulphate solution and keep the products of both reactions for reference in parts (b) and (c).
(b) Add about 1 cm³ of concentrated nitric acid (care) to 5 cm³ of iron(II) sulphate solution in a boiling tube. Heat the mixture to boiling and then add small quantities of sodium hydroxide solution, stirring and spotting on to universal indicator paper after each addition, until the liquid is alkaline.
(c) Heat 5 cm³ of iron(III) sulphate solution to boiling in a boiling tube and

then bubble sulphur dioxide into the hot liquid for 30 seconds. (Carry out this operation in a fume cupboard.) Note the colour of the resultant liquid and then add sodium hydroxide solution, with stirring, until the mixture is alkaline.

Questions

1 Describe and name the precipitates obtained in (a). Write chemical equations and ionic equations for the reactions which occurred.
2 What did you see on adding the concentrated nitric acid in (b)? Give the name of this product.
3 What was the final precipitate in (b)? How could you tell?
4 What type of change did the iron(II) ions undergo in (b)? What was the function of the nitric acid?
5 What happened to the iron(III) ions in (c)? Explain how you reached your conclusion.

20.7.a Some reactions involving lead(II) ions

Lead compounds are poisonous: wash your hands when you have completed this experiment.

PROCEDURE (a) Add 5 cm³ of magnesium sulphate solution to an equal volume of lead(II) nitrate solution in a test tube and record the result.
(b) Repeat (a) using potassium iodide solution in place of the magnesium sulphate solution.
(c) Repeat (a) using sodium chloride solution in place of the magnesium sulphate solution. In this case tip the resultant mixture into a boiling tube and heat, with shaking, until the liquid boils. If the solid does not dissolve add small quantities of distilled water, boiling after each addition, until it does. Allow the tube to cool and note what happens.
(d) To 5 cm³ of lead(II) nitrate solution in a test tube add small quantities of sodium hydroxide solution, stirring after each addition, until the tube is full. Tip the liquid into a second tube to ensure that mixing is complete.

Questions

1 Describe what you saw in (a) and name the products formed. Write a chemical equation and an ionic equation to represent the reaction.
2 Answer question 1 for part (b).
3 Answer question 1 for part (c).
4 What type of reactions take place in (a), (b) and (c)?
5 Describe and explain what happened in (d).

20.7.b To compare the properties of lead(II) oxide and lead(IV) oxide

PROCEDURE Carry out each of the following tests using separate samples of lead(II) oxide and lead(IV) oxide.
(a) Heat a spatula measure of the oxide in an ignition tube and test for the evolution of oxygen.
(b) Add half a spatula measure of the oxide to 10 cm³ of dilute nitric acid in a boiling tube and heat, with shaking, until the liquid boils.

(c) Add half a spatula measure of the oxide to $5 \, cm^3$ of concentrated hydrochloric acid in a test tube. Smell the product *cautiously* and apply suitable tests to identify it. Wash the contents of the tube down the sink as soon as possible.

Questions

1   Which of the oxides gave off oxygen in (a)? How did you identify the gas?
2   Which of the oxides dissolved in the dilute nitric acid in (b)? Name the products formed and write an equation for the reaction.
3   Which oxide reacted to give a gas in (c)? What was the gas and how did you identify it? Did the oxide behave as a normal basic oxide in this experiment? Give a reason for your answer.

20.7.c   To investigate the properties of 'red lead oxide'

PROCEDURE   (a) Heat a spatula measure of 'red lead oxide' in an ignition tube and test for the evolution of oxygen.
(b) Add half a spatula measure of the oxide to $10 \, cm^3$ of dilute nitric acid in a boiling tube and heat, with shaking, to boiling. Filter the mixture and divide the filtrate into two parts. Leave the filter paper and residue in the funnel.
(c) To one portion of the filtrate in a test tube add a few drops of potassium iodide solution. To the second portion in another test tube add small quantities of sodium hydroxide solution, stirring after each addition, until the tube is full. Tip the liquid into a second tube to ensure that mixing is complete.
(d) Add a few drops of concentrated hydrochloric acid to the residue on the filter paper and identify the gas evolved.

Questions

1   Was oxygen given off in (a)?
2   Did all, part, or none of the oxide dissolve in dilute nitric acid?
3   What can you deduce about the nature of the filtrate used in (c)? Explain your reasoning clearly.
4   What are your conclusions about the residue used in (d)? Again, explain your reasoning.
5   Did the 'red lead oxide' behave as if it were a mixture of two substances? If it did, name the substances and explain how you reached your conclusion.

20.8.a   To prepare two copper(I) compounds ★

PROCEDURE   (a) Mix $5 \, cm^3$ of Fehling's solution A with $5 \, cm^3$ of Fehling's solution B in a boiling tube to obtain a royal blue solution containing complex copper(II) ions. Add a spatula measure of glucose and boil the mixture, with shaking, until no further change occurs. Filter off the resultant precipitate of copper(I) oxide.
(b) Mix $5 \, cm^3$ of concentrated hydrochloric acid with $5 \, cm^3$ of 0.5 M copper(II) chloride solution in a boiling tube and add a few copper turnings. Boil the mixture, with shaking, until no further change occurs

and then allow it to cool. Pour the *liquid* into 100 cm³ of cold, freshly boiled distilled water in a 250 cm³ beaker, stir well and filter off the resultant precipitate of copper(I) chloride at the pump.

Questions

1 Describe the precipitate obtained in (a). What colour is the corresponding copper(II) compound?
2 What type of change did the copper(II) ions undergo during the reaction?
3 What was the purpose of the glucose?
4 Describe the precipitate in (b). What colour is the corresponding copper(II) compound?
5 What was the reducing agent in (b)?

20.1   The reactivity series

If the metals are arranged in order of their reactivity towards oxygen, the *reactivity series* is obtained. The same order of reactivity is not necessarily adhered to in *all* reactions but it does not vary a great deal. The series is therefore very useful for predicting the way in which various metals are likely to react with a given substance.

The relative positions of the metals in the series may be determined by comparing the ease with which they burn in air (Experiment 20.1.a) but it is not easy to decide on the correct order for metals which have similar reactivities. It is better to allow two metals to compete for oxygen by heating one of them with the oxide of the other. If the first metal is higher in the reactivity series it will take oxygen from the second metal and a reaction will occur; if the first metal is lower in the series it will be unable to remove oxygen from the second metal and no reaction will take place (see Experiment 20.1.b). These ideas are summarised in Figure 20.1.

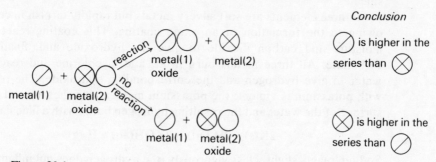

Figure 20.1

Hydrogen can be allocated a place in the series in a similar way by investigating the reaction between metals and water (hydrogen oxide). Metals which are near the top of the series react violently with cold water; those lower down only react when heated in steam, and any which are below hydrogen do not react at all. Hydrogen reduces the oxides of this last group to the metals (Experiments 20.1.c, 20.1.d and 20.1.e). Aluminium is less reactive with water than it is with metal oxides for reasons which are discussed in 20.4. Details of all the reactions are given with the properties of the metals later in the chapter. It should be stressed that although the experiments in this chapter place hydrogen at approximately the same level as iron in the *reactivity* series, its position in the *electrochemical* series is below that of lead (see 11.1).

A metallic oxide can be reduced to the metal by heating it with another metal which is higher in the reactivity series as in Experiment 20.1.b, but this would be a very expensive method to use on a large scale. Carbon is a much cheaper reducing agent and both zinc and iron are obtained industrially by heating their oxides with it.

The thermal stability of metallic compounds

When the thermal stabilities of metallic hydroxides, carbonates and nitrates are compared (Experiments 18.2.b, 16.4.c and 17.4.d) it is found that, as a general rule, the higher the metal is in the reactivity series, the more stable are its compounds when heated. This is an important relationship and is well worth remembering.

The ways in which the chemical properties of a metal and its compounds may be related to the position of the metal in the reactivity series are summarised in Table 20.1.

20.2   Potassium, sodium and lithium – the alkali metals

These three metals are found in group I of the periodic table and all have one electron in the outermost shells of their atoms. They are too reactive to occur free. Potassium and sodium are found mainly as chlorides, both in the sea and in salt deposits formed where seas have dried up. Sodium nitrate (Chile saltpetre) also occurs naturally. Lithium is found in complex silicates.

All three elements are soft silvery metals but rapidly tarnish in cold air, owing to the formation of an oxide coating. This coating reacts with moisture and carbon dioxide to give the hydroxide and, finally, the carbonate. All three metals burn in air and in chlorine and react with water to give hydrogen and the corresponding hydroxide. The reaction with potassium is violent: the potassium melts and rushes around on the surface of the water and the resulting hydrogen burns with a lilac flame.

$$2K(s) + 2H_2O(l) \rightarrow 2KOH(aq) + H_2(g)$$

Sodium reacts slightly less vigorously (the hydrogen does not inflame) and lithium less vigorously still. With acids, potassium and sodium explode and lithium reacts very violently.

An explanation of the reactivity trend in the alkali metals

The reactivity of each metal is related to the ease with which its atoms lose their one outermost electron to form positive ions. If you look at Table 20.2 you will see that the most reactive element, potassium, has the largest atoms of the three. The outermost electron in a potassium atom is

Table 20.1 Chemical properties and the reactivity series

	Action of cold air	Action of water	Action of dilute hydrochloric and sulphuric acids	Reduction of oxides	Nature of hydroxides	Effect of heat on carbonates	Effect of heat on nitrates
K	rapidly attacked	violent in cold	explode	not reduced by carbon or hydrogen	strongly basic: stable to heat	stable	give nitrite and oxygen
Na	rapidly attacked	violent in cold	explode	not reduced by carbon or hydrogen	strongly basic: stable to heat	stable	give nitrite and oxygen
Ca	attacked	attacked in cold	give hydrogen with decreasing vigour as series is descended	not reduced by carbon or hydrogen	strongly basic: stable to heat	give oxide and carbon dioxide with increasing ease as series is descended	give oxide, nitrogen dioxide and oxygen with increasing ease as series is descended
Mg	little action: protective oxide coating formed	react with steam	give hydrogen with decreasing vigour as series is descended	not reduced by carbon or hydrogen	weakly basic or amphoteric: on heating give oxide and water with increasing ease as series is descended	give oxide and carbon dioxide with increasing ease as series is descended	give oxide, nitrogen dioxide and oxygen with increasing ease as series is descended
Al	little action: protective oxide coating formed	react with steam	give hydrogen with decreasing vigour as series is descended	not reduced by carbon or hydrogen	weakly basic or amphoteric: on heating give oxide and water with increasing ease as series is descended	give oxide and carbon dioxide with increasing ease as series is descended	give oxide, nitrogen dioxide and oxygen with increasing ease as series is descended
Zn	little action: protective oxide coating formed	react with steam	give hydrogen with decreasing vigour as series is descended	reduced by hot carbon	weakly basic or amphoteric: on heating give oxide and water with increasing ease as series is descended	give oxide and carbon dioxide with increasing ease as series is descended	give oxide, nitrogen dioxide and oxygen with increasing ease as series is descended
Fe	rusts in moist air	no reaction	give hydrogen with decreasing vigour as series is descended	reduced by hot carbon	weakly basic or amphoteric: on heating give oxide and water with increasing ease as series is descended	give oxide and carbon dioxide with increasing ease as series is descended	give oxide, nitrogen dioxide and oxygen with increasing ease as series is descended
Pb	little action	no reaction	give hydrogen with decreasing vigour as series is descended	reduced by hot carbon or hydrogen	weakly basic or amphoteric: on heating give oxide and water with increasing ease as series is descended	give oxide and carbon dioxide with increasing ease as series is descended	give oxide, nitrogen dioxide and oxygen with increasing ease as series is descended
Cu	little action	no reaction	no reaction	reduced by hot carbon or hydrogen	weakly basic or amphoteric: on heating give oxide and water with increasing ease as series is descended	give oxide and carbon dioxide with increasing ease as series is descended	give oxide, nitrogen dioxide and oxygen with increasing ease as series is descended

further from the nucleus than that in a sodium atom and therefore is less strongly held by the nucleus. On the other hand, lithium atoms are smaller than sodium atoms and thus the outermost electron in each one is more firmly held by the nucleus. For this reason, lithium is the least reactive of these three elements. Would you expect the group I metal, rubidium (atomic radius 0.216 nm) to be more reactive or less reactive than potassium?

Metal	Atomic radius/nm
Lithium	0.133
Sodium	0.157
Potassium	0.203

Table 20.2

Hydrides

When the alkali metals are heated in hydrogen the corresponding hydrides are formed. Electrolysis of these compounds in the molten state yields hydrogen at the *anode*, showing that they contain H^- ions (see also 7.4).

Oxides and hydroxides

The oxides are ionic solids, formed by burning the elements in a limited air supply. They react violently with water to give solutions of the corresponding hydroxides.

e.g. $$Na_2O(s) + H_2O(l) \rightarrow 2NaOH(aq)$$

The oxides and hydroxides are all strongly basic and, with the exception of lithium hydroxide, stable to heat.

Salts

Salts of the alkali metals contain colourless K^+, Na^+ or Li^+ ions. All the simple ones are soluble in water and, apart from the carbonates, may be prepared by the titration method. Lithium salts are less thermally stable than those of sodium and potassium.

Uses of the metals and their compounds

The alkali metals and their compounds are widely used: the following paragraphs give a selection of these uses.

Metallic *potassium* has few uses but its *chloride, nitrate* and *sulphate* are important fertilisers. *Potassium hydroxide* is used in soap making and *potassium carbonate* in making hard glass.

Sodium is used to make the petrol additive tetraethyl-lead(IV), as a coolant in nuclear reactors, and in street lamps. Its *hydroxide* is used in making sodium carbonate, soap, paper, glass and artificial silk. *Sodium carbonate* is used in making paper and glass and as a water softener while the *hydrogencarbonate* is present in baking powder and health salts.

Sodium nitrate is an important fertiliser. *Sodium sulphate* is used in its *anhydrous* form in the paper and glass industries and in its *hydrated* form is known as Glauber's salt (a laxative). *Sodium chloride* is used in preparing food and for the manufacture of sodium, chlorine, sodium hydroxide and domestic bleach.

Lithium fluoride is used as a welding flux and *lithium chloride* as a dehumidifier in air conditioners.

Manufacture of sodium

Sodium, in common with other very reactive metals, cannot be obtained by chemical reduction of its oxide. Instead it is manufactured by the electrolysis of its fused chloride in the *Downs cell* (Figure 20.2.1).

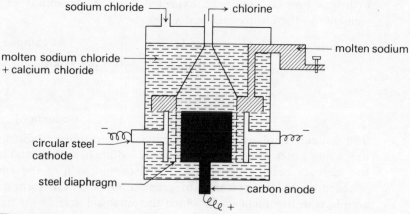

Figure 20.2.1 The Downs cell

The electrolyte is molten sodium chloride containing calcium chloride to lower its melting point. Once the process has been started the very high current used (up to 30 000 amp) keeps the electrolyte in the liquid state. The diagram shows how the products are collected.

Chlorine is liberated at the carbon anode and is an important by-product of the process. Molten sodium is produced at the steel cathode; it must be kept away both from the air and from the chlorine (why?).

Changes   At anode   At cathode

$$2Cl^-(l) \rightarrow 2Cl(g) + 2e^- \dashrightarrow 2e^- + 2Na^+(l) \rightarrow 2Na(l)$$

$$Cl_2(g) \qquad\qquad\qquad 2Hg(l)$$

$$\qquad\qquad\qquad\qquad\qquad 2Na/Hg(l)$$

(oxidation)   (reduction)

Manufacture of sodium hydroxide

In this process brine is electrolysed in a *mercury cathode cell* (Figure 20.2.2). Chlorine is liberated at carbon anodes and is piped off as shown. The use of a mercury cathode results in the discharge of sodium ions, *not*

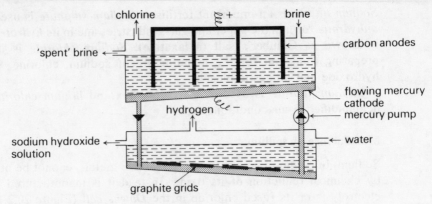

chlorine

brine

'spent' brine

carbon anodes

flowing mercury cathode
mercury pump

hydrogen

sodium hydroxide solution

water

graphite grids

Figure 20.2.2 Mercury cathode cell

hydrogen ions as would be expected from the positions of the two elements in the electrochemical series.

| *Changes* | At anode | | At cathode |

$$2Cl^-(aq) \rightarrow 2Cl(g) + 2e^- \qquad \rightarrow 2e^- + 2Na^+(aq) \rightarrow 2Na(s)$$

$$Cl_2(g)$$

$$2Hg(l)$$

$$2Na/Hg(l)$$

(oxidation)   (reduction)

The sodium dissolves in the mercury and the resulting sodium amalgam drops into water where it reacts to form sodium hydroxide and hydrogen, the mercury being regenerated. The decomposition of the amalgam is speeded up by the presence of graphite grids. These form a galvanic couple with the amalgam in which the amalgam acts as the anode.

$$2Na/Hg(l) + 2H_2O(l) \rightarrow 2NaOH(aq) + 2Hg(l) + H_2(g)$$

The mercury is pumped up to the top cell and used again. The sodium hydroxide solution is evaporated to dryness and some of the hydrogen burned in chlorine to make hydrogen chloride. The excess chlorine has many uses of its own (19.2).

Manufacture of sodium carbonate

The increasing demand for chlorine has resulted in the production of vast quantities of sodium hydroxide. Some of this sodium hydroxide is converted into sodium carbonate by reacting its aqueous solution with carbon dioxide produced by heating limestone in a limekiln (see page 302).

20.3   Calcium and magnesium

Calcium and magnesium are found in group II of the periodic table and their atoms all have two electrons in their outermost shells. The elements are too reactive to occur free but are widely found in the form of carbonates and sulphates. In addition, many rocks contain calcium phosphate. The atoms of these metals are smaller than those of the corresponding group I elements and so their outermost electrons are more firmly held by the nucleus. This, coupled with the fact that *two* electrons must be removed from each atom to leave an ion with a stable octet, means that the group II metals are less reactive than their group I neighbours. Calcium, being lower in the group than magnesium, has larger atoms and is generally the more reactive of the two (why?).

Both metals are silvery in appearance but soon tarnish in air owing, initially, to the formation of a coating of oxide. They burn in air and chlorine. Calcium reacts fairly vigorously with cold water, producing hydrogen and a suspension of the slightly soluble calcium hydroxide (Experiment 20.1.c).

$$Ca(s) + 2H_2O(l) \rightarrow Ca(OH)_2(aq) + H_2(g)$$

The reaction between magnesium and water is very slow because a layer of almost insoluble magnesium hydroxide is formed on the metal. Magnesium does, however, burn in steam (Experiment 20.1.d) to form hydrogen and magnesium *oxide* (the temperature is too high for the existence of the hydroxide). Both metals dissolve in dilute hydrochloric and sulphuric acids to liberate hydrogen but the reaction between calcium and dilute sulphuric acid is slowed down by the formation of a coating of insoluble calcium sulphate.

Oxides and hydroxides

The oxides are white *refractory* solids (i.e. have very high melting points) with giant ionic lattices. They combine with water to form the corresponding hydroxides, the reaction being violent in the case of calcium oxide (Experiment 20.3.a).

$$CaO(s) + H_2O(l) \rightarrow Ca(OH)_2(s)$$

The oxides and hydroxides of both metals are basic, but considerably less so than those of the corresponding group I metals. The hydroxides decompose to give the oxides and water on being heated strongly.

e.g. $$Ca(OH)_2(s) \rightarrow CaO(s) + H_2O(g)$$

Calcium hydroxide solution (lime water) is used to test for carbon dioxide.

Salts

These all contain colourless Ca^{2+} or Mg^{2+} ions and may be prepared by the standard methods described in 5.4. They are less thermally stable than the corresponding salts of potassium and sodium.

Uses of the metals and their compounds

Metallic *calcium* is used as a reducing agent in the manufacture of uranium. *Calcium oxide* (quicklime) is an important industrial base and is used to make calcium hydroxide, glass and fertilisers. *Calcium hydroxide* (slaked lime) is a cheap industrial alkali and is used to reduce soil acidity, to make glass and to soften water. *Calcium carbonate* (limestone, chalk) is used in the manufacture of calcium oxide, iron, cement and toothpaste, and for decoration in the form of marble. *Calcium sulphate* is used in the manufacture of plaster, plaster of Paris and paper while *anhydrous calcium chloride* is used as a drying agent.

Magnesium is used in light alloys, flashlights, flares, as a sacrificial anode to protect steel from corrosion, and in the manufacture of titanium. *Magnesium oxide* is sometimes used as a furnace lining. A suspension of *magnesium hydroxide* (Milk of Magnesia) reduces stomach acidity and *magnesium sulphate-7-water* (Epsom salt) is a laxative.

Hardness of water

A **'hard' water** is one which does not readily form a lather with soap.

The most common causes of hardness are dissolved calcium and magnesium ions (Experiment 20.3.b) from the corresponding sulphates and hydrogencarbonates. These ions react with soap (sodium octadecanoate) to give a precipitate of insoluble calcium or magnesium octadecanoate (scum). All of the calcium and magnesium ions must be removed from solution as scum before the soap can produce a lather. This, of course, wastes soap.

e.g. $2C_{17}H_{35}COONa(aq) + Ca^{2+}(aq) \rightarrow (C_{17}H_{35}COO)_2Ca(s) + 2Na^+(aq)$
$\qquad$ soap $\qquad\qquad\qquad\qquad\qquad\qquad$ scum

Soapless detergents, made from petroleum, have molecules which are similar in form and action to those of soap but their calcium and magnesium salts are *soluble* in water and so do not form scum (21.10).

HOW WATER BECOMES HARD $\quad$ Magnesium sulphate, which is soluble in water, occurs naturally in some parts of Britain. Gypsum ($CaSO_4 \cdot 2H_2O$) is widely distributed and although it is only slightly soluble in water, sufficient dissolves to affect the action of soap.

Rain water dissolves a little carbon dioxide as it falls through the air and considerably more as it soaks through the soil, where the gas is produced by living organisms. You should know already that water containing carbon dioxide dissolves calcium carbonate to give a solution of calcium hydrogencarbonate (Experiment 16.4.a). Magnesium carbonate is similarly affected.

e.g. $\qquad\qquad CaCO_3(s) + H_2O(l) + CO_2(g) \rightleftharpoons Ca(HCO_3)_2(aq)$

Because of this reaction, chalk, limestone and rocks containing magnesium carbonate will gradually dissolve in rain water, giving rise to potholes and underground caverns and rendering the water hard.

The slow evaporation of drops of dilute calcium hydrogencarbonate solution hanging from cavern roofs causes the above process to reverse and leaves minute deposits of calcium carbonate behind. As a result of this, *stalactites* form, growing down towards the floor. Where drops of the solution fall to the floor and evaporate, *stalagmites* grow upwards towards the roof.

Stalactites and stalagmites in Campanet Caves, Spain

THE REMOVAL OF HARDNESS FROM WATER There are two main
methods of removing hardness from water (i.e. 'softening' the water). The
calcium and magnesium ions which are present in solution may either be
converted to their carbonates (which are insoluble in water and therefore
cannot react with soap) or else they can be completely removed.

1   *Precipitation of calcium and magnesium carbonates*

(a) Boiling: this converts calcium and magnesium hydrogencarbonates to
the corresponding carbonates, steam and carbon dioxide (16.4) and
reverses the process by which the ions entered the water.

$$Ca(HCO_3)_2(aq) \xrightarrow{\text{heat}} CaCO_3(s) + H_2O(g) + CO_2(g)$$

Any calcium or magnesium sulphate present in solution is unaffected by boiling.

> Hardness which can be removed simply by boiling is referred to as **temporary hardness**. Hardness which cannot be removed by boiling is called **permanent hardness**.

(b) Clark's process: a *calculated* quantity of calcium hydroxide is added to reservoirs to soften the water. This method removes *temporary* hardness only.

$$Ca(HCO_3)_2(aq) + Ca(OH)_2(aq) \rightarrow 2CaCO_3(s) + 2H_2O(l)$$

You should realise that if excess calcium hydroxide were used it would produce hardness of its own, because it contains calcium ions.

(c) The addition of sodium carbonate: calcium and magnesium carbonates are precipitated when sodium carbonate is dissolved in hard water. Both temporary and permanent hardness are removed in this way.

e.g. $$Ca(HCO_3)_2(aq) + Na_2CO_3(aq) \rightarrow CaCO_3(s) + 2NaHCO_3(aq)$$

$$MgSO_4(aq) + Na_2CO_3(aq) \rightarrow MgCO_3(s) + Na_2SO_4(aq)$$

Sodium carbonate-10-water is used as *washing soda* and is the main ingredient in bath salts.

2 Removal of calcium and magnesium ions from water

(a) Distillation: this process removes all dissolved solids but is, of course, expensive. It is used in some countries to provide drinking water from sea water (4.1).

(b) Ion exchange: some substances remove dissolved ions from hard water and replace them with ions of their own, the process being known as *ion exchange*. One such substance is 'Permutit' ($Al_2Na_2SiO_4 \cdot xH_2O$): this exchanges its sodium ions for any calcium and magnesium ions in the water and thus removes both temporary and permanent hardness. If Permutit is given the formula Na_2X, for simplicity, the reactions may be represented by equations such as

$$CaSO_4(aq) + Na_2X(s) \rightarrow CaX(s) + Na_2SO_4(aq).$$

When all of the sodium ions in the Permutit have been exchanged the original compound can be regenerated by running concentrated sodium chloride solution through it.

$$CaX(s) + 2NaCl(aq) \rightarrow Na_2X(s) + CaCl_2(aq)$$

The calcium chloride solution which is formed at the same time is washed away.

There are some man-made ion exchange resins which replace dissolved cations with $H^+(aq)$ ions and others which replace dissolved anions with $OH^-(aq)$ ions. If tap water is passed through *both* types of resin most of these $H^+(aq)$ and $OH^-(aq)$ ions combine to form water molecules and the resulting product is very pure 'de-ionised' water.

DISADVANTAGES OF HARD WATER Hard water (a) wastes soap and precipitates scum, (b) deposits 'fur' (calcium carbonate) in kettles, pipes and boilers, thereby causing wastage of fuel, and (c) interferes with some manufacturing processes, e.g. in the wool industry.

ADVANTAGES OF HARD WATER Hard water (a) does not dissolve lead from the water pipes in older houses (lead is a dangerous cumulative poison), (b) provides calcium, which is essential for teeth and bones, (c) seems to cut down the incidence of heart disease, and (d) is better for some manufacturing processes, such as brewing.

20.4 Aluminium

Aluminium is found in group III of the periodic table and therefore its atoms have three electrons in their outermost shells. It is the most abundant metal in the earth's crust and occurs as its oxide in bauxite, emery, ruby, emerald and sapphire, as a complex fluoride in cryolite (Na_3AlF_6), and in the form of complex silicates in many rocks and clays.

Aluminium is a silvery metal which, in spite of its high position in the reactivity series, is stable in air. This is because it becomes coated with a thin layer of oxide which protects it from further attack. If this layer is removed by amalgamation with mercury the metal reacts rapidly with cold air and becomes coated with feathery growths of oxide (Experiment 20.4.a). Aluminium burns in air and in chlorine and gives its oxide and hydrogen when strongly heated in steam. It readily dissolves in dilute hydrochloric acid, liberating hydrogen, but dilute sulphuric acid has little effect on it and dilute nitric acid renders it passive. It dissolves in sodium hydroxide solution, producing hydrogen and a solution of sodium aluminate.

$$2Al(s) + 2NaOH(aq) + 6H_2O(l) \rightarrow 2Na[Al(OH)_4](aq) + 3H_2(g)$$

Oxide and hydroxide

Aluminium oxide is a white refractory solid with a giant ionic lattice. Unlike the oxides of the more reactive metals, it does not combine with water to give the hydroxide. Both the oxide and hydroxide are amphoteric: they will dissolve in acids to give aluminium salts and in strong alkalis to give aluminates.

e.g.
$$Al(OH)_3(s) + 3HCl(aq) \rightarrow AlCl_3(aq) + 3H_2O(l)$$
base

$$Al(OH)_3(s) + NaOH(aq) \rightarrow Na[Al(OH)_4](aq)$$
acid

Aluminium hydroxide decomposes on heating to give the oxide and water.

Salts

The common hydrated salts contain colourless Al^{3+} ions. They are soluble in water and may be prepared from the hydroxide by the insoluble base method. (N.B. Aluminium carbonate does not exist.) Because of the difficulty in removing all three electrons from the outermost shell of an aluminium atom, the anhydrous salts show some covalent character. For example, anhydrous aluminium chloride sublimes when heated and is hydrolysed by water (5.6). It cannot be made by heating crystals of the hydrated salt but must be prepared from the elements by direct combination, using the method of Experiment 19.2.b.

Uses of aluminium and its compounds

The *metal* is used to make a wide range of food containers for which it can be rendered more attractive by anodising and dyeing (page 209); it is also used in overhead electric cables and in light alloys. The *oxide* is an abrasive and the *hydroxide* is used as a mordant to make dyes adhere to cloth. *Aluminium sulphate* is used in paper-making, sewage treatment and water purification.

Manufacture of aluminium

As yet there is no way of extracting aluminium from clay. The usual method of obtaining the metal is by the electrolysis of a solution of aluminium oxide in molten cryolite, Na_3AlF_6. The oxide has too high a melting point to be electrolysed in the molten state on its own.

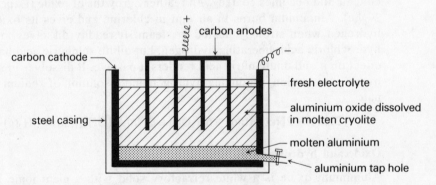

Figure 20.4 The manufacture of aluminium

Pure aluminium oxide is obtained from bauxite (impure $Al_2O_3 \cdot 2H_2O$) and is dissolved in cryolite at 1173 K (900 °C) in the cell shown in Figure 20.4. The high current used (up to 125 000 amp) keeps the electrolyte molten. Aluminium ions are discharged at the graphite cathode which acts as a lining for the cell, and oxide ions are discharged at the carbon anodes. The oxygen so produced gradually oxidises the anodes to carbon dioxide. Fresh aluminium oxide is added from time to time and the molten aluminium which collects at the bottom of the cell is tapped off. Can you think why the Na^+ and F^- ions from the cryolite are not discharged?

Changes	At anode	At cathode

$$6O^{2-}(l) \rightarrow 6O(g) + 12e^- \quad\quad 12e^- + 4Al^{3+}(l) \rightarrow 4Al(l)$$
$$\downarrow$$
$$3O_2(g)$$

(oxidation)   (reduction)

20.5   Zinc

Zinc is in the last group of the centre block of the periodic table but it has none of the special properties associated with this block (see page 404). It occurs naturally as zinc blende (zinc sulphide) and calamine (zinc carbonate).

It is a lustrous bluish-white metal which is fairly stable in air owing to the formation of a protective oxide coating. However, this oxide coating is less of a protection than that on the surface of aluminium. The metal burns in air and in chlorine and gives its oxide and hydrogen when strongly heated in steam. When pure it dissolves only slowly in dilute hydrochloric and sulphuric acids but the reactions may be speeded up by adding copper(II) sulphate solution to set up galvanic couples on the metal surface. Like aluminium it dissolves in hot aqueous solutions of potassium and sodium hydroxides, producing hydrogen and a solution of potassium or sodium zincate.

e.g. $Zn(s) + 2NaOH(aq) + 2H_2O(l) \rightarrow Na_2[Zn(OH)_4](aq) + H_2(g)$

Oxide and hydroxide

Zinc oxide is a white ionic solid which turns yellow on heating but reverts to white on cooling. Both the oxide and the hydroxide are amphoteric, like those of aluminium, and will dissolve in acids to give zinc salts and in strong alkalis to give zincates (see page 347). Zinc hydroxide decomposes on heating to form the oxide and water. Unlike aluminium hydroxide it dissolves in aqueous ammonia solution by forming a complex ion.

Salts

These contain the colourless Zn^{2+} ion. The common ones, apart from the carbonate and sulphide, are soluble in water and all may be prepared by standard methods. Like the corresponding aluminium compound, anhydrous zinc chloride must be made by direct combination of the elements to avoid hydrolysis.

Uses of zinc and its compounds

Zinc is used for galvanising iron (page 215), for roofing, to make the casings of dry batteries and to make alloys (e.g. brass). *Zinc oxide* is used in ointments and *zinc carbonate* in calamine lotion for the treatment of sunburn.

20.6 Iron

Iron is one of the transition metals which make up the centre block of the periodic table. It is the second most abundant metal in the earth's crust and occurs naturally in iron pyrites (fool's gold, FeS_2), haematite (Fe_2O_3), magnetite (Fe_3O_4), and siderite ($FeCO_3$). It is essential for the production of haemoglobin in the blood and chlorophyll in plants.

Iron is a white, lustrous metal which rusts in moist air; the rust does not protect it from further corrosion. The metal burns in oxygen and chlorine and gives hydrogen and iron(II) diiron(III) oxide (magnetic iron oxide) when heated in steam (Experiment 20.1.d). If this oxide is heated in hydrogen it is reduced back to iron (Experiment 20.1.e). Thus the reaction is reversible and the metal and hydrogen are approximately equally reactive towards oxygen.

$$3Fe(s) + 4H_2O(g) \rightleftharpoons Fe_3O_4(s) + 4H_2(g)$$

The metal dissolves fairly slowly in dilute hydrochloric and sulphuric acids, giving hydrogen and the corresponding iron(II) salt; concentrated nitric acid renders it passive.

Special properties of the transition metals

The transition metals differ from most of the main group metals of the periodic table in several ways, the most important being that they have more than one possible oxidation number, form coloured ions in solution, act as catalysts and form many complex ions. A detailed treatment of these properties is not possible at this level but you should realise that they are all due to the fact that in transition metal atoms the electron shell next to the outermost one is not full. Electrons can be removed from or added to this shell, a process which is not possible in atoms of the main group elements. In zinc atoms the shell *is* full and this explains why zinc does not share the special properties of the transition metals.

Oxides and hydroxides

There are three oxides of iron. Iron(II) oxide, FeO, is a black, unstable solid which inflames spontaneously in air to give the more stable reddish-brown iron(III) oxide, Fe_2O_3. Both of these compounds are basic oxides. The third oxide, iron(II) diiron(III) oxide (magnetic iron oxide), Fe_3O_4, is formed when iron is heated in air or steam and behaves as if it were a compound of the other two, i.e. ($FeO + Fe_2O_3$)(s). One of its naturally occurring forms, lodestone, was used as a compass by the early mariners.

Iron(II) hydroxide is dirty green in colour and iron(III) hydroxide is reddish-brown. They are formed as gelatinous precipitates when sodium hydroxide solution is added to solutions containing iron(II) or iron(III) ions. Both are basic and decompose on heating.

Salts

Iron(II) salts are generally less stable than the corresponding iron(III) salts. Solutions of the former contain pale green $Fe^{2+}(aq)$ ions while those of the latter contain yellow $Fe^{3+}(aq)$ ions. The conversion of iron(II) compounds to iron(III) compounds involves removal of electrons and therefore is an example of oxidation, the reverse process being reduction.

$$Fe^{2+} \underset{\text{reduction}}{\overset{\text{oxidation}}{\rightleftharpoons}} Fe^{3+} + e^-$$

Both anhydrous iron(II) chloride and anhydrous iron(III) chloride are hydrolysed by water and thus cannot be prepared by heating crystals of the hydrated salts. Direct combination of iron and chlorine gives iron(III) chloride (Experiment 19.2.b): to make anhydrous iron(II) chloride the iron must be heated in a stream of hydrogen chloride.

Uses of iron

The uses of iron are numerous and well-known. *Cast iron* is cheap but brittle and so it can only be used where little stress will be experienced (e.g. in domestic boilers and bunsen burner bases). *Wrought iron* is almost pure iron. It is softer and more flexible than cast iron and is used to make such things as ornamental gates (much 'wrought iron' nowadays is actually mild steel). *Steels* are alloys of iron with carbon and other elements, such as chromium and manganese. The distinctive properties of the different types of steel depend upon the nature of these elements and the proportions in which they are present, e.g. stainless steel contains about 15% of chromium, together with some nickel.

The manufacture of iron

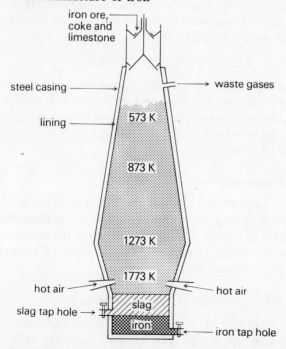

Figure 20.6.1 The blast furnace

Iron is low enough in the reactivity series for its oxide to be reduced chemically. The reaction is carried out in a blast furnace (Figure 20.6.1) and takes place in several stages.

At the top of the furnace a mixture of iron ore (e.g. haematite, Fe_2O_3), coke and limestone is fed in by means of a double cup and cone device to prevent the escape of gas. A blast of hot air at the base of the furnace causes the coke to burn at white heat to form carbon dioxide.

$$C(s) + O_2(g) \rightarrow CO_2(g)$$

As it rises through the furnace the carbon dioxide is reduced by more hot coke to carbon monoxide.

$$CO_2(g) + C(s) \rightarrow 2CO(g)$$

In the upper part of the furnace, at 1023 K (750 °C), the carbon monoxide reduces the iron(III) oxide to give iron in the form of a porous solid.

$$Fe_2O_3(s) + 3CO(g) \rightleftharpoons 2Fe(s) + 3CO_2(g)$$

As the iron reaches the lower part of the furnace it melts and flows down to the bottom. The gas which reaches the top of the furnace contains about one-third carbon monoxide by volume and is piped off and burned to heat the air for the hot-air blast.

The purpose of the limestone is to remove the infusible sandy impurities (mainly silicon(IV) oxide). Firstly the limestone decomposes at a temperature of 1123 K (850 °C) to form calcium oxide.

$$CaCO_3(s) \rightarrow CaO(s) + CO_2(g)$$

The calcium oxide (basic) then reacts with the silicon(IV) oxide (acidic) to form a *fusible* slag (calcium silicate) which flows to the bottom of the furnace and floats on the molten iron.

$$CaO(s) + SiO_2(s) \rightarrow CaSiO_3(l)$$

The iron and slag are tapped off periodically and the furnace continually recharged. The resultant iron (cast iron) is brittle because it contains up to 4% carbon, together with elements such as sulphur and phosphorus. A typical blast furnace produces over 1 000 000 kg of iron in 24 hours and uses up 6 000 000 kg of air.

The conversion of iron into steel

There are several ways of carrying out this conversion but all of them involve the removal of impurities from cast iron by oxidation and the subsequent addition of carbon and other elements in the correct proportions to give the required type of steel. Two methods of removing the impurities will be considered here.

THE SIEMENS-MARTIN OPEN HEARTH PROCESS

In this process, a mixture of cast iron, steel scrap and iron(III) oxide (which provides some of the oxygen needed) is melted in a shallow reverberatory furnace, the roof of which is specially shaped to reflect the

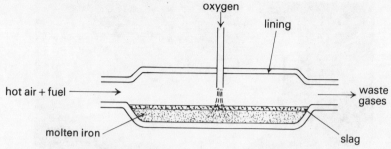

Figure 20.6.2 The Siemens–Martin open hearth process

heat supplied by burning gas or oil down on to the charge. Non-metallic impurities form oxides which escape as gases or combine with the basic lining of the furnace to form slag. Oxygen is injected by means of a water-cooled lance to speed up the process. Over half of the world's steel is produced by this method.

THE LINZ-DONAWITZ (LD) PROCESS

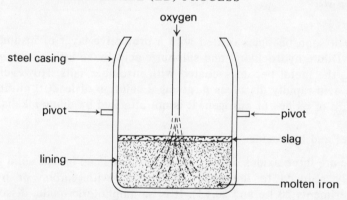

Figure 20.6.3 The Linz–Donawitz process

The LD converter looks something like a giant concrete mixer. It is tilted to receive a charge of scrap steel and molten cast iron and is then returned to the vertical. Oxygen is blown on to the surface of the charge by means of a water-cooled oxygen lance to oxidise the impurities in the same way as in the open hearth process. Vast quantities of oxygen (referred to as 'tonnage oxygen') are used. This method is rapid, but it cannot recycle as much scrap steel as the open hearth process.

20.7   Lead

Lead is the last of the group IV elements and its atoms have four electrons in their outermost shells. Its most important naturally occurring compound is galena, PbS.

The metal is soft and silvery but its surface rapidly tarnishes in air and takes on a grey appearance. It combines only slowly with oxygen and chlorine, even when heated, and is hardly attacked by pure water. The presence of dissolved oxygen causes it to dissolve slowly but in hard

Making steel

water it soon becomes coated with a protective layer of insoluble lead salts. Dilute hydrochloric and sulphuric acids have little action because, again, the metal becomes coated with insoluble salts. However, dilute nitric acid rapidly dissolves it, giving a solution of lead(II) nitrate and a mixture of oxides of nitrogen. It is not attacked by dilute alkalis.

Oxides and hydroxide

There are three oxides of lead. Lead(II) oxide is an orange solid which is readily reduced to the metal by heating with carbon or hydrogen (Experiments 20.1.e and 20.1.f). It is an amphoteric oxide, dissolving in those dilute acids which have soluble lead(II) salts and also in solutions of strong alkalis to give plumbates(II).

e.g.
$$PbO(s) + 2HNO_3(aq) \rightarrow Pb(NO_3)_2(aq) + H_2O(l)$$
$$PbO(s) + 2NaOH(aq) + H_2O(l) \rightarrow Na_2[Pb(OH)_4](aq)$$

Lead(II) oxide forms 'red lead oxide' when heated in air to about 723 K (450 °C) but at higher temperatures – above 773 K (500 °C) – the reaction reverses.

$$6PbO(s) + O_2(g) \rightleftharpoons 2Pb_3O_4(s)$$

Lead(IV) oxide is a brown powder which is a powerful oxidising agent. For example, it oxidises concentrated hydrochloric acid to chlorine while itself being reduced to lead(II) chloride.

$$PbO_2(s) + 4HCl(aq) \rightarrow PbCl_2(s) + 2H_2O(l) + Cl_2(g)$$

On heating, lead(IV) oxide gives off oxygen and leaves a residue of lead(II) oxide.

$$2PbO_2(s) \rightarrow 2PbO(s) + O_2(g)$$

The third oxide of lead, dilead(II) lead(IV) oxide (red lead oxide), Pb_3O_4, is a red solid which behaves as if it were a mixture of lead(II) and lead(IV) oxides, i.e. $(2PbO + PbO_2)(s)$. For example, on heating it gives off oxygen and leaves a residue of lead(II) oxide,

$$2Pb_3O_4(s) \xrightarrow{\text{heat}} 6PbO(s) + O_2(g),$$

and when heated with dilute nitric acid it partly dissolves, forming a solution of lead(II) nitrate and leaving a residue of lead(IV) oxide.

$$Pb_3O_4(s) + 4HNO_3(aq) \rightarrow 2Pb(NO_3)_2(aq) + PbO_2(s) + 2H_2O(l)$$

Lead(II) hydroxide is a white amphoteric solid which, like the corresponding oxide, dissolves in those dilute acids which have soluble lead(II) salts and also in aqueous solutions of strong alkalis to form plumbates(II).

Salts

Apart from the nitrate and ethanoate, the common lead(II) salts are insoluble in water and may be prepared by double decomposition. Lead(II) chloride dissolves in hot water but is deposited as crystals when the solution cools again.

Uses of lead and its compounds

Lead has been known since ancient times because, being low in the reactivity series, it is readily extracted from its ores. It is used in roofing (owing to its resistance to corrosion), in car batteries, for shielding against X-rays and radioactive materials, and in alloys such as solder, typemetal and pewter. *Lead(II) oxide* is used in making glass, and *'red lead oxide'* in making paints for exterior surfaces. *Lead(II) nitrate* is used in the laboratory preparation of nitrogen dioxide because it contains no water of crystallisation (why is this an advantage?). Tetraethyl-lead(IV) is added to petrol as an 'anti-knock' compound (see page 251).

20.8 Copper

Copper is found in the centre block of the periodic table and like iron it shows the special properties associated with transition metals (see 20.6). It occurs naturally as the free metal in a few areas and also as copper pyrites ($CuFeS_2$), copper glance (Cu_2S), and malachite ($CuCO_3 \cdot Cu(OH)_2$).

It is a fairly soft, reddish-brown metal but its surface slowly tarnishes on exposure to the atmosphere, owing to the formation of a coating of green basic sulphate (or basic chloride near the coast). It combines with oxygen and chlorine when heated but does not react with water or steam. Copper is below hydrogen in the electrochemical series and therefore cannot displace the gas from dilute acids. It does, however, dissolve in dilute nitric acid, which acts as an oxidising agent and is itself reduced to nitrogen oxide (17.2).

$$3Cu(s) + 8HNO_3(aq) \rightarrow 3Cu(NO_3)_2(aq) + 4H_2O(l) + 2NO(g)$$

Concentrated nitric acid is reduced by copper to nitrogen dioxide.

$$Cu(s) + 4HNO_3(aq) \rightarrow Cu(NO_3)_2(aq) + 2H_2O(l) + 2NO_2(g)$$

Oxides and hydroxide

Copper(I) oxide is a red solid, formed when an alkaline solution of copper(II) ions is reduced by glucose in the presence of a stabilising agent such as the tartrate ions found in Fehling's solution (Experiment 20.8.a).

Copper(II) oxide is formed as a black solid when many copper(II) compounds are heated. Both copper(II) oxide and copper(II) hydroxide (which is a pale blue solid) are basic and dissolve in dilute acids to give copper(II) salts. Copper(II) hydroxide is not amphoteric but it does dissolve in ammonia solution to give a royal blue solution owing to the formation of complex tetraamminecopper(II) ions $[Cu(NH_3)_4]^{2+}$. The hydroxide readily decomposes on heating to give copper(II) oxide and water.

Salts

Most copper(I) salts are unstable in the presence of water. However, copper(I) chloride can be made from copper(II) chloride by heating it with copper in the presence of concentrated hydrochloric acid. When the solution is diluted, the copper(I) chloride is obtained as a white precipitate (Experiment 20.8.a).

$$CuCl_2(aq) + Cu(s) \rightarrow 2CuCl(s)$$

The conversion of copper(II) compounds to copper(I) compounds involves the addition of electrons and therefore is an example of reduction, the reverse change being oxidation. In the above reaction the reducing agent is metallic copper.

The simple copper(II) salts are soluble in water and may be prepared by standard methods. They readily decompose to give copper(II) oxide on heating.

Uses of copper and its compounds

Because it occurs free in nature and its ores are readily reduced, *copper* was one of the first metals discovered by man. It is used in roofing, as an electrical conductor, and to make water pipes and boilers. It can be alloyed with zinc to make brass, with tin to make bronze and with nickel to make British 'silver' coins; nine carat gold is almost two-thirds copper. *Copper(I) oxide* is used to colour glass red and *copper(II) sulphate* is present in many fungicides and timber preservatives.

20.9 Differences between metals and non-metals

You will appreciate by now that metals and non-metals have certain characteristic properties. These are summarised in Table 20.9 but you should refer to the relevant chapters for more details.

	Properties	Metals	Non-metals
PHYSICAL	State at room temperature and pressure	Solid (except mercury)	Gas or volatile liquid; solid if a giant structure
	m.p.	High	Low (unless a giant structure)
	Density	High	Low
	Flexibility	Flexible	Brittle
	Thermal conductivity	Good	Poor
	Electrical conductivity	Good	Poor (except graphite)
CHEMICAL	Ions formed	Positive	Negative
	Reaction with acids	The more electro-positive metals give salt and hydrogen, unless the acid is a powerful oxidising agent	Give non-metal oxide with hot, concentrated oxidis-ing acids
	Electron attracting power	Low	High
	Oxides	Ionic solids, basic; if they react with water, they form alkalis	Covalent, acidic, often react with water to give acids; mostly volatile
	Hydroxides	Bases (alkalis if soluble in water)	Acids
	Chlorides	Generally ionic, usually stable to water	Covalent, react with water
	Hydrides	Those of the more electropositive metals are ionic solids (contain H^- ion)	Covalent liquids or gases

Table 20.9 The characteristic properties of metals and non-metals

Questions on Chapter 20

1 The relative positions of the elements rubidium (Rb), beryllium (Be) and bismuth (Bi) in the electrochemical or metal activity series are given below, with the positions of some more familiar elements. The formulae of three chlorides are also given.

Position in series	Chloride
rubidium	RbCl
sodium	
magnesium	
beryllium	$BeCl_2$
iron	
lead	
hydrogen	
bismuth	$BiCl_3$
copper	

Make use of this information to answer the following questions about rubidium, beryllium and bismuth. You may assume that these do not show variable oxidation number.

(a) Write formulae for the hydroxides for these three elements. Which one of the hydroxides will be soluble in water?

(b) Give the formulae of the solid compounds you would expect to remain when samples of the three hydroxides in (a) are heated.

(c) How, and under what conditions, would you expect rubidium and beryllium to react with water? Write equations for these reactions. State, with reasons, whether any reaction between bismuth and water is to be expected.

(d) What would you expect to happen when beryllium and bismuth are separately treated with dilute hydrochloric acid?

(e) Which one of the three elements could be readily obtained by heating its oxide in a stream of hydrogen?   (C)

2 Describe the reactions (if any) of the metals calcium, copper, iron, with (i) air, (ii) water, (iii) dilute hydrochloric acid. If there is a reaction, state the necessary conditions and give the *names* and *formulae* of the products.

Using these reactions, and giving your reasons, arrange the three metals in order of *decreasing* chemical activity.   (O & C)

3 State and explain what is *observed* when aqueous solutions of each of the following pairs of substances are mixed.

(a) Iron(III) chloride and sodium hydroxide.

(b) Calcium chloride and sodium carbonate.

(c) Lead(II) nitrate and potassium iodide.

Describe how you would employ the reactions (a) and (c), where appropriate, as essential steps when attempting the following preparations. All necessary practical details should be given.

   (i) The preparation of iron(III) oxide from aqueous iron(III) chloride.

   (ii) The preparation of lead(II) iodide from solid lead(II) carbonate.   (W)

4 Explain the following reactions, giving equations where possible.

(a) When sodium hydroxide solution is added to zinc sulphate solution there is a white precipitate. With excess of sodium hydroxide solution this precipitate redissolves.

(b) When sodium hydroxide solution is added to ammonium sulphate and the mixture warmed, a gas is evolved which turns red litmus blue. If this gas is passed into aqueous copper(II) sulphate, a deep blue colour is produced.

(c) When a gas jar containing equal volumes of carbon monoxide and carbon dioxide is opened under a solution of sodium hydroxide, the liquid eventually rises about half-way up the jar. (O & C)

5 In many places calcium hydrogencarbonate is present in tap water. Explain why this happens.

Describe what happens when such water is used (a) in a boiler, and (b) with soap.

Explain how hard water can be softened by the process of ion exchange. (Reference to *one* process is sufficient.) (O & C)

6 Give the name and formula of a common ore of aluminium.

Describe the chemistry of the extraction of aluminium from the ore.

Describe how, and under what conditions aluminium reacts with (a) sodium hydroxide solution, (b) air, (c) iron(III) oxide. (SUJB)

7 (a) Give an account of the chemical reactions taking place in the blast furnace during the manufacture of iron from haematite (impure iron(III) oxide, Fe_2O_3).

(b) Iron

|
| dilute hydrochloric acid
↓

Compound A (pale green in solution)

|
| chlorine
↓

Compound B (yellow in solution)

|
| sodium hydroxide solution
↓

Compound C (brown)

|
| heat
↓

Compound D (brown)

(i) Give the names and formulae of A, B, C and D, which are all compounds of iron, and write equations for any *two* of the reactions.

(ii) Calculate the maximum mass of compound D that could be obtained from 5.60 g of iron.

(iii) If a solution of compound B is shaken with powdered iron, a solution of compound A is obtained. Write an *ionic* equation for this reaction. (C)

8 Starting from crystals of lead(II) nitrate, which are soluble in water, describe how you would prepare reasonably pure samples of the following substances: (a) lead(II) chloride, which is almost insoluble in water, (b) lead(II) oxide, (c) nitrogen dioxide, (d) lead(II) carbonate, which is insoluble in water. (O & C)

9 A blue powder is said to be a compound of formula $Cu(OH)_2 \cdot 2CuCO_3$. State, with equations, how you would expect this compound to react (a) when heated, (b) when treated with dilute nitric acid.

Describe fully how you would determine the percentage by mass of copper in the powder. If the formula given is correct, what would this percentage be? (C)

10 For *each* of the elements magnesium and sulphur, give *two* physical properties and *three* chemical properties that illustrate the general differences between metals and non-metals. (C)

21 Organic chemistry

Originally the term 'organic chemistry' referred to the chemistry of substances found in living organisms and it was thought that the cells of these organisms possessed some 'vital force' without which such substances could not be made. However, as scientific techniques improved, many of these substances, together with similar compounds not associated with the living world, were synthesised in the laboratory. All organic compounds have one element in common and nowadays we regard organic chemistry to be the chemistry of the compounds of this element. What is the element? Perform Experiment 21.1.a and you will find out.

Caution: many organic compounds are inflammable. When you carry out the experiments described in this chapter take care to avoid the risk of fire and *be ready* to deal with any accidents.

Experiments

21.1.a The action of heat on organic compounds

PROCEDURE   (a) Heat a spatula measure of sugar in an ignition tube until no further change takes place. Note the appearance of the residue and then repeat the experiment using other materials of plant or animal origin, e.g. bread, potato, rice, nuts, chopped leaves.
(b) Scrape the residue from each of the above experiments into a test tube, add an equal volume of dry copper(II) oxide and mix the solids well by means of a glass rod. Heat the mixture as described in Experiment 16.2.a.

Questions

1 What appeared to be formed in each experiment in (a)?
2 Was the presence of this substance confirmed by (b)? Explain your reasoning.
3 Which element did all the organic compounds which you tested appear to have in common?

21.1.b To study some physical properties of the alkanes

The alkanes form a series of compounds which can be represented by the general formula C_nH_{2n+2}, where n is a whole number. Thus the first member has the formula CH_4, the second member C_2H_6, and so on.

PROCEDURE   Using the information given in Table 21.1.b, plot graphs of boiling point against n and density against n. In both graphs n should be plotted along the horizontal axis.

n	b.p./K (°C)	Density in the liquid state/g cm^{-3}
1	112 (−161)	0.42
2	184(−89)	0.57
3	229(−44)	0.59
4	272(−1)	0.60
5	309(36)	0.63
6	341(68)	0.66
7	372(99)	0.68
8	399(126)	0.70
9	424(151)	0.72
10	447(174)	0.73

Table 21.1.b

Questions

1   As n increases, do the boiling points increase in a regular manner or do they follow no pattern at all?
2   In what way do the densities of the alkanes vary with n?
3   Can you suggest an explanation for these results?

21.2.a   Some chemical properties of the alkanes ★

PROCEDURE   (a) Use a burning splint to ignite 2–3 drops of hexane, C_6H_{14}, in a watch glass. Hold a beaker of cold water in the flame.
(b) Collect a test tube full of North Sea Gas (which is mainly methane, CH_4) over water. Add 2 cm³ of lime water to the tube and apply a flame to the gas. When combustion ceases cork the tube and shake it.
(c) Add 5 drops of a solution of bromine in tetrachloromethane to two small tubes, each half-filled with hexane (the tetrachloromethane is an inert solvent and takes no part in the reaction). Place one of these tubes in a cupboard and position the other one 20 cm away from a 100 watt lamp. Shake the tubes at intervals.
(d) Add a few drops of potassium manganate(VII) solution, previously acidified with an equal volume of dilute sulphuric acid, to a small tube half-filled with hexane and observe the result.
(e) Repeat part (d) but use sodium carbonate solution in place of the dilute sulphuric acid.

Questions

1   What colour was the flame in (a)? Name two products of the combustion of hexane and explain how you came to your conclusion.
2   Do you think that the combustion of methane was more complete or less complete than that of hexane? What evidence was there for your answer? Can

you think *why* combustion should be more complete for one alkane than for the other?

3 Write an equation for the *complete* combustion of methane.

4 What did you observe in (c)? Does the reaction between bromine and an alkane appear to be photosensitive?

5 What would you have expected to happen in (d) and (e) if the hexane had been oxidised? Was the hexane easily oxidised or resistant to oxidation?

21.3.a Some chemical properties of ethene★

PROCEDURE Collect four test tubes and one gas jar full of ethene as described on page 429. Stopper the test tubes, close the gas jar with a cover slip and then proceed as follows.

(a) Repeat part (b) of the previous experiment using a test tube full of ethene in place of the methane.

(b) Add 1 cm³ of a solution of bromine in tetrachloromethane to a test tube full of ethene and shake the tube.

(c) Shake a test tube full of ethene with 1 cm³ of potassium manganate(VII) solution, previously acidified with an equal volume of dilute sulphuric acid.

(d) Repeat part (c) but use sodium carbonate solution in place of the dilute sulphuric acid.

(e) Invert a gas jar full of bromine vapour over the jar of ethene, remove the cover slips and observe what happens.

Questions

1 What did you deduce from your observations in (a)? Write an equation for the complete combustion of ethene.

2 Compare the results of (b), (c) and (d) with those obtained for the corresponding reactions of alkanes. Which are the more reactive, alkanes or alkenes?

3 What did you observe in (e)? Describe the appearance of the product. Why did you not see this same product in (b)?

21.4.a Some chemical properties of ethyne

Part (e) of this experiment must not be performed by pupils.

PROCEDURE Add 2–3 lumps of calcium dicarbide to 5 cm³ of distilled water in a test tube, quickly fit a bung and delivery tube and collect four test tubes full of the resulting gas, ethyne, over water in the usual way. Treat the ethyne in the same way as the ethene in parts (a)–(d) of the previous experiment.

(e)★★ Pour dilute hydrochloric acid into a gas jar to a depth of about 2 cm and add 2–3 spatula measures of bleaching powder to produce chlorine. When the jar is full of the gas, add a lump of calcium dicarbide at arm's length.

Questions

1 What did you observe in parts (a)–(d)?

2 Do you think that ethyne is a saturated compound or an unsaturated compound? Give reasons for your answer.

3 Ethyne has the formula C_2H_2. Suggest a structural formula for it.

4 Describe what you saw in part (e). What were the products of the reaction?

21.6.a The fractional distillation of crude oil★

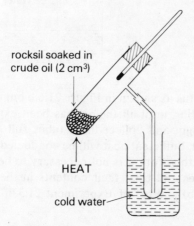

rocksil soaked in crude oil (2 cm³)

HEAT

cold water

Figure 21.6.a

PROCEDURE Set up the apparatus shown in Figure 21.6.a; the filter tube should be of size 125×16 mm and the collecting tubes of size 75×12 mm. Heat the crude oil gently and collect the distillate in four fractions according to boiling point: (a) up to 70°C, (b) 70–120°C, (c) 120–170°C, (d) 170–220°C. Pour the fractions into separate crucibles and note their colour and viscosity (thickness). Test each fraction for inflammability; if any fraction does not ignite, add a *small* tuft of cotton wool to act as a wick and test again. Record your results as shown in Table 21.6.a.

b.p./°C	Colour	Viscosity	Inflammability	Type of flame
Up to 70 70–120 120–170 170–220				

Table 21.6.a

Questions

1 Does the table show any trends in colour, viscosity, inflammability or type of flame as the boiling point increases?

2 Describe the appearance of the residue in the tube. Can you suggest a large scale use for this substance?

21.6.b The effect of heat on alkanes★

Alkanes burn if they are heated in air but what happens if air is absent?

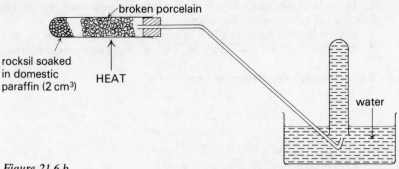

Figure 21.6.b

PROCEDURE   Set up the apparatus as shown in Figure 21.6.b but do not place the collecting tube in position until all the air has been expelled. Heat the broken porcelain strongly and collect three tubes full of gas (**caution:** beware of sucking back). Sufficient heat will be conducted along the tube to vaporise the paraffin, therefore it is not necessary to heat the rocksil directly. Stopper the tubes and treat their contents in the same way as the ethene in parts (a), (b) and (d) of Experiment 21.3.a.

Questions

1   Did the paraffin simply distil over, or was it decomposed in the high temperature region of the tube? Give a reason for your answer.
2   What can you say about the nature of the gas produced? Explain how you reached your conclusion.
3   How might this reaction be useful on a large scale?

21.7.a   Tests for sugars and starch

This experiment involves the use of Fehling's solution, a royal blue liquid containing copper(II) ions which can be reduced to give a red precipitate of copper(I) oxide. The precipitate is sometimes yellow at first but turns red on standing.

PROCEDURE   Carry out tests (a) and (b) with 1% solutions of the following substances: glucose, fructose, maltose, starch, and sucrose.
(a) Boil 5 cm³ portions of each solution with 5 cm³ of Fehling's solution.
(b) Add 1–2 drops of a dilute solution of iodine in potassium iodide to a 5 cm³ portion of each solution.
    Record your results as shown in Table 21.7.a.

Substance	Positive result with Fehling's solution	Positive result with iodine solution
	√ or ×	√ or ×

Table 21.7.a

Questions

1 Which substances would you class as reducing agents?
2 What did you observe in (b)? Could this reaction be used to test for a specific substance?

21.7.b Breaking down molecules of sucrose and starch

PROCEDURE   (a) Dissolve a few crystals of sucrose in 10 cm³ of distilled water in a boiling tube. Add 4 cm³ of dilute sulphuric acid and then place the tube in a beaker of boiling water for 5 minutes. After cooling, neutralise the acid with anhydrous sodium carbonate (test with universal indicator paper) and boil a 5 cm³ portion with Fehling's solution. Finally, boil the remaining solution down to about one-quarter of its original volume and keep the residual liquid for Experiment 21.7.c.

(b) Place 10 cm³ of a 1% aqueous solution of starch and 1 cm³ of concentrated hydrochloric acid in a boiling tube and heat the tube in a beaker of boiling water. At five-minute intervals, transfer a few drops of the mixture by means of a dropping pipette to a test tube containing 1 cm³ of a dilute solution of iodine in potassium iodide. Maintain the level of the solution in the tube with distilled water if necessary and continue heating and testing with fresh samples of the iodine solution until no starch remains. Cool the residual liquid and then neutralise it, test a portion with Fehling's solution and concentrate the remainder as in (a).

(c) Add about 1 cm³ of saliva to 10 cm³ of a 1% aqueous starch solution in a boiling tube. Stir well and allow the mixture to stand at room temperature, testing with iodine solution as in (b) until no starch remains. Finally, carry out a Fehling's test on a portion of the residual liquid and concentrate the remainder as before.

Questions

1 The sucrose and starch were hydrolysed (i.e. they reacted with the water). What do you think was the function of the acids and the saliva and which of them was the most effective? Give a reason for your answer.
2 What type of compounds were produced by these reactions?
3 Is there any evidence as to whether the molecules of sucrose and starch were broken down into simpler molecules or built up into more complex ones?
4 In what way is the hydrolysis of starch an important part of your life?

21.7.c To identify the products of the hydrolysis of sucrose and starch

PROCEDURE   Using a very fine capillary tube, place a small spot (diameter no more than 5 mm) of a 1% solution of glucose on a 15 cm × 10 cm piece of filter paper as shown in Figure 21.7.c. Rinse the capillary tube well and repeat the procedure using 1% solutions of fructose and maltose and the concentrated solutions from the previous experiment. After allowing the spots to dry, fasten the edges of the filter paper together by means of paper clips and lower the resulting cylinder into a gas jar containing 1 cm depth of solvent. (A suitable solvent may be made by mixing together 15 cm³ of propan-2-ol, 5 cm³ of anhydrous

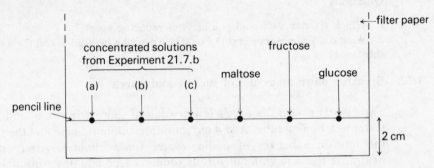

Figure 21.7.c

ethanoic acid and 5 cm³ of distilled water.) Close the jar with a cover slip and leave it for about 18 hours, then remove the filter paper, dry it in the air and develop it by dipping it quickly into a mixture of 5 volumes of a 2% solution of phenylamine in propanone, 5 volumes of a 2% solution of diphenylamine in propanone and 1 volume of an 85% aqueous solution of phosphoric(v) acid. Allow the filter paper to dry in the air and then heat it in an oven at 100 °C for a few minutes. Note the positions of the spots on the filter paper.

Questions

1   Draw a diagram of the filter paper, showing the final positions of the spots.
2   What do the results of this experiment tell you about the hydrolysis of sucrose and starch?

21.8.a   The fermentation of glucose ★ ★

The previous two experiments have shown that starch can be broken down into various sugars. Can the sugar molecules be broken down even further?

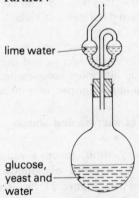

Figure 21.8.a

PROCEDURE   Dissolve 50 g of glucose in 500 cm³ of lukewarm water in a 2 dm³ flask. Add about 10 g of dried yeast, stir well and place the fermentation trap in position as shown in Figure 21.8.a. Leave the mixture in a warm place for a few days and then filter it into a 1 dm³ round-bottomed flask. Fractionally distil the filtrate in the apparatus used for

Experiment 1.2.k and collect the liquid which boils below 80°C. Smell the product and try to ignite a little of it in an evaporating basin.

Questions

1 Describe the appearance of the mixture during fermentation.
2 Which gas was given off during the fermentation? How could you tell?
3 What can you deduce about the nature of the liquid which distilled over below 80°C? Explain your reasoning.

21.8.b Some properties of ethanol

PROCEDURE   (a) Ignite 3–4 drops of ethanol on a watch glass by means of a lighted splint.
(b) Pour 5 cm³ of ethanol into a test tube and add a 5 mm cube of sodium. (**Caution:** do not handle the sodium.) Fit the tube with a bung and delivery tube and collect the resulting gas over water in the usual way (see Figure 4.4.a). Test this gas with a flame. (If any sodium remains in the tube, add more ethanol to dissolve it; *do not* tip the metal into the sink.) Half-fill a 250 cm³ beaker with tap water, heat it to boiling and then turn the gas down so that the water is just simmering. Pour the residual liquid from the test tube into a watch glass and place this on top of the beaker. Evaporate the ethanol, remembering that it is inflammable, and examine the solid which remains.
(c) Add a *small* amount of phosphorus pentachloride (**caution:** avoid skin contact) to 1 cm³ of ethanol in a dry test tube. Breathe gently across the top of the tube and test the vapour with a piece of moist universal indicator paper.
(d) To 2 cm³ of potassium dichromate(VI) solution in a test tube add 1 cm³ of concentrated sulphuric acid (**care**) and 1 cm³ of ethanol. Note any colour change and cautiously smell the product.

Questions

1 What would you expect to be formed by the complete combustion of ethanol? Write an equation for this reaction.
2 Do you think that the equation accurately represents the reaction which took place in (a)? Give a reason for your answer.
3 Which gas was evolved in (b)? Describe the solid product which remained on the watch glass and write an equation for the reaction between sodium and ethanol. (Hint: the reaction is similar to that between sodium and water.)
4 Describe what you saw in (c). Name one of the products obtained and explain how you identified it.
5 What did you observe in (d)? Were the dichromate(VI) ions oxidised or reduced?
6 From your answer to question 5, decide whether the ethanol was oxidised or reduced. Describe the smell of the product and find out its name.

21.9.a Some properties of ethanoic acid

PROCEDURE   (a) Carry out Experiments 5.2.a, 5.2.b and 5.2.c using dilute ethanoic acid as the only acid.

(b) Carry out part (b) of the previous experiment using *anhydrous* ethanoic acid (**caution:** corrosive, harmful vapour) in place of the ethanol. Do *not* evaporate the resulting solution.

(c) Carry out part (c) of the previous experiment using *anhydrous* ethanoic acid in place of the ethanol.

Questions

1 Is dilute ethanoic acid a strong or a weak acid? Give reasons for your answer.

2 Sum up the acidic properties of dilute ethanoic acid, giving a chemical equation and an ionic equation to illustrate each property. Are there any reactions where this acid does not behave in the expected manner?

3 What happened in (b) and (c)? What do these results tell you about the structure of the ethanoic acid molecule?

21.10.a To study the reaction between organic acids and alcohols

PROCEDURE (a) Place 1 cm³ of ethanol and an equal volume of anhydrous ethanoic acid in a test tube and carefully add a few drops of concentrated sulphuric acid. *Warm* the mixture, with shaking, over a *low* bunsen flame for about a minute. (Remember that the liquid is highly inflammable: if it approaches boiling point remove the tube from the flame.) Pour the contents of the tube into a 100 cm³ beaker half-filled with tap water and smell the mixture cautiously.

(b) Repeat (a) using 2–3 drops of pentanol in place of the ethanol.

(c) Repeat (a) using 1 cm³ of methanol in place of the ethanol and a spatula measure of 2-hydroxybenzoic acid (salicylic acid) in place of the ethanoic acid.

Questions

1 To which class of compounds do the organic products belong?

2 Describe the smell of each of the products.

3 Write an equation for the reaction which occurred in (a) and name the organic product.

4 What was the purpose of the concentrated sulphuric acid in these experiments?

21.10.b The laboratory preparation of soap

This experiment involves the hydrolysis of a naturally occurring ester (castor oil) in the presence of sodium hydroxide solution.

PROCEDURE Pour 2 cm³ of castor oil and 10 cm³ of 4 M sodium hydroxide solution (**caution:** corrosive) into a 100 cm³ beaker. Boil the mixture gently for 10 minutes, stirring continuously and topping up with distilled water as required, then allow it to cool. Add 2 spatula measures of sodium chloride (soap is insoluble in sodium chloride solution), stir well, filter off the solid and wash it with a little cold distilled water. Place a sample of the solid in a test tube and shake it with warm distilled water, then with warm tap water.

Questions

1 Did the product behave like soap? Explain your answer.

2 The hydrolysis of an ester to give an acid and an alcohol is a reversible reaction:

$$\text{ester} + \text{water} \rightleftharpoons \text{acid} + \text{alcohol}.$$

Why did the addition of sodium hydroxide solution favour the forward reaction?

3 What type of compound is soap?

4 Why does the behaviour of soap with distilled water differ from that with tap water?

21.10.c The preparation of a soapless detergent

PROCEDURE Slowly pour $2\,\text{cm}^3$ of concentrated sulphuric acid, with stirring, into a test tube containing $1\,\text{cm}^3$ of olive oil. Add $5\,\text{cm}^3$ of distilled water, stir and decant the liquid from the solid product. Wash the solid with two further $5\,\text{cm}^3$ portions of distilled water and test the residue with distilled water and tap water as in the previous experiment.

Questions

1 Did the product behave like a detergent? Explain your answer.

2 In what way does the behaviour of a soapless detergent differ from that of soap?

21.11.a What have proteins in common?

This experiment should be carried out in a well-ventilated laboratory.

PROCEDURE Mix half a spatula measure of milk powder with twice its own volume of soda lime. Gently heat the mixture in an ignition tube and test the gases given off with moist universal indicator paper. Repeat the experiment using other substances with high protein content, such as gelatine and cheese.

Questions

1 Which gas was given off? How could you tell?

2 Which elements were common to all the proteins which you tested?

21.12.a The polymerisation of phenylethene (styrene) ★ ★

Read the section on addition polymerisation on page 446 before attempting this experiment.

PROCEDURE Add $0.1\,\text{g}$ of di(dodecanoyl) peroxide (lauroyl peroxide) to $5\,\text{cm}^3$ of styrene in a boiling tube and set up the apparatus shown in Figure 18.4.a. Heat the tube in a beaker of boiling water for 30 minutes, allow it to cool and then pour the contents into a $100\,\text{cm}^3$ beaker half-filled with ethanol. Push the polystyrene into a lump by means of a glass rod, pour off the ethanol and dry the solid on a filter paper. Gently heat the solid in a crucible until it softens, leave it to cool and then heat it again.

Questions

1 What was the purpose of the di(dodecanoyl) peroxide in the experiment?
2 Describe the appearance of the polystyrene.
3 Does polystyrene appear to be a thermoplastic or a thermosetting plastic? Give a reason for your answer.
4 List some everyday uses of polystyrene.

21.12.b An experiment with poly(methyl 2-methylpropenoate) (Perspex)★★

This experiment must be carried out in a fume cupboard.

PROCEDURE   (a) Place about 5 g of Perspex chips in a 150×19 mm filter tube and set up the apparatus shown in Figure 21.6.a. (There is no need for rocksil or a thermometer.) Gently heat the tube until the residue starts to blacken.
(b) Heat 5 cm³ of the distillate from part (a) with 0.1 g of di(dodecanoyl) peroxide as in Experiment 21.12.a but do not allow the temperature of the water in the beaker to exceed 80°C (the liquid in the tube boils at 100°C). When the appearance of the mixture changes, stir it with a glass rod until it solidifies and then remove the tube from the beaker and examine the product.

Questions

1 What happened when the Perspex chips were heated in part (a)? Do you think that the product consisted of smaller or larger molecules than the original material? How did you draw this conclusion?
2 What do you think was formed in part (b)?

21.12.c Making nylon★

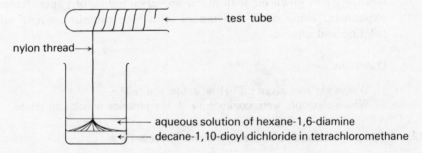

Figure 21.12.c

PROCEDURE   Place 10 cm³ of a 5% solution of decane-1,10-dioyl dichloride in tetrachloromethane (**caution:** toxic vapour) in a 100 cm³ beaker and carefully pour on to this 10 cm³ of a 5% aqueous solution of hexane-1,6-diamine to which a few drops of 2 M sodium hydroxide solution have previously been added. Using tongs, pull out the film of nylon which forms at the interface of the two liquids. Wrap the resulting thread round a test tube and rotate the tube to wind more thread out of the beaker (see

Figure 21.12.c). When a reasonable amount has been collected, wash the nylon with a mixture of equal volumes of water and ethanol and leave it to dry. Heat a sample of the dry polymer in a crucible in a fume cupboard until it melts, allow it to cool and then heat it again.

Questions

1 Describe the appearance of the nylon. Do you think it was pure?
2 Does nylon appear to be a thermoplastic or a thermosetting plastic? Give a reason for your answer.
3 List some everyday uses of nylon.

21.1 Organic chemistry – an introduction

Organic chemistry is the chemistry of carbon compounds, excluding carbonates and the oxides of carbon. Because carbon atoms can join together in so many different ways to form rings and chains of atoms, the number of organic compounds which exist is very large indeed. However, these compounds can be divided into classes, or 'families' in which all of the members have similar chemical properties, and this makes the study of organic chemistry considerably simpler than it would otherwise be. The families of compounds are referred to as *homologous series*.

What are the characteristics of a homologous series of compounds? Consider the simplest series of all, the alkanes, which are the major constituents of petroleum. The molecular formulae of the first four members are given below, together with graphic formulae, showing the arrangement of the atoms within the molecules.

methane, CH_4 ethane, C_2H_6

propane, C_3H_8 butane, C_4H_{10}

As you can see, all four compounds have the general formula C_nH_{2n+2} and each one differs from its neighbours by the group —CH_2—. Investigation of the data given in Experiment 21.1.b shows that the densities and boiling points of the alkanes increase fairly regularly as successive —CH_2— groups are added to the molecules and it is found that other physical properties, such as viscosity and melting point, also show a gradual change with increase in relative molecular mass. Since their molecules are so similar you will not be surprised to learn that the alkanes have similar chemical properties and methods of preparation. These points are summarised in the following definition.

> A **homologous series** is a group of compounds in which
> 1   all members possess the same general formula,
> 2   each member differs from the next by the group —CH_2—,
> 3   the physical properties of members change gradually as the relative molecular mass changes,
> 4   all members have similar chemical properties and
> 5   all members can be made by general methods.

Isomerism

Consider again the structure of a molecule of butane, C_4H_{10}. Its atoms can be arranged in two different ways,

$$\text{either} \quad H-\underset{\underset{H}{|}}{\overset{\overset{H}{|}}{C}}-\underset{\underset{H}{|}}{\overset{\overset{H}{|}}{C}}-\underset{\underset{H}{|}}{\overset{\overset{H}{|}}{C}}-\underset{\underset{H}{|}}{\overset{\overset{H}{|}}{C}}-H \quad \text{or} \quad H-\underset{\underset{H}{|}}{\overset{\overset{H}{|}}{C}}-\underset{\underset{\underset{H-\underset{\underset{H}{|}}{\overset{\overset{}{|}}{C}}-H}{|}}{|}}{\overset{\overset{H}{|}}{C}}-\underset{\underset{H}{|}}{\overset{\overset{H}{|}}{C}}-H.$$

These two compounds have the same molecular formula but different structures and hence different names. (The second one is called methylpropane because it has a methyl group, —CH_3, attached directly to a propane carbon chain.) The different structures of the two molecules give the compounds different physical properties but their chemical properties are the same because they are both alkanes. Such compounds are examples of *isomers*.

> **Isomerism** is the existence of two or more compounds with the same molecular formula but different structures, the different forms being called **isomers**.

Isomerism is not confined to members of the same homologous series. For example, ethanol and methoxymethane (dimethyl ether)* both have the molecular formula C_2H_6O, but their structures are as shown. The two compounds are very different chemically because they belong to different homologous series.

$$H-\underset{\underset{H}{|}}{\overset{\overset{H}{|}}{C}}-\underset{\underset{H}{|}}{\overset{\overset{H}{|}}{C}}-O-H \qquad\qquad H-\underset{\underset{H}{|}}{\overset{\overset{H}{|}}{C}}-O-\underset{\underset{H}{|}}{\overset{\overset{H}{|}}{C}}-H$$

ethanol   methoxymethane

* The naming system for organic compounds has been changed in recent years. For reference, the older names of the more common compounds are given in brackets after the modern names the first time that such compounds are mentioned.

N.B. You must remember that these structures are really three-dimensional, each carbon atom being surrounded *tetrahedrally* by four other atoms. Also free rotation of the atoms about the bonds can occur. Thus there are only *two* isomers of dichloroethane, $C_2H_4Cl_2$:

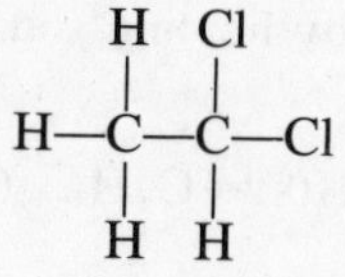

$$
\begin{array}{ccc}
& \text{H} & \text{Cl} \\
| & | \\
\text{H}-\text{C}-\text{C}-\text{Cl} \\
| & | \\
& \text{H} & \text{H}
\end{array}
\quad \text{and} \quad
\begin{array}{ccc}
& \text{H} & \text{H} \\
| & | \\
\text{H}-\text{C}-\text{C}-\text{H}. \\
| & | \\
& \text{Cl} & \text{Cl}
\end{array}
$$

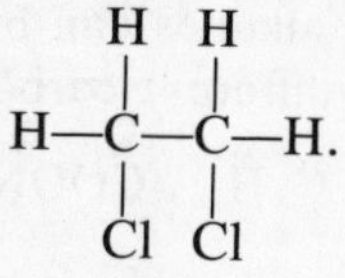

1,1-dichloroethane   1,2-dichloroethane

Structures such as those shown below are, in fact, all the same (i.e. 1,2-dichloroethane) but you will need to make models of the molecules to understand this point properly.

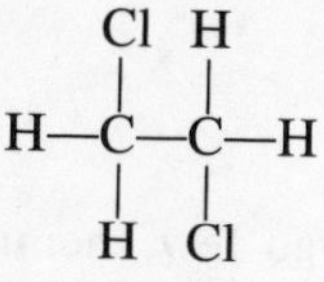

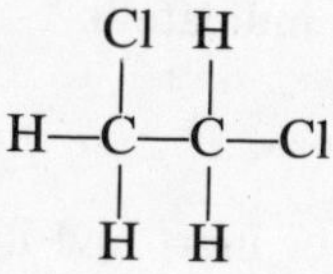

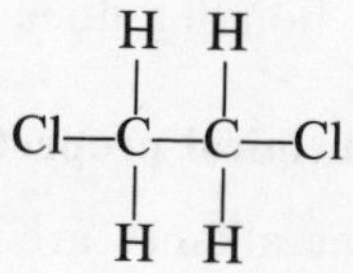

$$
\begin{array}{cc}
\text{Cl} & \text{H} \\
| & | \\
\text{H}-\text{C}-\text{C}-\text{H} \\
| & | \\
\text{H} & \text{Cl}
\end{array}
\qquad
\begin{array}{cc}
\text{Cl} & \text{H} \\
| & | \\
\text{H}-\text{C}-\text{C}-\text{Cl} \\
| & | \\
\text{H} & \text{H}
\end{array}
\qquad
\begin{array}{cc}
\text{H} & \text{H} \\
| & | \\
\text{Cl}-\text{C}-\text{C}-\text{Cl} \\
| & | \\
\text{H} & \text{H}
\end{array}
$$

21.2   The alkanes

The alkanes are *hydrocarbons*, i.e. compounds containing hydrogen and carbon only. They form a homologous series of general formula C_nH_{2n+2} and, as can be seen from the structures on page 425, their molecules contain only single covalent bonds.

> Compounds which have molecules containing only single covalent bonds are said to be **saturated**.

The names, formulae and physical state at room temperature of the first few alkanes are given in Table 21.2.

Name	Formula	Physical state at room temperature
Methane	CH_4	Gas
Ethane	C_2H_6	Gas
Propane	C_3H_8	Gas
Butane	C_4H_{10}	Gas
Pentane	C_5H_{12}	Liquid
Hexane	C_6H_{14}	Liquid
Heptane	C_7H_{16}	Liquid
Octane	C_8H_{18}	Liquid

Table 21.2 The first eight alkanes

Laboratory preparation

Methane may be obtained by heating a mixture of anhydrous sodium ethanoate (sodium acetate) and soda lime. (Soda lime acts like sodium hydroxide but has a higher melting point; it is made by reacting sodium hydroxide solution with calcium oxide.) The methane is collected over water.

$$CH_3COONa(s) + NaOH(s) \rightarrow CH_4(g) + Na_2CO_3(s)$$

Other alkanes can be prepared by heating soda lime with the sodium salts of different carboxylic acids.

$$C_nH_{2n+1}COONa(s) + NaOH(s) \rightarrow C_nH_{2n+2}(g) + Na_2CO_3(s)$$

Physical properties

The simplest alkanes are colourless gases or liquids and variations in their melting point, density etc. have already been discussed (see page 425); those with longer carbon chains (e.g. octadecane, $C_{18}H_{38}$) are solids. All are insoluble in water but readily dissolve in organic solvents. (What type of bonding does this indicate?)

Chemical properties

The alkanes are rather inert and undergo few chemical reactions.

COMBUSTION Alkanes burn in a plentiful air supply to form carbon dioxide and water.

e.g. $$CH_4(g) + 2O_2(g) \rightarrow CO_2(g) + 2H_2O(g)$$

In a limited air supply carbon monoxide, or even carbon (soot), may be produced because of incomplete combustion. This can be dangerous if, for example, a gas water heater is operated in a badly ventilated room or a car engine is run in a closed garage (North Sea Gas is mainly methane and petrol is a mixture of hydrocarbons).

REACTION WITH CHLORINE AND BROMINE Methane reacts with chlorine to give a mixture of products according to the equations given below. Bromine reacts similarly, but more slowly. The reactions of both chlorine and bromine with alkanes are photosensitive (speeded up by light).

$$CH_4(g) + Cl_2(g) \rightarrow HCl(g) + CH_3Cl(g)$$
chloromethane

$$CH_3Cl(g) + Cl_2(g) \rightarrow HCl(g) + CH_2Cl_2(l)$$
dichloromethane

$$CH_2Cl_2(l) + Cl_2(g) \rightarrow HCl(g) + CHCl_3(l)$$
trichloromethane
(chloroform)

$$CHCl_3(l) + Cl_2(g) \rightarrow HCl(g) + CCl_4(l)$$
tetrachloromethane
(carbon tetrachloride)

These reactions, in which hydrogen atoms in the methane molecules are successively replaced by chlorine atoms, are examples of *substitution reactions*.

A **substitution reaction** is a reaction in which one atom or group of atoms in a molecule is replaced by another.

Uses of the alkanes

1 As fuels (North Sea Gas, bottled gas, petrol, paraffin, fuel oil, etc.),
2 As the starting materials in the manufacture of a vast number of different substances, e.g. plastics, synthetic fibres, aerosol propellants, insecticides, paints and soapless detergents.

Manufacture of the alkanes

Alkanes are obtained on a large scale from petroleum and natural gas (21.6).

21.3 The alkenes

The alkenes form a homologous series of general formula $C_n H_{2n}$, the first member being ethene, C_2H_4. Their names are derived from those of the corresponding alkanes by replacing the ending *-ane* with *-ene*. Alkene molecules all contain a carbon-to-carbon double bond in which two carbon atoms share two pairs of electrons.

e.g.

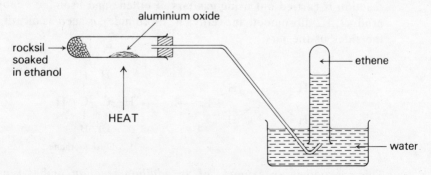

ethene (ethylene)

propene (propylene)

Compounds which have molecules containing double or triple covalent bonds are said to be **unsaturated**.

Laboratory preparation

rocksil soaked in ethanol

aluminium oxide

HEAT

ethene

water

Figure 21.3 The laboratory preparation of ethene

Ethene can be prepared by the catalytic dehydration of ethanol, C_2H_5OH, in the apparatus shown in Figure 21.3. The catalyst of aluminium oxide is heated strongly and sufficient heat is conducted down the tube to vaporise the ethanol. The ethene is collected over water.

$$C_2H_5OH(g) \rightarrow C_2H_4(g) + H_2O(g)$$

The dehydration may also be brought about by heating ethanol with concentrated sulphuric acid at 453 K (180 °C). The use of different alcohols will produce different alkenes.

$$C_nH_{2n+1}OH(g) \rightarrow C_nH_{2n}(g) + H_2O(g)$$

Physical properties

Ethene is a colourless gas with a sweetish smell and a density about the same as that of air. It is practically insoluble in water but readily dissolves in organic solvents. Trends in the physical properties of the alkenes are similar to those shown by the alkanes.

Chemical properties

A double covalent bond might be expected to be twice as strong as a single one. However, it is found that this is not so: one of the two bonds in the double bond is easily broken and, as a result of this, alkenes are much *more* reactive than the corresponding alkanes.

COMBUSTION   Like all hydrocarbons, alkenes burn in excess oxygen to form carbon dioxide and water.

e.g. $$C_2H_4(g) + 3O_2(g) \rightarrow 2CO_2(g) + 2H_2O(g)$$

When ethene burns in air the flame is smoky because insufficient oxygen is available for complete combustion. This is characteristic of all unsaturated hydrocarbons.

REACTION WITH CHLORINE AND BROMINE   When ethene is shaken with a solution of bromine in tetrachloromethane the reddish-brown colour of the bromine disappears almost immediately (Experiment 21.3.a). This is in marked contrast to the behaviour of the alkanes. If the reaction is carried out using gas jars of ethene and bromine *vapour*, the product, 1,2-dibromoethane (ethylene dibromide), is seen as oily drops on the sides of the jars.

1,2-dibromoethane

This reaction is an example of an *addition reaction* and is typical of unsaturated compounds.

> An **addition reaction** is a reaction in which two or more molecules react to give a single product.

Notice that the product is a *saturated* compound, the electrons which formed the weaker of the two bonds in the double bond being shared with the bromine atoms. Chlorine reacts in a similar way, but more vigorously, to give 1,2-dichloroethane.

REACTION WITH HYDROGEN CHLORIDE AND HYDROGEN BROMIDE These compounds both react with alkenes by addition but the reactions are much slower than with the corresponding halogens.

e.g.

$$\underset{H}{\overset{H}{>}}C=C\underset{H}{\overset{H}{<}} + H-Br \longrightarrow H-\underset{\underset{H}{|}}{\overset{\overset{H}{|}}{C}}-\underset{\underset{Br}{|}}{\overset{\overset{H}{|}}{C}}-H$$

bromoethane

HYDROGENATION At about 473 K (200 °C) and in the presence of a finely divided nickel catalyst, hydrogen adds on to alkenes to form alkanes.

e.g.

$$\underset{H}{\overset{H}{>}}C=C\underset{H}{\overset{H}{<}} + H-H \longrightarrow H-\underset{\underset{H}{|}}{\overset{\overset{H}{|}}{C}}-\underset{\underset{H}{|}}{\overset{\overset{H}{|}}{C}}-H$$

This reaction is not important in the case of ethene but is widely used in the margarine industry where hydrogenation of animal and vegetable oils (which contain carbon-to-carbon double bonds) 'hardens' them into solid fats.

N.B. Animal and vegetable oils are unsaturated esters (21.10), *not* alkenes.

REACTION WITH POTASSIUM MANGANATE(VII) SOLUTION An acidified solution of potassium manganate(VII) is decolourised by alkenes and an alkaline solution is turned green, finally depositing a brown precipitate (Experiment 21.3.a). In both cases the alkenes are oxidised to diols (i.e. compounds containing two —OH groups), the reactions being further examples of addition.

e.g.

$$\underset{H}{\overset{H}{>}}C=C\underset{H}{\overset{H}{<}} + H_2O + [O] \longrightarrow H-\underset{\underset{OH}{|}}{\overset{\overset{H}{|}}{C}}-\underset{\underset{OH}{|}}{\overset{\overset{H}{|}}{C}}-H$$

from potassium manganate(VII)

ethane-1,2-diol
(ethylene glycol)

Tests for unsaturation in organic compounds

Unsaturated organic compounds:
1 rapidly decolourise a solution of bromine in tetrachloromethane,
2 decolourise an acidified solution of potassium manganate(VII) and
3 turn an alkaline solution of potassium manganate(VII) green and then give a brown precipitate.

Uses of the alkenes

The alkenes can be converted into a wide range of products; for example, ethene is used:
1 to make plastics such as polythene (21.12),
2 to make ethanol (21.8),
3 to make 1,2-dibromoethane (a petrol additive) and
4 to make ethane-1,2-diol (used as an antifreeze and to make 'Terylene').

Manufacture of the alkenes

Alkenes are manufactured by the large-scale cracking of oils (21.6).

21.4 The alkynes

This homologous series of unsaturated hydrocarbons has the general formula $C_n H_{2n-2}$ and its members all possess a carbon-to-carbon triple bond in their molecules. The names of the alkynes are obtained by replacing the -*ane* ending of the corresponding alkane with -*yne*.

e.g.

$$H—C\equiv C—H$$

ethyne (acetylene)

$$H—\overset{\displaystyle H}{\underset{\displaystyle H}{C}}—C\equiv C—H$$ propyne

Only ethyne will be discussed in this book.

Laboratory preparation of ethyne

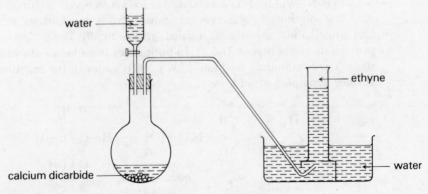

Figure 21.4 The laboratory preparation of ethyne

Ethyne is prepared in the laboratory by dropping water on to calcium dicarbide in the apparatus shown in Figure 21.4. It is collected over water.

$$CaC_2(s) + 2H_2O(l) \rightarrow Ca(OH)_2(aq) + C_2H_2(g)$$

Physical properties

Ethyne is a colourless gas, almost odourless when pure and slightly less dense than air. It is insoluble in water but readily dissolves in organic solvents. Because the gas readily explodes when compressed alone it is usually stored dissolved under pressure in propanone.

Chemical properties

The reactions of ethyne are similar to those of ethene but, because of the higher degree of unsaturation, addition reactions take place in two steps.

e.g.
$$H-C \equiv C-H + Br-Br \rightarrow$$

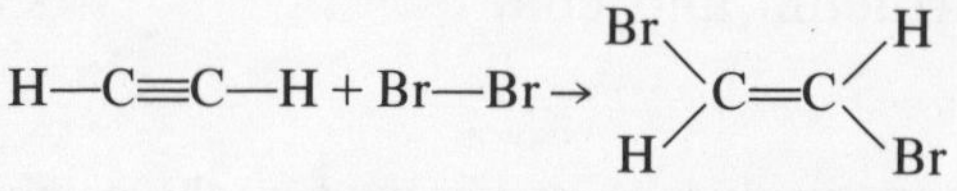

1,2-dibromoethene

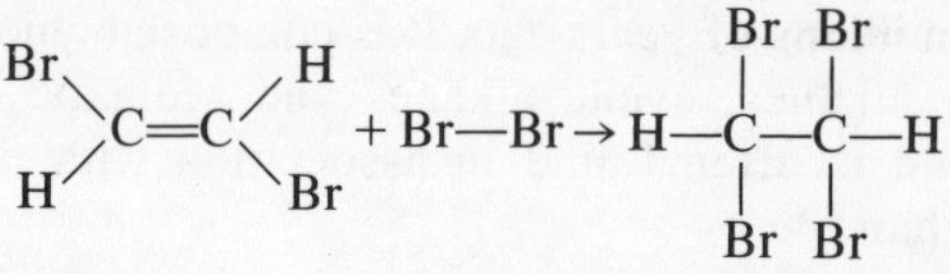

1,1,2,2-tetrabromoethane

Ethyne explodes when mixed with chlorine, giving steamy fumes of hydrogen chloride and black clouds of carbon (Experiment 21.4.a).

$$C_2H_2(g) + Cl_2(g) \rightarrow 2C(s) + 2HCl(g)$$

Uses of ethyne

1   In oxyacetylene blowpipes, used for cutting and welding steel.
2   To make a wide range of organic compounds, e.g. plastics, solvents and synthetic rubber.

Manufacture of ethyne

Ethyne, like most other hydrocarbons, is manufactured from petroleum.

21.5   Ring compounds

All of the compounds discussed so far in this chapter consist of molecules in which the carbon atoms are arranged in chains. However, it is possible for the ends of a chain to join up so that a ring of atoms is formed. Cyclohexane, C_6H_{12}, and benzene, C_6H_6, are two compounds whose molecules consist of rings.

As its name implies, cyclohexane behaves like an alkane. Although it fits the general formula C_nH_{2n}, it has no double bonds and thus is *not* an alkene.

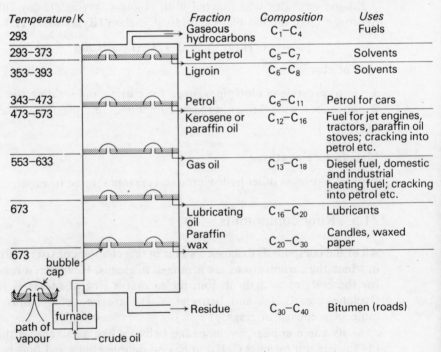

The structure of a benzene molecule is rather special. Each carbon atom uses three of its four outermost electrons to form single covalent bonds: the fourth electron is 'delocalised' and becomes the property of the ring as a whole, rather than being associated with a particular atom. This gives benzene unique properties; for example, it shows a marked reluctance to undergo addition reactions, even though the formula C_6H_6 suggests that it is an unsaturated compound.

21.6   Petroleum and coal

Petroleum

Petroleum was produced underground by the combined effects of heat, pressure and bacteria on the remains of marine animals and plants which died many millions of years ago. It is composed chiefly of hydrocarbons, particularly alkanes, cyclic alkanes and aromatic compounds such as benzene, and is often found in association with natural gas, which is mainly methane.

Liquid petroleum (crude oil) is fractionally distilled in tall fractionating towers. These are divided into compartments, each maintained at a

Temperature/K	Fraction	Composition	Uses
293	Gaseous hydrocarbons	C_1–C_4	Fuels
293–373	Light petrol	C_5–C_7	Solvents
353–393	Ligroin	C_6–C_8	Solvents
343–473	Petrol	C_6–C_{11}	Petrol for cars
473–573	Kerosene or paraffin oil	C_{12}–C_{16}	Fuel for jet engines, tractors, paraffin oil stoves; cracking into petrol etc.
553–633	Gas oil	C_{13}–C_{18}	Diesel fuel, domestic and industrial heating fuel; cracking into petrol etc.
673	Lubricating oil	C_{16}–C_{20}	Lubricants
	Paraffin wax	C_{20}–C_{30}	Candles, waxed paper
673			
	Residue	C_{30}–C_{40}	Bitumen (roads)

bubble cap

path of vapour   furnace   crude oil

Figure 21.6 The fractionation of liquid petroleum

particular temperature, and as the vapour passes upwards from one compartment to the next the appropriate fractions condense out. Each fraction contains groups of hydrocarbons with boiling points within a particular range, rather than pure samples of individual compounds (see Figure 21.6).

There is little demand for the heavier oils and they are usually converted to more useful products by *cracking* (Experiment 21.6.b). In industry the long-chained alkane molecules are broken down by heat in the presence of an aluminium oxide catalyst to give a number of much simpler alkanes and alkenes.

e.g.
$$C_{10}H_{22}(g) \xrightarrow[Al_2O_3]{heat} C_8H_{18}(g) + C_2H_4(g)$$

This equation shows one way in which decane could be cracked. You should be able to write the equations for several possible alternatives.

Petroleum is immensely important in today's world, as should be obvious from the selection of uses of the alkanes and alkenes given in 21.2 and 21.3.

Coal

Coal was formed in a similar way to petroleum by the underground compression of vegetable matter. It consists mainly of carbon, together with compounds containing hydrogen, oxygen, nitrogen and sulphur.

The disadvantages of using coal as a fuel are many. For example, it is difficult to obtain and its dust causes serious damage to miners' lungs, it must be transported by train and lorry rather than through pipes, and when it is burned it produces ash and pollutes the atmosphere with smoke and harmful gases such as sulphur dioxide.

However, the destructive distillation of coal gives useful products such as coal gas, coal tar, ammonia and coke. Coal gas and coke were important industrial and domestic fuels and coal tar was used as a source of many organic chemicals. In fact, until recent years, coal was an essential raw material for the organic chemicals industry. As the world's stocks of natural gas and oil dwindle it is possible that coal will once again feature as a major feed-stock for industry.

21.7 Carbohydrates

Carbohydrates are compounds composed of carbon, hydrogen and oxygen where the hydrogen and oxygen atoms are present in the same ratio as in water.

Examples of carbohydrates include sugars (e.g. glucose, $C_6H_{12}O_6$) and starch $(C_6H_{10}O_5)_n$ which is present in foods like potatoes and rice. The value of n in the formula of starch is somewhere in the region of 5000: molecules as large as this are referred to as *macromolecules*.

The structure of a glucose molecule is shown on the left in Figure 21.7.1 but, for simplicity, the version on the right will be used in equations.

Sugars like glucose which have molecules made up of single six-membered rings are referred to as *monosaccharides*. A molecule of glucose and one of its isomer fructose can join together and eliminate a water molecule to form sucrose (cane sugar, $C_{12}H_{22}O_{11}$). This is called a

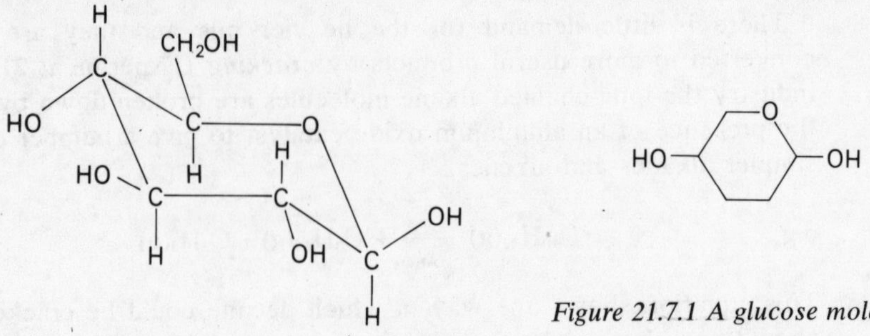

Figure 21.7.1 A glucose molecule

disaccharide because its molecules contain two six-membered rings. Another disaccharide, maltose, is obtained by the joining together of pairs of glucose molecules (Figure 21.7.2).

$$2C_6H_{12}O_6(aq) \rightleftharpoons C_{12}H_{22}O_{11}(aq) + H_2O(l)$$

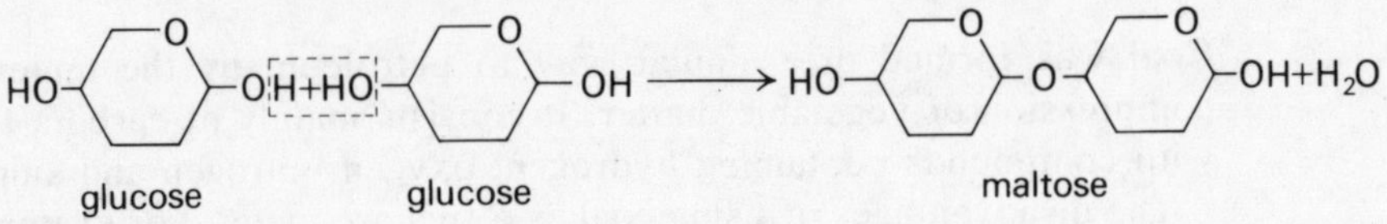

Figure 21.7.2

Starch is a *polysaccharide*, its molecules being formed by the joining together of many monosaccharide units.

Tests for sugars and starch

1 Glucose, fructose and maltose all reduce the royal blue complex copper(II) ions present in Fehling's solution to give a red precipitate of copper(I) oxide: they are referred to as *reducing sugars*. Sucrose is not a reducing sugar and therefore has no effect on Fehling's solution (Experiment 21.7.a).
2 Starch does not reduce Fehling's solution but gives a blue–black colour when added to an aqueous solution of iodine in potassium iodide.

The hydrolysis of sucrose and starch

Since glucose and fructose can combine together to form sucrose and water it would seem reasonable to expect that the hydrolysis of sucrose (i.e. its reaction with water) would simply reverse this process. This does, in fact, happen, the reaction being readily brought about by boiling an aqueous solution of sucrose with a catalyst of dilute sulphuric acid (Experiment 21.7.b).

$$C_{12}H_{22}O_{11}(aq) + H_2O(l) \rightleftharpoons C_6H_{12}O_6(aq) + C_6H_{12}O_6(aq)$$

sucrose(aq) + water $\rightleftharpoons$ glucose(aq) + fructose(aq)

Starch can also be hydrolysed but the products of the reaction depend upon the nature of the catalyst used. Amylase is an *enzyme* (i.e. a complex organic catalyst) found in saliva and at room temperature it catalyses the hydrolysis of the huge starch molecules into the much smaller ones of the disaccharide, maltose. However, if starch is boiled with a dilute acid catalyst it breaks down into the monosaccharide, glucose. Amylase is a more efficient catalyst than the acid in terms of the temperature at which it operates and the low concentration required, but it does not break the starch molecules down into such small units. The different products of the two reactions (Figure 21.7.3) may be identified by chromatography (Experiment 21.7.c).

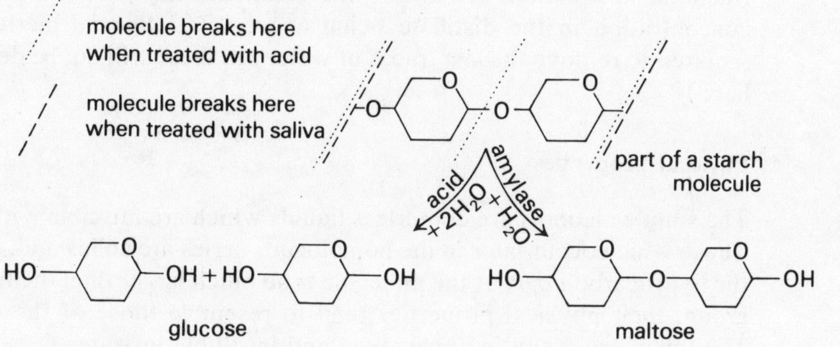

Figure 21.7.3

The enzyme-catalysed hydrolysis of the starch present in foods like potatoes and bread is, of course, an essential part of our digestive process and subsequent oxidation of the resultant sugars into carbon dioxide and water provides us with energy for movement and warmth (3.3). The reverse reactions take place during plant growth, where glucose is built up from carbon dioxide and water by photosynthesis and then converted to starch and cellulose.

21.8   The alcohols

The alcohols are members of a homologous series which has the general formula $C_nH_{2n+1}OH$, their molecules being similar to those of the alkanes but having a hydroxyl group in place of one of the hydrogen atoms. They are named by replacing the final -*e* of the corresponding alkane with -*ol*.

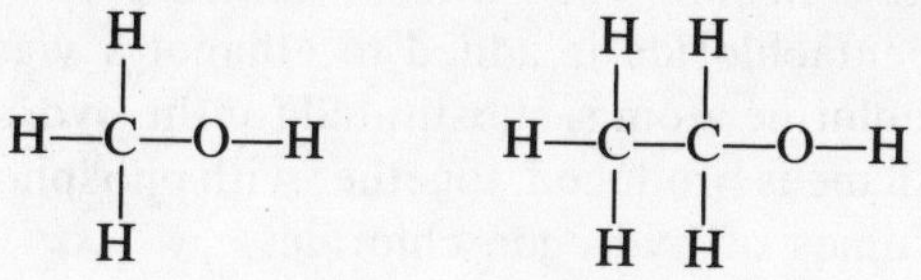

methanol   ethanol

Preparation of ethanol by fermentation

When a mixture of yeast and aqueous glucose solution is kept in a warm place *fermentation* occurs, giving ethanol and bubbles of carbon dioxide

(Experiment 21.8.a). The reaction is brought about by the enzyme zymase, which is present in the yeast.

$$C_6H_{12}O_6(aq) \rightarrow 2C_2H_5OH(aq) + 2CO_2(g)$$

> **Fermentation** is the slow decomposition of an organic substance brought about by micro-organisms and usually accompanied by the evolution of heat and a gas.

The yeast is killed when the proportion of ethanol reaches about 15%, so that any further concentration of the solution has to be carried out by fractional distillation. Even then the ethanol is not pure, its maximum concentration in the distillate being about 96%. (Special methods are required to remove the last traces of water but these will not be dealt with here.)

Physical properties

The simple alcohols are colourless liquids which are miscible with water. Those which occur later in the homologous series are solids and, because the hydrocarbon part of the molecule is so much larger than the hydroxyl group, their physical properties tend to resemble those of the alkanes. Thus they are waxy in appearance and insoluble in water.

Chemical properties (See Experiment 21.8.b)

COMBUSTION Ethanol readily burns in air with an almost colourless flame. In a plentiful supply of air the products are carbon dioxide and water.

$$C_2H_5OH(l) + 3O_2(g) \rightarrow 2CO_2(g) + 3H_2O(g)$$

REACTION WITH SODIUM Sodium effervesces quietly in ethanol, giving hydrogen and sodium ethoxide.

$$2C_2H_5OH(l) + 2Na(s) \rightarrow 2C_2H_5ONa(s) + H_2(g)$$

In this reaction ethanol behaves in a similar way to water and this implies that its molecules contain an —OH group.

REACTION WITH PHOSPHORUS PENTACHLORIDE When phosphorus pentachloride is added to ethanol a vigorous reaction occurs in which a chlorine atom is substituted for the hydroxyl group of the alcohol. Chloroethane is produced, together with phosphorus trichloride oxide and steamy fumes of hydrogen chloride.

$$C_2H_5OH(l) + PCl_5(s) \rightarrow C_2H_5Cl(g) + POCl_3(l) + HCl(g)$$

The production of steamy fumes of hydrogen chloride when phosphorus pentachloride is added to an organic compound indicates the presence of a hydroxyl group in the compound, provided that the compound is *dry*. (Why must it be dry?)

REACTION WITH ORGANIC ACIDS Alcohols react with organic acids in the presence of concentrated sulphuric acid to form compounds called *esters*. This topic is dealt with in 21.10.

DEHYDRATION When the vapour of an alcohol is passed over a heated aluminium oxide catalyst, or a liquid alcohol is heated with concentrated sulphuric acid at about 453 K (180°C), the corresponding alkene is produced (see page 430).

REACTION WITH ACIDIFIED POTASSIUM DICHROMATE(VI) SOLUTION On warming with this reagent, alcohols are oxidised to *aldehydes* (general formula $C_nH_{2n+1}CHO$) and then to *carboxylic acids* (21.9). For example, ethanol is converted to ethanal and then to ethanoic acid.

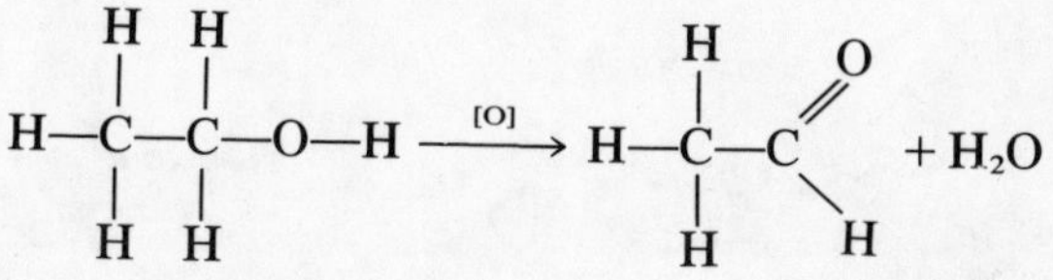

ethanal (acetaldehyde)

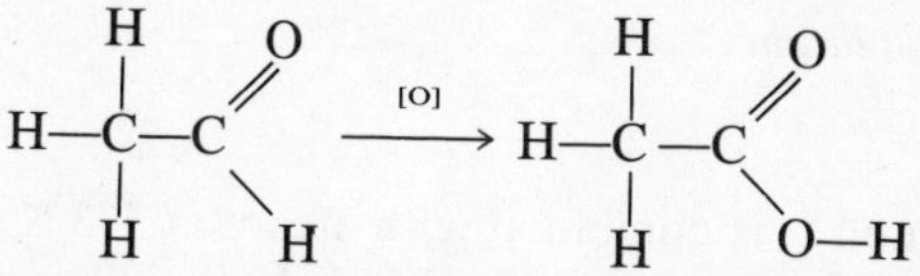

ethanoic acid (acetic acid)

The orange potassium dichromate(VI) solution turns green, showing that it has been reduced. This colour change is made use of in the 'breathalyser' to test whether a motorist has been consuming alcoholic drinks.

Uses of the alcohols

1 As solvents.
2 To make other organic compounds, such as synthetic fibres, organic acids and esters.
3 As fuels.
4 In beer, wines and spirits.
N.B. The ethanol used in school laboratories is normally industrial methylated spirit, which consists of 95% ethanol together with 5% methanol to make it undrinkable (methanol is very poisonous) and render it exempt from excise duty. 'Ordinary' methylated spirit contains about 90% ethanol and 9% methanol plus small quantities of unpleasant-tasting substances and a purple dye.

Manufacture of ethanol

1 Ethene, which is obtained from the cracking of oils, can be hydrated to

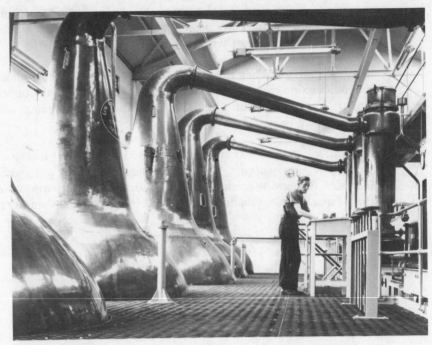

Whisky distillation

give ethanol: it is combined with steam at 573 K (300°C) and 6.5×10^6 Pa (65 atm) in the presence of a phosphoric(v) acid catalyst.

$$CH_2\!\!=\!\!CH_2(g) + H_2O(g) \longrightarrow CH_3CH_2OH(g)$$

2   Beer is made by fermentation of the starch in barley. Various enzymes in the barley and yeast break the starch down into glucose and this is converted to ethanol and carbon dioxide as in the laboratory preparation (page 437). Wines are made in a similar way by fermenting the sugars in grapes (other fruits and vegetables may be used but then the product is not strictly a wine). Spirit drinks are obtained by distillation of the more dilute solutions produced by fermentation. During fermentation it is important to exclude air because oxidising bacteria may enter and convert the ethanol to ethanoic acid (vinegar).

21.9   The carboxylic acids

This homologous series has the general formula $C_nH_{2n+1}COOH$ and its members are named by replacing the *-e* in the name of the corresponding alkane with *-oic*. Note that it is the *total* number of carbon atoms in the carbon chain and not just those in the hydrocarbon group which gives the acid its name.

e.g.

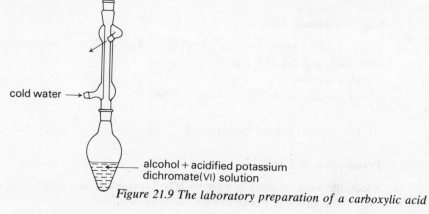

methanoic acid
(formic acid)

ethanoic acid
(acetic acid)

propanoic acid
(propionic acid)

Preparation

Figure 21.9 The laboratory preparation of a carboxylic acid

cold water →

alcohol + acidified potassium
dichromate(VI) solution

Carboxylic acids are made by oxidation of the corresponding alcohols. This reaction may be carried out using acidified potassium dichromate(VI) solution in the apparatus shown in Figure 21.9. The reflux condenser ensures that the aldehyde which is first formed is returned to the flask and oxidised further to the carboxylic acid. The condenser is then placed in the normal position and the acid distilled off.

e.g.
$$CH_3CH_2OH(l) \xrightarrow{2[O]} CH_3COOH(l) + H_2O(l)$$

An alternative procedure is to pass a mixture of alcohol vapour and air over a heated platinum catalyst.

Physical properties

The simple carboxylic acids are colourless liquids which are completely miscible with water. As is the case with the alcohols, the later members of the series are waxy solids and are insoluble in water (why?).

Chemical properties

REACTION WITH PHOSPHORUS PENTACHLORIDE AND WITH SODIUM *Anhydrous* ethanoic acid gives steamy fumes of hydrogen chloride when reacted with phosphorus pentachloride and reacts quietly with sodium to produce hydrogen, thus confirming the presence of an —OH group in its molecule.

e.g.
$$2CH_3COOH(l) + 2Na(s) \rightarrow 2CH_3COONa(s) + H_2(g)$$
$$CH_3COOH(l) + PCl_5(s) \rightarrow CH_3COCl(l) + POCl_3(l) + HCl(g)$$

REACTION WITH ALCOHOLS The combination of anhydrous carboxylic acids with alcohols in the presence of concentrated sulphuric acid gives esters (21.10).

ACID PROPERTIES Aqueous solutions of the carboxylic acids show all of the typical properties of acids mentioned in 5.2. However, they are only weak acids because most of the molecules in their solutions are in the unionised state.

Uses of ethanoic acid

1 To make a large number of organic chemicals, e.g. synthetic fibres, varnishes, aspirin and esters.
2 As a solvent.
3 As vinegar (a 4–5% aqueous solution).

21.10   Esters and detergents

Preparation of esters

Alcohols react reversibly with acids to form water and compounds called *esters*, the process being known as *esterification*:

$$acid + alcohol \rightleftharpoons ester + water.$$

The reaction is very slow at room temperature but may be speeded up by heating and by the addition of a catalyst of concentrated sulphuric acid (Experiment 21.10.a). In addition to acting as a catalyst, the concentrated sulphuric acid also removes water from the mixture, thus minimising the backward reaction and increasing the yield of ester. The ester can be separated from the other liquids by fractional distillation.

e.g. $CH_3CO|OH(l) + H|OC_2H_5(l) \rightleftharpoons CH_3COOC_2H_5(l) + H_2O(l)$

ethanoic acid + ethanol $\rightleftharpoons$ ethyl ethanoate   + water
                                                            (ethyl acetate)

Notice how the name of the ester is derived from the two parts of its molecule, the *ethyl* group, $-C_2H_5$ (i.e. ethane minus a hydrogen atom), and the *ethanoate* group, CH_3COO-. Similarly the ester $CH_3COOC_3H_7$ is called *propyl ethanoate*. What is the name and structure of the compound which has the formula $CH_3COOC_5H_{11}$?

ethyl ethanoate   propyl ethanoate

Physical properties of esters

The simple esters are colourless liquids which are immiscible with water and have characteristic pleasant smells. (They are responsible for the fragrance of flowers and the flavour of fruits.)

Chemical properties of esters

The only chemical property of the esters which will be considered here is their hydrolysis to give an acid and an alcohol.

e.g. $CH_3COOC_2H_5(l) + H_2O(l) \rightleftharpoons CH_3COOH(l) + C_2H_5OH(l)$

As you can see, this reaction is the reverse of esterification. It is carried out by boiling the ester with water and a catalyst of dilute sulphuric acid. (Why is *dilute* acid used for hydrolysis and *concentrated* acid used for esterification?)

Sodium hydroxide solution may also be used to catalyse the hydrolysis of an ester but in this case the reaction goes to completion. Why is it not reversible this time? The answer is that the sodium hydroxide neutralises the acid as soon as it is formed, thus removing it from the mixture and preventing the back reaction.

e.g. $CH_3COOC_2H_5(l) + NaOH(aq) \rightarrow CH_3COONa(aq) + C_2H_5OH(l)$

sodium ethanoate

Detergents

Detergents (i.e. cleaning agents) fall into two main types, *soapy detergents* and *soapless detergents*. We normally just refer to these as 'soaps' and 'detergents' respectively.

Soaps are the sodium or potassium salts of long-chained carboxylic acids (e.g. octadecanoic acid (stearic acid), $C_{17}H_{35}COOH$) and their molecules are shaped rather like long-tailed tadpoles (see Figure 21.10.1). The 'tail' is the hydrocarbon radical, which is *hydrophobic* (water-hating) and is not attracted to water molecules: it can, however, enter droplets of non-polar liquids like oil. The 'head' of the soap molecule is the ionic —COO⁻ part, which is *hydrophilic* (water-loving): it is attracted by water molecules but not by oil or grease.

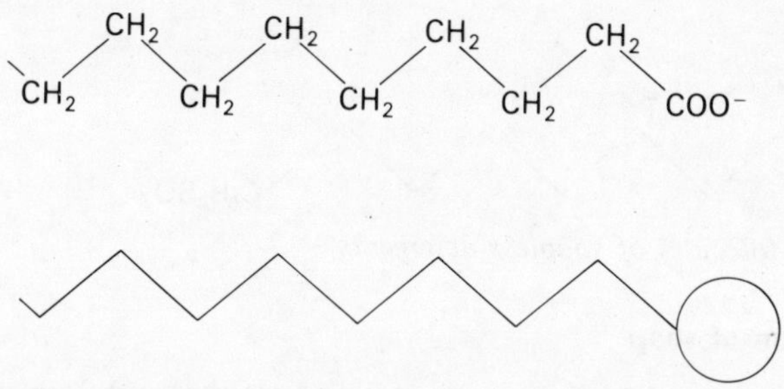

Figure 21.10.1 Molecules of soap

The presence of soap in water lowers the intermolecular forces between the water molecules and thus reduces the surface tension. This allows the water to spread out and wet a surface rather than forming small droplets on it.

When greasy material, such as a piece of cloth, is placed in soap solution, the hydrophobic tails of the soap molecules enter the grease but the hydrophilic heads stay in the water. Figure 21.10.2 shows how the grease is broken into small particles and carried off into the solution; the negatively charged heads of the soap molecules which surround the grease particles cause them to repel one another so that they cannot join together to form large globules.

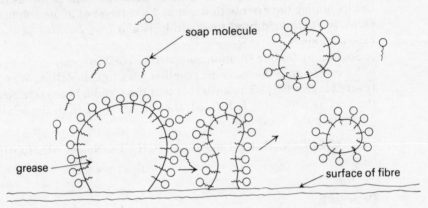

Figure 21.10.2 How soap removes grease from cloth

Soapless detergents are made from petroleum and consist of molecules which are similar in form and action to those of soap (see Figure 21.10.3). They have two main advantages when compared with soaps: firstly, unlike soaps, they are not made from animal fats and vegetable oils which could otherwise be used for food and, secondly, their calcium and magnesium salts are *soluble* in water and thus do not form scum.

Figure 21.10.3 Molecules of soapless detergents

The preparation of soap

Animal fats and vegetable oils are esters of alcohols such as propane-1,2,3-triol (glycerol), $CH_2(OH)CH(OH)CH_2(OH)$, and carboxylic acids with fairly long hydrocarbon chains, e.g. octadecanoic acid (stearic acid), $C_{17}H_{35}COOH$. If these esters are boiled with sodium hydroxide solution they are hydrolysed to give the alcohol and the sodium salt of the acid.

The sodium salt of a long-chained carboxylic acid is, of course, soap (Experiment 21.10.b).

e.g.
$$C_{17}H_{35}COOCH_2$$
$$C_{17}H_{35}COOCH \quad + \quad 3NaOH \quad \rightarrow 3C_{17}H_{35}COONa \quad + \quad CHOH$$
$$C_{17}H_{35}COOCH_2 \qquad\qquad\qquad\qquad\qquad\qquad\qquad CH_2OH$$

| ester (fat) | + | sodium hydroxide | → | sodium salt (soap) | + | alcohol (glycerol) |

The alkaline hydrolysis of an ester is known as *saponification* (from the Latin *sapo*, meaning 'soap').

21.11 Proteins

Proteins are substances made by living cells to build up animal and plant tissue. When they are heated with soda-lime, proteins evolve ammonia (Experiment 21.11.a) and this suggests that there is some basic unit, containing nitrogen, which is common to all of them. It has been found that proteins are built up from *aminoacids*, molecules of which contain an amino group (—NH₂) and a carboxyl group (—COOH). The structure of these molecules is as shown, R often being a hydrocarbon group.

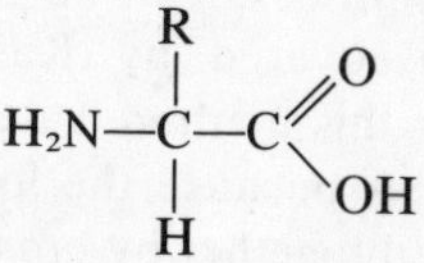

$$H_2N-\underset{\underset{H}{|}}{\overset{\overset{R}{|}}{C}}-C\overset{O}{\underset{OH}{}}$$

The relative molecular masses of proteins vary from a few thousands up to several millions and thus protein molecules, like those of starch and cellulose, are macromolecules. The way in which hundreds, or even thousands, of aminoacid molecules link together to make a protein is illustrated below. Since R can be any one of 20 or more different groups you can appreciate that the structures of proteins must be very complex indeed.

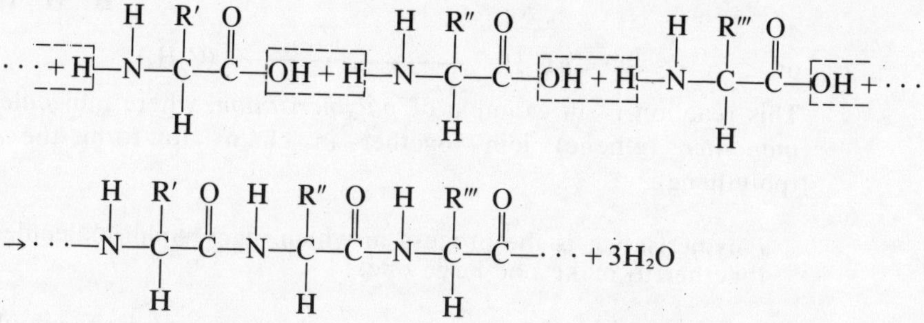

$$\cdots + H-N-\underset{\underset{H}{|}}{\overset{\overset{R'}{|}}{C}}-C-OH + H-N-\underset{\underset{H}{|}}{\overset{\overset{R''}{|}}{C}}-C-OH + H-N-\underset{\underset{H}{|}}{\overset{\overset{R'''}{|}}{C}}-C-OH + \cdots$$

$$\rightarrow \cdots -N-\underset{\underset{H}{|}}{\overset{\overset{R'}{|}}{C}}-C-N-\underset{\underset{H}{|}}{\overset{\overset{R''}{|}}{C}}-C-N-\underset{\underset{H}{|}}{\overset{\overset{R'''}{|}}{C}}-C- \cdots + 3H_2O$$

The $-\overset{O}{\overset{||}{C}}-\overset{H}{\overset{|}{N}}-$ grouping by which the different aminoacid molecules join together is known as a *peptide linkage*.

The reverse of the above process (i.e. the hydrolysis of proteins) can be brought about by the action of enzymes or by heating with dilute acids or alkalis. As with starch, it is possible to identify the products of this reaction by means of chromatography.

21.12 Man-made macromolecules

You have seen that naturally occurring macromolecules such as those of starch and proteins are built up by the linking together of large numbers of much smaller molecules. In recent years, chemists have copied this process to make the many plastics and synthetic fibres which nowadays we take for granted. Look around you at your clothes, furniture and personal possessions and you will see just how much our daily lives depend on these man-made macromolecules.

Addition polymerisation

When ethene is heated to 473 K (200°C) at a pressure of 1000 atm and in the presence of a trace of oxygen as catalyst, large numbers of molecules join together to form polythene. This reaction was discovered by accident in 1933 when a crack developed in a piece of apparatus containing ethene at a high pressure and the ethene came into contact with the oxygen of the air. Nowadays the process can be carried out at atmospheric pressure using catalysts discovered by Karl Ziegler and Giulio Natta. The polythene made by this method is denser and more rigid than that made by the older process because the long carbon chains of the polythene molecules are joined together by *cross-linking* (i.e. bonds acting between the chains).

The relative molecular mass of polythene is between 10 000 and 40 000 but its empirical formula is the same as that of ethene, as can be seen from the equation for its formation.

or $\quad n\,C_2H_4 \longrightarrow (C_2H_4)_n$

This reaction is an example of *polymerisation*, where molecules of the *monomer* (ethene) join together in chains to form the *polymer* (polythene).

> **Polymerisation** is the process in which many small molecules join together to make one large one.

As you can see, one of the bonds in the carbon-to-carbon double bond of each ethene molecule breaks and the molecules undergo an addition reaction. For this reason the process is called *addition polymerisation*. Other substances which can be made to undergo addition polymerisation include phenylethene (Experiment 21.12.a), chloroethene and methyl

2-methylpropenoate (Experiment 21.12.b). Notice that all of the monomers contain a carbon-to-carbon double bond.

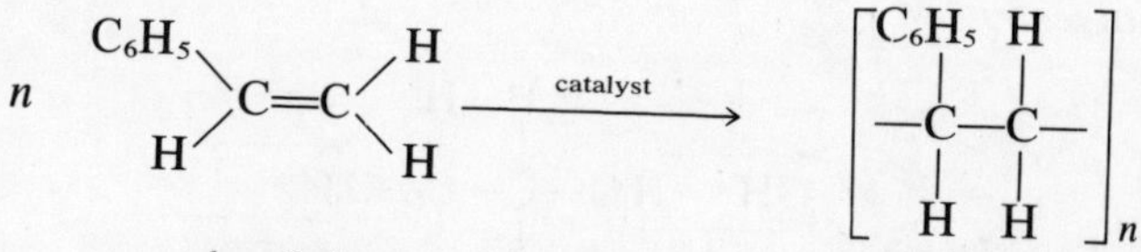

phenylethene
(styrene)

poly(phenylethene)
(polystyrene)

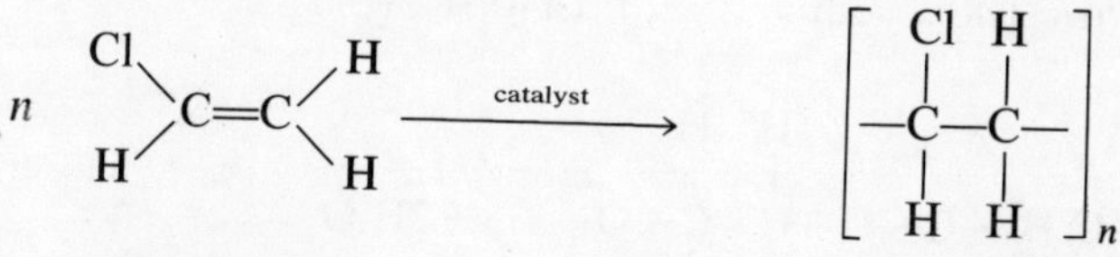

chloroethene
(vinyl chloride)

poly(chloroethene)
(polyvinylchloride, PVC)

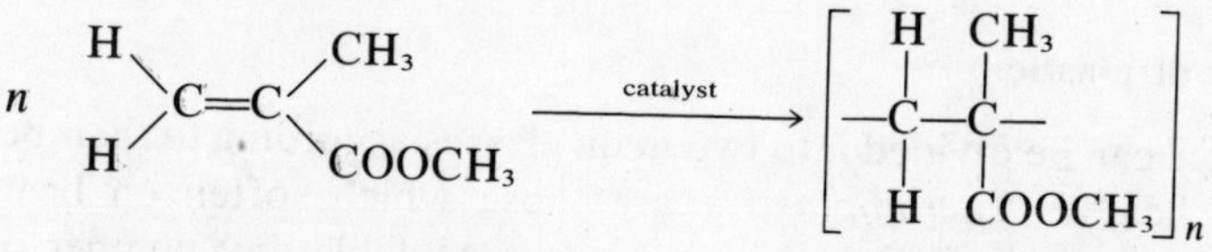

methyl 2-methylpropenoate
(methyl methacrylate)

poly(methyl 2-methylpropenoate)
(polymethylmethacrylate, Perspex)

Polymerisation is often reversible. For example, Perspex can easily be depolymerised into the monomer by heating, and then repolymerised by adding a catalyst (Experiment 21.12.b). **Caution:** you should not heat other polymers without consulting your teacher because some of them give off very poisonous vapours.

Condensation polymerisation

This differs from addition polymerisation in that small molecules such as water or hydrogen chloride are eliminated when the monomer units link together. Usually the process involves two or more different types of monomer, in which case the product is called a *co-polymer*.

One example of a condensation polymer is nylon, which is made from hexane-1,6-diamine and decane-1,10-dioyl dichloride. Notice that nylon molecules, like protein molecules, contain the peptide linkage.

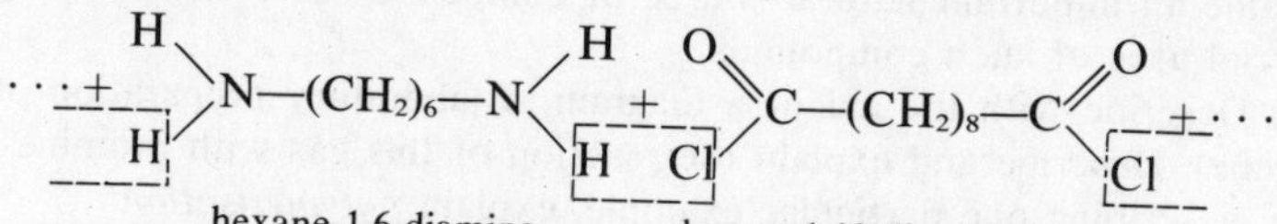

hexane-1,6-diamine
(hexamethylenediamine)

decane-1,10-dioyl dichloride
(sebacoyl chloride)

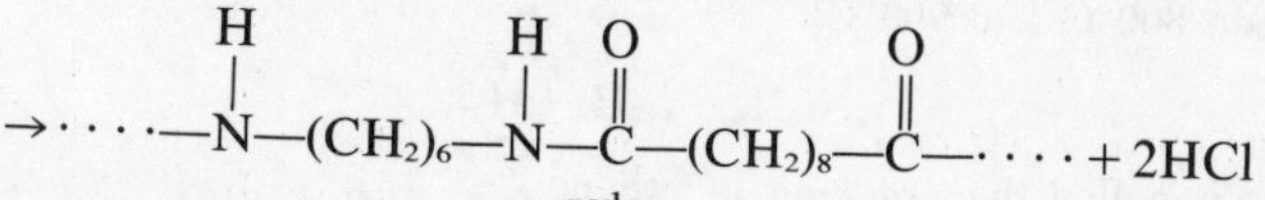

nylon

Terylene is made in a similar manner from benzene-1,4-dicarboxylic acid and ethane-1,2-diol. The process involves reaction between hydroxyl and carboxyl groups and therefore is a multi-stage esterification. Terylene is thus a *polyester*.

benzene-1,4-dicarboxylic
acid (terephthalic acid)

ethane-1,2-diol
(ethylene glycol)

Terylene

Types of plastic

Plastics can be divided into two main classes according to their behaviour when heated. *Thermoplastics* are those which soften on heating and harden on cooling, the process being repeatable any number of times. They are not usually very resistant to heat. *Thermosetting plastics* soften on being heated the first time after manufacture and thus can be moulded. However, this heating causes cross-linking to occur between the long-chained macromolecules and results in the setting up of a rigid three-dimensional network which cannot be softened by subsequent reheating. Polystyrene is a thermoplastic and bakelite a thermosetting plastic.

Questions on Chapter 21

1 (a) Which homologous series of organic compounds can be represented by the following general formulae?
Series A, $C_n H_{2n+2}$; series B, $C_n H_{2n}$; series C, $C_n H_{2n+1}COOH$; series D, $C_n H_{2n+1}OH$:
(b) Give the name and structural formula of *one* compound in *each* series.
(c) Describe reactions by which (i) a *named* compound of series B can be converted to a compound of series A; (ii) a *named* compound of series D can be converted to a compound of series B.
(d) Name an important natural source of compounds of series A and give *two* industrial uses of such compounds.   (C)
2 (a) Describe with the aid of a diagram, a laboratory preparation of ethene (ethylene). Describe and explain the reaction of this gas with bromine water.
(b) By describing one particular example, explain *polymerisation*.
(c) Outline the manufacture of ethanol from ethene.   (O & C)
3 In the USA, ethene (ethylene) is manufactured by heating ethane to temperatures between 800°C and 900°C:

$$C_2H_6 \rightarrow C_2H_4 + H_2.$$

The process is called the *cracking* of ethane.

In the cracking of the hydrocarbon, C_3H_8, several reactions occur, one of which is represented by the equation

$$C_3H_8 \rightarrow C_3H_6 + H_2.$$

(a) Give the names of the hydrocarbons C_3H_8 and C_3H_6.

(b) Give the names and general formulae of the homologous series to which C_3H_8 and C_3H_6 belong.

(c) Give the structural formulae for C_3H_8 and C_3H_6, using lines to represent covalent bonds between atoms.

(d) When C_3H_8 is cracked, another reaction occurs in which a mixture of equal volumes of two other gaseous hydrocarbons is formed. Give the formulae of these two hydrocarbons.

(e) Give two possible industrial uses of the hydrogen formed as a by-product of the manufacture of ethene.

(f) Explain how ethane and ethene differ in their reactions with chlorine.   (C)

4   Proteins and starches are both large molecules containing thousands of atoms. They are important constituents of a balanced diet.

Chemically speaking, what is the main difference between proteins and starches and what do these substances give on hydrolysis?

How would you show that hair is a protein and not a carbohydrate?   (SCE)

5   Name the process used to obtain an aqueous solution of ethanol from a sugar, for example, glucose, $C_6H_{12}O_6$. Describe how you would carry out this preparation.

State, without further description, the method you would use to obtain a sample of reasonably pure ethanol from the aqueous solution and indicate the physical bases of this separation.

Write down the *structural* formula of ethanol.

Explain the following, stating the chemical reactions which take place:

(a) ethanol burns in air with a pale blue flame,

(b) when sodium is added to pure ethanol a colourless gas is evolved,

(c) when a mixture of equal volumes of ethanol and ethanoic acid is warmed with a few drops of concentrated sulphuric acid, a vapour with a pleasant smell is obtained.   (AEB)

6   A neutral organic liquid W of relative molecular mass 46 gives, on dehydration, a colourless gas X having the empirical formula CH_2. X reacts (i) with hydrogen to give a gas Y of relative molecular mass 30 and (ii) with bromine to give a liquid Z of relative molecular mass 188.

(a) Give the names and molecular formulae for compounds X, Y, Z and W.

(b) Write an equation for the dehydration of W and describe how this reaction can be carried out in the laboratory.

(c) Describe one other reaction of W (excluding its combustion in air or oxygen).

(d) 0.46 g of W is vaporised by heating to 100°C at a pressure of 760 mm Hg. Calculate the volume occupied by the vapour under these conditions. (Molar volume = $22.4 \, dm^3 \, mol^{-1}$ at s.t.p.)   (C)

7   (a) An organic acid A was found to contain 48.7% of carbon, 8.1% of hydrogen and 43.2% of oxygen by mass. Compound A reacted with ethanol to form an ester B having a relative molecular mass of 102. Calculate the molecular formula for A and write structural formulae for both A and B.

(b) Give briefly the conditions necessary for the efficient formation of an ester from an organic acid. Write equations for the reactions of ethanoic acid with (i) sodium hydroxide solution, and (ii) phosphorus trichloride *or* phosphorus pentachloride.   (C)

8   Describe briefly how you would obtain in the laboratory a small sample of *five* of the following. You should in each case indicate the materials and apparatus

required and the conditions for the reaction to take place. It is not necessary to obtain a pure sample, but you should give one property of the product which would confirm that you had probably obtained some of it.

(a) A soapless detergent, (b) soap, (c) an ester, (d) ethanol, (e) an unsaturated hydrocarbon such as ethylene (ethene), (f) a polymer.   (L)

22 Chemical analysis

Analysis is a branch of chemistry which has many important applications. Apart from its more obvious uses, such as in the identification of rocks and minerals, it is also employed in police work, agriculture, food technology, medicine and many other fields. There are two types of analysis, *qualitative analysis*, in which the various elements in a substance are identified, and *quantitative analysis*, where their relative masses are found.

22.1 Qualitative analysis

The tests in this section have been kept as simple as possible so that you will *understand* what you are doing instead of blindly following a recipe. To avoid the use of complex reagents the scheme has been restricted to the detection of the following ions: NH_4^+, K^+, Na^+, Li^+, Ca^{2+}, Al^{3+}, Zn^{2+}, Fe^{2+}, Fe^{3+}, Pb^{2+}, Cu^{2+}, CO_3^{2-}, HCO_3^-, S^{2-}, SO_3^{2-}, SO_4^{2-}, NO_3^-, Cl^-, Br^- and I^-.

Unless otherwise instructed, use a spatula measure of solid for each of the tests. Carry out all of the preliminary tests and then confirm your results with a further test where possible. If no anion is detected in the preliminary tests the salt is probably a sulphate and *further test 1* (page 453) should be performed.

1 Preliminary tests

	Experiment	Observation	Conclusion
1	Note appearance of solid.	(a) Blue or bright green (b) Pale green (c) Yellow or brown	Possibly Cu^{2+} Possibly Fe^{2+} Possibly Fe^{3+}
2	Heat in an ignition tube.	(a) O_2 (glowing splint relit), no brown fumes of NO_2	Nitrate of K or Na
		(b) O_2 (glowing splint relit), + brown fumes of NO_2	Nitrate (not of K or Na)
		(c) CO_2 (lime water milky)	CO_3^{2-} or HCO_3^-
		(d) Sublimation	Possibly NH_4^+
		(e) Steam	Water of crystallisation
		(f) Yellow when hot, white when cold	ZnO *residue*
		(g) Red when hot, yellow when cold	PbO *residue*

	Experiment	Observation	Conclusion
3	Add dil. HCl. Warm if no reaction.	(a) Immediate effervescence (CO_2 – lime water milky) (b) SO_2 (smell, potassium dichromate(VI) paper green) (c) H_2S (smell, lead(II) ethanoate paper black)	CO_3^{2-} or HCO_3^- (Go on to FT2*) SO_3^{2-} S^{2-}
4	Add 2 cm³ conc. H_2SO_4. Warm *gently* if no reaction. Do not pour hot mixture into water. Add a little Cu powder if no reaction.	(a) Same gases as in test 3 (b) HCl (colourless, fuming, acidic gas) (c) $HBr + Br_2$ (colourless, fuming, acidic gas + reddish-brown vapour) (d) I_2 (black solid + purple vapour) + SO_2 + H_2S (e) NO_2 (acidic brown fumes)	Anions as in 3 Cl^- (Go on to FT3) Br^- (Go on to FT3) I^- (Go on to FT3) NO_3^- (Go on to FT4)

Prepare a solution of the substance by shaking a spatula measure of it with 10 cm³ of distilled water and heating to boiling if necessary. If the solid is insoluble, replace the water with dil. HCl and repeat the procedure. (If the solid still will not dissolve it is probably a lead salt and dil. HNO_3 must be used.) Top up the test tube with distilled water, tip the liquid into another tube to ensure thorough mixing and use about 3 cm³ of it for tests 5 and 6.

	Experiment	Observation	Conclusion
5	Add NaOH solution until *just* alkaline (test with indicator paper), then fill up the tube and tip its contents into a second tube. Heat if no precipitate.	(a) White precipitate, insoluble in excess alkali ($Ca(OH)_2$) (b) White precipitate, soluble in excess alkali ($Al(OH)_3$, $Zn(OH)_2$ or $Pb(OH)_2$). Keep solution for FT5 (c) Pale blue precipitate ($Cu(OH)_2$) (d) Dirty-green precipitate ($Fe(OH)_2$) (e) Red-brown precipitate ($Fe(OH)_3$) (f) NH_3 (smell, alkaline gas)	Ca^{2+} (Go on to test 7) Al^{3+}, Zn^{2+} or Pb^{2+} Cu^{2+} Fe^{2+} Fe^{3+} NH_4^+
6	Repeat test 5 using NH_3 solution in place of the NaOH solution.	(a) White precipitate, insoluble in excess alkali ($Ca(OH)_2$, $Al(OH)_3$ or $Pb(OH)_2$) (b) White precipitate, soluble in excess alkali ($Zn(OH)_2$) (c) Pale blue precipitate ($Cu(OH)_2$), giving royal blue solution with excess alkali (d) Dirty-green precipitate ($Fe(OH)_2$) (e) Red-brown precipitate ($Fe(OH)_3$)	Ca^{2+}, Al^{3+} or Pb^{2+} (Go on to FT5) Zn^{2+} Cu^{2+} Fe^{2+} Fe^{3+}

*FT is used here as an abbreviation for 'further test'.

	Experiment	Observation	Conclusion
7	Dip a platinum wire into conc. HCl in a watch glass and then heat the *end* of the wire in a flame. Repeat if necessary until the wire does not colour the flame. Then dip the wire into the acid again, touch it on the powdered solid and replace it in the flame. Clean the wire after use.	(a) Lilac flame (red through blue glass) (b) Golden yellow flame (c) Crimson flame (d) Brick-red flame (e) Green-blue flame	K^+ Na^+ Li^+ Ca^{2+} Cu^{2+}

2 Further tests

	Experiment	Observation	Conclusion
1	Dissolve in distilled water (dil. HCl if insoluble in water). Acidify with dil. HCl and add a little $BaCl_2$ solution.	Thick white precipitate (of $BaSO_4$)	SO_4^{2-}
2	If preliminary test 3(a) is positive, dissolve original solid in water and add 3 drops of universal indicator.	(a) Original solid insoluble (b) Purple colour (c) Green colour, turning purple on boiling	CO_3^{2-} CO_3^{2-} HCO_3^-
3	Dissolve in distilled water (dil. HNO_3 if insoluble in water). Acidify with dil. HNO_3 and add a few drops of $AgNO_3$ solution.	(a) Thick white precipitate (of AgCl), soluble in excess ammonia solution (b) Pale cream precipitate (of AgBr), slightly soluble in ammonia solution (c) Yellow precipitate (of AgI), insoluble in ammonia solution	Cl^- Br^- I^-
4	Dissolve in distilled water and add freshly made $FeSO_4$ solution. Pour 1 cm^3 of conc. H_2SO_4 *carefully* into tilted tube.	(a) Insoluble in water (b) Brown ring at junction of layers	NO_3^- absent NO_3^-

Experiment	Observation	Conclusion
5 Pass H_2S into *solution* from preliminary test 5(b).	(a) No precipitate (b) White precipitate (of ZnS) (c) Black precipitate (of PbS)	Al^{3+} Zn^{2+} Pb^{2+}

3 Tests for oxidising and reducing agents

The usual method of testing for an oxidising agent is to mix it with a substance which is easily oxidised (i.e. a reducing agent) and which gives a visible change when the reaction takes place. Similarly, a suspected reducing agent is added to an oxidising agent which undergoes a visible change when reduced.

Experiment	Result	Explanation
1 *Oxidising agents* (a) Test with a moist starch–potassium iodide paper.	Paper goes blue–black	I^- ions are oxidised to iodine which then reacts with the starch
(b) Bubble H_2S into a solution of the substance.	Yellow precipitate	S^{2-} ions oxidised to S
(c) Warm with conc. HCl.	Cl_2 (smell, bleaches moist indicator paper)	Cl^- ions oxidised to Cl_2
2 *Reducing agents* (a) Add potassium manganate(VII) solution, acidified with dil. H_2SO_4.	Purple solution is decolourised	Manganate(VII) ions are reduced to pale pink manganese(II) ions
(b) Add potassium dichromate(VI) solution, acidified with dil. H_2SO_4.	Orange solution turns green	Dichromate(VI) ions are reduced to green chromium(III) ions
(c) Add a solution of an iron(III) salt, e.g. iron(III) chloride.	Yellow solution turns pale green	Fe^{3+} ions reduced to Fe^{2+} ions (test with NaOH solution as described earlier)

22.2 Quantitative analysis

Once the components of a substance have been identified their relative proportions may be found by quantitative analysis. *Gravimetric analysis*, involving a series of weighings, may be carried out but the technique is too difficult for use at this level. A quicker and easier method is that of *volumetric analysis*, where a solution of the substance under investigation is titrated with a standard solution of a suitable reagent. (A standard solution is one whose concentration is known.) In this book we shall only consider titrations between acids and alkalis.

Concentrations of solutions

Although the concentration of a solution can be expressed in $g\ dm^{-3}$, these units are not very useful for the purpose of a titration. Compounds react in mole ratios and so it is more usual to quote concentrations in $mol\ dm^{-3}$. A convenient abbreviation which is often employed for $mol\ dm^{-3}$ is the letter 'M': thus a 0.1 M solution of a given substance has a concentration of 0.1 mole of solute per dm^3 of solution.

Procedure

The apparatus required for a titration is a pipette for delivering a fixed volume (usually 20 or 25 cm^3) of the alkali, a burette for delivering a variable volume of the dilute acid, and a 150 cm^3 conical flask in which the reaction takes place. Before beginning the experiment, rinse the pipette with distilled water and then with the alkali, making sure that the whole of the inside surface is wet. Rinse the burette, including the jet, with distilled water and then with the dilute acid. Then fill the burette with the acid, open the tap and allow the acid to run out until all air bubbles have been expelled from the jet. Wash the flask with distilled water only.

Suck the alkali into the pipette (with the jet well below the surface of the liquid to avoid entry of air) until the level is above the mark. Close the end with the forefinger and release the pressure slightly until the bottom of the meniscus comes level with the mark. Make sure that the eye is level with the meniscus and that the jet of the pipette is just below the surface of the alkali during this operation. Finally, transfer the alkali to the conical flask, holding the pipette vertically with its jet in contact with the side of the flask, as shown in Figure 22.2. A drop of solution will remain in the pipette but must *not* be forced out because its presence is allowed for when the instrument is calibrated.

Add 2–3 drops of indicator (e.g. screened methyl orange) to the alkali and then read the level of the acid in the burette. Again, make sure that the eye is level with the bottom of the meniscus when the reading is taken. Run 1 cm^3 portions of acid into the alkali, swirling the flask after each

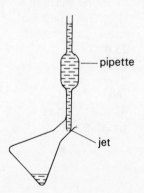

Figure 22.2 The use of a pipette

addition, until the indicator shows that the acid is just in excess. Note the new burette reading and subtract the initial reading from this to give the rough volume, to the nearest 1 cm³, of the acid required to neutralise the alkali. Empty the flask, rinse it with distilled water and repeat the whole process, this time adding the acid in a continuous stream until the volume is within 1 cm³ of the end-point. Carry out the final additions dropwise, with swirling, until the solution is exactly neutral. Perform further accurate titrations until two results within 0.1 cm³ of one another are obtained.

Treatment of results

The following example shows how a typical set of results should be treated. Notice the similarity between the method of setting out the final calculation and that given in the example on page 170.

Example 1   To find the concentration of a sample of hydrochloric acid

25 cm³ portions of standard sodium carbonate solution, made by dissolving 1.31 g of the anhydrous salt in 250 cm³ of distilled water, were titrated with dilute hydrochloric acid until two results within 0.1 cm³ of one another were obtained. The indicator used was screened methyl orange.

Titration	Rough	1	2
Burette reading 2	24–25	24.6	24.9
Burette reading 1	0	0.3	0.6
Volume of acid/cm³	24–25	24.3	24.3

Average accurate result = 24.3 cm³

$$Na_2CO_3(aq) + 2HCl(aq) \rightarrow 2NaCl(aq) + H_2O(l) + CO_2(g)$$
$$\text{mol } Na_2CO_3(aq):\text{mol } HCl(aq) = 1:2 \qquad \ldots A$$

$$\text{g } Na_2CO_3 \, dm^{-3} = 1.31 \times 4$$
$$\text{mol } Na_2CO_3 \, dm^{-3} = \frac{1.31}{106} \times 4 \left(\begin{array}{l} \text{molar mass of} \\ Na_2CO_3 = 106 \, g \, mol^{-1} \end{array} \right)$$

$$\text{mol } Na_2CO_3 \quad 25.0 \, cm^{-3} = \frac{1.31}{106} \times 4 \times \frac{25.0}{1000}$$

$$\text{mol } HCl \qquad 24.3 \, cm^{-3} = \frac{1.31}{106} \times 4 \times \frac{25.0}{1000} \times 2 \qquad \ldots \text{from A}$$

$$\text{mol } HCl \qquad dm^{-3} \quad = \frac{1.31}{106} \times 4 \times \frac{25.0}{1000} \times 2 \times \frac{1000}{24.3}$$

$$= \underline{0.102}$$

When you become familiar with the method of setting out you should have no difficulty in starting at the third line, thus saving both time and space. This point is illustrated in the next example.

Example 2

When dilute sulphuric acid was titrated with standard 0.094 M sodium hydroxide solution it was found that 26.8 cm^3 of the acid was required to neutralise 25.0 cm^3 of the alkali. Calculate the mass concentration of the acid. (Relative atomic masses: H = 1, S = 32, O = 16.)

$$H_2SO_4(aq) + 2NaOH(aq) \rightarrow Na_2SO_4(aq) + 2H_2O(l)$$

mol NaOH(aq) : mol H$_2$SO$_4$(aq) = 2 : 1 = 1 : $\frac{1}{2}$ ... A

mol NaOH	25.0 cm^{-3}	$= 0.094 \times \dfrac{25.0}{1000}$
mol H$_2$SO$_4$	26.8 cm^{-3}	$= 0.094 \times \dfrac{25.0}{1000} \times \dfrac{1}{2}$ ... from A
mol H$_2$SO$_4$	dm^{-3}	$= 0.094 \times \dfrac{25.0}{1000} \times \dfrac{1}{2} \times \dfrac{1000}{26.8}$
g H$_2$SO$_4$	dm^{-3}	$= 0.094 \times \dfrac{25.0}{26.8} \times \dfrac{1}{2} \times 98$
		$= \underline{4.30}$

Questions on Chapter 22

1   Calculate the masses of (a) potassium hydroxide in 250 cm^3 of 0.1 M solution, (b) anhydrous potassium carbonate in 500 cm^3 of 0.5 M solution, (c) sulphuric acid in 100 cm^3 of 0.05 M solution.

2   Find the masses of (a) magnesium oxide required to neutralise 500 cm^3 of 2 M nitric acid, (b) anhydrous sodium carbonate required to neutralise 250 cm^3 of 2 M hydrochloric acid, (c) sodium hydroxide required to neutralise 100 cm^3 of 1 M sulphuric acid.

3   From the following titration results, calculate the concentrations of the two acids.

(a) 25.0 cm^3 of 0.1 M potassium hydroxide solution neutralised 23.4 cm^3 of dilute sulphuric acid.

(b) 25.0 cm^3 of 0.05 M sodium carbonate solution neutralised 28.7 cm^3 of dilute nitric acid.

4   (a) 25.0 cm^3 of sodium hydroxide solution neutralised 26.2 cm^3 of 0.05 M sulphuric acid.

(b) 25.0 cm^3 of potassium carbonate solution neutralised 22.9 cm^3 of 0.1 M hydrochloric acid.

Find the concentrations of the two alkalis.

5   The following statement summarises the results of a quantitative exercise involving an alkali and a standard acid solution:

0.12 M aqueous nitric acid was titrated against 25.0 cm^3 of aqueous sodium hydroxide contained in a conical flask. 22.5 cm^3 of the acid was required to react completely with the base.

(a) (i) What is meant by a *standard* acid solution?

(ii) Write a balanced equation for the reaction between the nitric acid and the sodium hydroxide.

(b) (i) Explain briefly why the volumes of the acid and the alkali are stated to one decimal place.

(ii) Name a piece of apparatus which is suitable for measuring the nitric acid.

(iii) In addition to making careful observations, how did the experimenter satisfy himself that the correct volume of acid was 22.5 cm^3?

(iv) Name one indicator, other than litmus or universal indicator in liquid or paper form, which would be suitable in the titration, and state the colour change.

(c) (i) Calculate the number of moles of nitric acid, HNO_3, involved in the complete reaction.

(ii) Calculate the concentration of the sodium hydroxide solution.

(d) Arrange the following liquids in order of increasing pH value, beginning with the solution of lowest pH: 1 M aqueous ammonia, 1 M nitric acid, 1 M ethanoic acid, water, and 1 M sodium hydroxide. (NI)

6 For each of the following pairs of substances, describe a *chemical* test by which you could distinguish between them. State the results of the test on both substances in the pair.

(a) Manganese(IV) oxide, MnO_2, and iron(II) sulphide, FeS.

(b) Carbon monoxide and hydrogen.

(c) Sodium hydrogencarbonate and anhydrous sodium carbonate.

(d) Ethene, C_2H_4, and ethane, C_2H_6.

(e) Potassium chloride and potassium iodide. (AEB)

7 Reactions of two metals X and Y are described below.

(a) X will not dissolve in dilute hydrochloric acid, but will dissolve in dilute nitric acid. Two tests were carried out on the resulting colourless solution.

(i) With sodium hydroxide solution there was a white precipitate which redissolved when excess alkali was added.

(ii) With potassium iodide solution there was a yellow precipitate.

(b) Y reacts with cold water giving a substance which is only slightly soluble in water. If carbon dioxide is passed through some of the clear solution obtained, a white precipitate is seen.

Explain these reactions and, giving your reasons, *name* the two metals X and Y.
(O & C)

8 (a) When a white powder A was heated in a test tube, a gas B which turned lime water milky was released leaving a yellow residue C in the tube; C was soluble in dilute nitric acid without effervescence producing a colourless solution D. When hot dilute hydrochloric acid was added to a hot portion of solution D, the mixture slowly became cloudy on standing and eventually deposited a white crystalline solid E; under similar conditions sulphuric acid produced an immediate white precipitate F. Name the substances A, B, C, D, E and F and write equations for the chemical reactions; explain why E is deposited slowly.

(b) Describe a confirmatory test which you could carry out for the cation of the original substance A. (O)

9 Explain and comment on the following observations.

(a) A strip of metal foil is heated in the air: it does not burn, but it changes in colour and becomes black. It is then put into dilute sulphuric acid and warmed; the solution turns slightly blue. When excess ammonia solution is added, it turns a deep blue.

(b) (i) On adding a black powder to hydrogen peroxide solution a colourless gas is given off. This black powder itself does not change.

(ii) The same powder gives off a greenish gas when warmed with concentrated hydrochloric acid. With excess acid the black powder is slowly dissolved.

(c) A colourless crystalline salt effervesces with dilute acid and gives an intense yellow colour in a bunsen flame. When left on a watch glass in the air it gradually loses mass. (O & C)

10 From the information given below, identify A, B, C and D explaining the observations and giving equations wherever possible.

(a) Addition of a colourless liquid A to copper causes the liberation of a colourless gas that turns brown on exposure to the atmosphere. The brown gas dissolves in aqueous sodium hydroxide.

(b) A white solid B reacts with water, evolving much heat. If enough water is added the resultant solid dissolves. When carbon dioxide is passed into this solution, a white precipitate is obtained.

(c) A black solid C dissolves in hydrochloric acid, evolving a foul-smelling gas which forms a black precipitate when bubbled into lead nitrate solution. An aqueous solution of the gas slowly turns cloudy on standing.

(d) A pure colourless liquid D, when mixed with water, gives a non-conducting solution. On warming a mixture of D with ethanoic acid and a few drops of concentrated sulphuric acid, a vapour with a fruity smell is obtained.   (AEB)

Answers

Chapter 4 5 (b) 200 cm³ of oxygen, 800 cm³ of nitrogen.

Chapter 5 7 (f) 287, (g) 57.4 g.

Chapter 6 4 80% of copper in each sample.
6 (b) (i) $X = {}^{4}_{2}He$, (ii) $X = {}^{32}_{16}S$.
8 (d) Cu_2O.
9 (a) 104 g, 208 g, (c) XO_2.
10 (a) 93, (b) 145.
11 (a) 13, (b) 14, (c) 2.8.3, (d) 51, (e) 122, (f) 51, (g) 20, (h) 20, (i) 2.8.8.2, (j) 35, (k) 17, (l) 2.8.7.

Chapter 8 5

Particle	Mass number	Number of protons	Number of neutrons	Number of electrons
Li atom	7	3	4	3
Li⁺ ion	7	3	4	2
¹²C isotope	12	6	6	6
¹³C isotope	13	6	7	6
N³⁻ ion	14	7	7	10
Ne atom	20	10	10	10
S²⁻ ion	32	16	16	18

Chapter 9 1 (a) 2.4 g, (b) 8 g, (c) 51 g, (d) 60 g.
2 (a) $\frac{1}{8}$, (b) $\frac{1}{2}$, (c) $\frac{1}{20}$, (d) $\frac{1}{50}$.
3 (a) 3.01×10^{22}, (b) 1.20×10^{23}, (c) 3.01×10^{23}, (d) 3.01×10^{22}.
4 (a) Na_2SO_4, (b) KI, (c) $FeSO_4$, (d) $Na_2S_2O_3 \cdot 5H_2O$.
5 (a) C_6H_6, (b) N_2H_4, (c) $C_6H_{12}O_6$, (d) P_4O_{10}.
6 (a) 35% N, 5% H, 60% O, (b) 28% Fe, 24% S, 48% O, (c) 20% Ca, 80% Br, (d) 25.5% Cu, 12.8% S, 25.7% O, 36.1% H_2O.
7 7.2 g.
8 16 g.
9 42.5 g.

10 30.8 g.

12 LiCl.

13 1.

14 Ammonium sulphate.

15 (a) 1 g of oxygen combines with 12.9 g of lead, (b) 6.5 g, (c) PbO_2, (d) white lead, 80% Pb.

16 (a) 461, 233, (b) 0.01, (c) 2.3 g, 4.6 g, 2.3 g, 2.3 g, (e) neutral, neutral, neutral, acidic, (f) alkaline.

17 Magnesium carbonate, 47.6 cm³, 40 cm³, 37.7 cm³.

18 9 g.

19 (a) 2 g, (b) 16 g, (c) 1, (d) $\frac{1}{2}$.

Chapter 10 **1** (a) 15 cm³, 10 cm³, (b) 5 cm³, 10 cm³, (c) 20 cm³, 10 cm³.

3 142, 88, 40.

4 (a) 1.79 dm³ of carbon dioxide, (b) 1.12 dm³ of hydrogen.

5 (a) 64, (b) 40, (c) 28.

6 (a) CH_2, (b) 42, (c) C_3H_6.

7 (a) 11.2 dm³, (b) 56 dm³, (c) 1.12 dm³, (d) 2.24 dm³.

8 (a) 16 g, (b) 16 g, (c) 3.4 g, (d) 4.25 g.

9 N_2O.

10 $8NH_3(g) + 3Cl_2(g) \rightarrow 6NH_4Cl(s) + N_2(g)$.

11 (a) 32, (b) 4 hydrogen atoms, (c) SiH_4.

12 Magnesium.

13 0.90 dm³, 0.72 g.

14 (a) 1.33 g, (b) C_3H_8, (c) Ca_3P_2, 4.8 dm³.

15 (a) $\frac{1}{8}$, (b) 56, 84, (c) C_4H_8, (d) 1, (h) 60 cm³, 40 cm³.

16 (i) Carbon, nitrogen, (ii) 52, (iii) 2 molecules carbon dioxide, 1 molecule nitrogen, (iv) C_2N_2.

Chapter 11 **5** (a) 38600, (b) 48250, (c) 2413, (d) 24125.

6 (a) (i) 16 g, (ii) 4.5 g, (iii) 54 g, (b) (i) 5.6 dm³, (ii) 5.6 dm³.

9 32 g, 5.6 dm³, 11.2 dm³.

Chapter 12 **3** 33.3 g (100 g water)⁻¹.

4 (b) (ii) 135 g (100 g water)⁻¹, (iii) 37.9 g, (c) (ii) 7.

5 (b) 81°C, (c) 23 g of potassium chlorate(v) precipitates out, (d) 735 g, no.

7 (a) (i) 106 g, (ii) 37°C, 1 M, (b) (ii) 6.

Chapter 13 **1** (b) (i) 255 cm³, (ii) 155 cm³, (iii) 35 cm³, (d) 2.2 g, (e) 5 g, (f) 73 g dm⁻³, (g) 25 cm³.

2 11 cm³ min⁻¹.

3 (b) (i) 2.24 dm³.

6 100 g.

Chapter 15 **1** 719 kJ.

2 (a) 55.4 kJ, (b) 57.5 kJ.

6 (b) (i) 5.5°C, (ii) 10°C, (c) 37 cm³, (e) (i) 20 cm³, (ii) 40 cm.

Chapter 16 **4** (b) 76, CS_2.

Chapter 17 **2** (c) $1.2 \, dm^3$.
  5 (b) 82 g of calcium nitrate, 85 g of sodium nitrate.

Chapter 20 **7** (b) (ii) 8 g.
  9 55.3%.

Chapter 21 **6** (a) W = ethanol, C_2H_5OH; X = ethene, C_2H_4:
  Y = ethane, C_2H_6; Z = 1,2-dibromoethane, $C_2H_4Br_2$;
  (d) $306 \, cm^3$.

7 (a) A = C_2H_5COOH,

Chapter 22 **1** (a) 1.4 g, (b) 34.5 g, (c) 0.49 g.
  2 (a) 20 g, (b) 26.5 g, (c) 8 g.
  3 (a) 0.053 M, (b) 0.087 M.
  4 (a) 0.10 M, (b) 0.046 M.
  5 (c) (i) 0.0027, (ii) 0.108 M.

Appendix

The periodic table

Groups   I   II

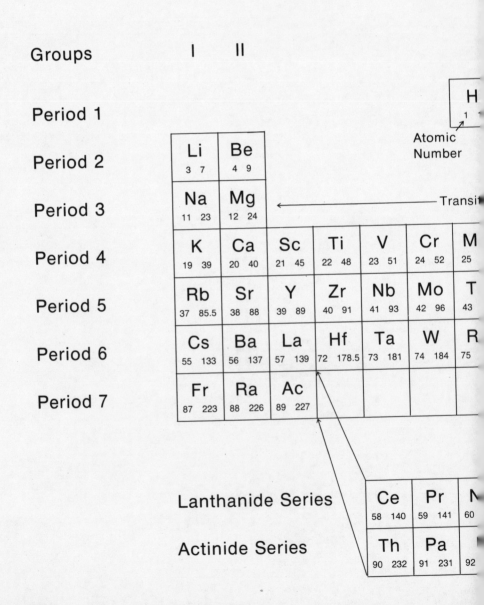

	I	II					
Period 1							H 1
Period 2	Li 3 7	Be 4 9					
Period 3	Na 11 23	Mg 12 24					
Period 4	K 19 39	Ca 20 40	Sc 21 45	Ti 22 48	V 23 51	Cr 24 52	M 25
Period 5	Rb 37 85.5	Sr 38 88	Y 39 89	Zr 40 91	Nb 41 93	Mo 42 96	T 43
Period 6	Cs 55 133	Ba 56 137	La 57 139	Hf 72 178.5	Ta 73 181	W 74 184	R 75
Period 7	Fr 87 223	Ra 88 226	Ac 89 227				

Atomic Number

Transit→

Lanthanide Series   Ce
58 140   Pr
59 141   N
60

Actinide Series   Th
90 232   Pa
91 231   92

| | | III | IV | V | VI | VII | O |

	III	IV	V	VI	VII	O

He
4

↑ Relative Atomic Mass

Metals ──────────────────→

			III	IV	V	VI	VII	O
			B 5 11	**C** 6 12	**N** 7 14	**O** 8 16	**F** 9 19	**Ne** 10 20
			Al 13 27	**Si** 14 28	**P** 15 31	**S** 16 32	**Cl** 17 35.5	**Ar** 18 40
e 56	**Co** 27 59	**Ni** 28 59	**Cu** 29 63.5	**Zn** 30 65	**Ga** 31 70	**Ge** 32 73	**As** 33 75	**Se** 34 79	**Br** 35 80	**Kr** 36 84
u 101	**Rh** 45 103	**Pd** 46 106	**Ag** 47 108	**Cd** 48 112	**In** 49 115	**Sn** 50 119	**Sb** 51 122	**Te** 52 128	**I** 53 127	**Xe** 54 131
s 190	**Ir** 77 192	**Pt** 78 195	**Au** 79 197	**Hg** 80 201	**Tl** 81 204	**Pb** 82 207	**Bi** 83 209	**Po** 84 210	**At** 85 210	**Rn** 86 222

| n 147 | **Sm** 62 150 | **Eu** 63 152 | **Gd** 64 157 | **Tb** 65 159 | **Dy** 66 162.5 | **Ho** 67 165 | **Er** 68 167 | **Tm** 69 169 | **Yb** 70 173 | **Lu** 71 175 |
| p 237 | **Pu** 94 242 | **Am** 95 243 | **Cm** 96 247 | **Bk** 97 249 | **Cf** 98 251 | **Es** 99 254 | **Fm** 100 253 | **Md** 101 256 | **No** 102 254 | **Lr** 103 257 |

Table of relative atomic masses

Element	Symbol	Atomic number	Relative atomic mass	Element	Symbol	Atomic number	Relative atomic mass
Actinium	Ac	89	227	Gold	Au	79	197
Aluminium	Al	13	27	Hafnium	Hf	72	178.5
Americium	Am	95	243	Helium	He	2	4
Antimony	Sb	51	122	Holmium	Ho	67	165
Argon	Ar	18	40	Hydrogen	H	1	1
Arsenic	As	33	75	Indium	In	49	115
Astatine	At	85	210	Iodine	I	53	127
Barium	Ba	56	137	Iridium	Ir	77	192
Berkelium	Bk	97	249	Iron	Fe	26	56
Beryllium	Be	4	9	Krypton	Kr	36	84
Bismuth	Bi	83	209	Lanthanum	La	57	139
Boron	B	5	11	Lawrencium	Lr	103	257
Bromine	Br	35	80	Lead	Pb	82	207
Cadmium	Cd	48	112	Lithium	Li	3	7
Caesium	Cs	55	133	Lutetium	Lu	71	175
Calcium	Ca	20	40	Magnesium	Mg	12	24
Californium	Cf	98	251	Manganese	Mn	25	55
Carbon	C	6	12	Mendelevium	Md	101	256
Cerium	Ce	58	140	Mercury	Hg	80	201
Chlorine	Cl	17	35.5	Molybdenum	Mo	42	96
Chromium	Cr	24	52	Neodymium	Nd	60	144
Cobalt	Co	27	59	Neon	Ne	10	20
Copper	Cu	29	63.5	Neptunium	Np	93	237
Curium	Cm	96	247	Nickel	Ni	28	59
Dysprosium	Dy	66	162.5	Niobium	Nb	41	93
Einsteinium	Es	99	254	Nitrogen	N	7	14
Erbium	Er	68	167	Nobelium	No	102	254
Europium	Eu	63	152	Osmium	Os	76	190
Fermium	Fm	100	253	Oxygen	O	8	16
Fluorine	F	9	19	Palladium	Pd	46	106
Francium	Fr	87	223	Phosphorus	P	15	31
Gadolinium	Gd	64	157	Platinum	Pt	78	195
Gallium	Ga	31	70	Plutonium	Pu	94	242
Germanium	Ge	32	73	Polonium	Po	84	210

Element	Symbol	Atomic number	Relative atomic mass	Element	Symbol	Atomic number	Relative atomic mass
Potassium	K	19	39	Tantalum	Ta	73	181
Praseodymium	Pr	59	141	Technetium	Tc	43	99
Promethium	Pm	61	147	Tellurium	Te	52	128
Protactinium	Pa	91	231	Terbium	Tb	65	159
Radium	Ra	88	226	Thallium	Tl	81	204
Radon	Rn	86	222	Thorium	Th	90	232
Rhenium	Re	75	186	Thulium	Tm	69	169
Rhodium	Rh	45	103	Tin	Sn	50	119
Rubidium	Rb	37	85.5	Titanium	Ti	22	48
Ruthenium	Ru	44	101	Tungsten	W	74	184
Samarium	Sm	62	150	Uranium	U	92	238
Scandium	Sc	21	45	Vanadium	V	23	51
Selenium	Se	34	79	Xenon	Xe	54	131
Silicon	Si	14	28	Ytterbium	Yb	70	173
Silver	Ag	47	108	Yttrium	Y	39	89
Sodium	Na	11	23	Zinc	Zn	30	65
Strontium	Sr	38	88	Zirconium	Zr	40	91
Sulphur	S	16	32				

Note: (1) The relative atomic masses are based on a scale where the mass of $^{12}C = 12$.

(2) For many of the radioactive elements the relative atomic mass given is the mass number of the most stable or the most common isotope.

Index

Page numbers in italics refer to experiments.

Acetaldehyde, *see* Ethanal
Acetic acid, *see* Ethanoic acid
Acetylene, *see* Ethyne
Acidity of a base, 80
Acids, 76, 81, 82
  anhydrides, 344, 355, 358
  carboxylic, *421*, 440
  characteristic ion, *70*, 81
  effect of water, *66*, *70*, 74, 80
  effect on indicators, *63*, 72
  oxides, *334*, 344
  properties of, *62*, *64*, *73*, *76*, *80*, 393
  salts, *69*, 79
  strength of, *71*, 82
  (*see also* names of individual acids)
Addition polymerisation, *423*, 446
Addition reaction, 431
Adsorption, 24, *291*, 299
Air
  combustion in, *37*, 43
  composition of, *39*, 43, 45
  liquefaction of, 345
  pollution, 46, 356
  reactions of metal with, *30*, *37*, 43,
    382, 393
  rusting in, *40*, 44
  solubility in water, *48*, 55
Alcohols, *420*, 437
Alkali metals, explanation of
  reactivity trend, 392
Alkalis, 74, 81
  characteristic ion, *70*, 81
  effect of water, *67*, *70*, 75, 80
  effect on indicators, *63*, 72
  properties of, *67*, *74*, *76*, 81
  strength of, *71*, 82
  (*see also* names of individual alkalis)
Alkanes, *414*, 427
Alkenes, *416*, 429
Alkynes, *416*, 432
Allotropy, 298
  of carbon, *290*, 297
  of phosphorus, 330
  of sulphur, *336*, 348

Aluminates, *334*, 401
Aluminium, *31*, *382*, *384*, *388*, 401
  compounds, *111*, *112*, 122, 401
  manufacture, 402
  uses of metal and compounds, 402
Aminoacids, 445
Ammonia
  manufacture, 324
  molecular shape, 138
  preparation of, *311*, 321
  properties of, *312*, 322
  solution, *312*, 323
  tests for, *311*, *312*, 322
  uses of, 324
Ammonium
  ion, 136
  salts, *315*, 321, 326
Ammonium chloride, *18*, *255*, 261, 326
  separation from salt, *18*, 25
Ammonium hydroxide, *312*, 323
Amphoteric, 83
  hydroxides, *334*, 347
  oxides, *334*, 345
Analysis
  qualitative, 451
  quantitative, 454
  volumetric, 454
Animal charcoal, *291*, 299
Anions, 205
Anode, 205
Anodising, 209
Anti-knock, 251
Argon, 46, 120
Atom, 100
  evidence for, *86*, 93
Atomic hydrogen, 60
Atomic mass, *88*, 95
  relative, 105, *152*, 158, 464, 466
Atomic number, 103, 464, 466
Atomic size, *88*, 95
Atomic structure, 102
Atomic theory, 99
Atomic volume, 118
Atomicity, 181

Avogadro constant, 159
Avogadro principle, *175*, 179

Baking powder, 302
Bases, 74, 82
  properties of, *65, 66*, 74
  solubility in water, 77
Basic oxides, *333*, 344
Basic salts, 80
Basicity, 79
Beer, 440
Benzene, 433
Blast furnace, 405
Bleaching, *340*, 355, *368*, 376
Boiling point
  effect of impurity on, *20*, 26
  effect of pressure on, 26
  of a pure liquid, *20*, 26
Bond-breaking, 278
Bond energy, 279
Bond-making, 278
Bonding
  coordinate, 136
  covalent, *125*, 132
  dative, 136
  electrovalent, *125*, 127
  ionic, *125*, 127
Boyle's law, 177
Bromides, *see* Halides
Bromine, *see* Halogens
Brønsted–Lowry theory, 82
Brownian motion, *89*, 96
Bunsen burner, *13*, 22

Calcite crystal, *87*, 94
Calcium, *383*, 396
  compounds, *386*, 397
  uses of metal and compounds, 397
Calorific values, 282
Carbohydrates, *418*, 435
Carbon
  allotropes, *290*, 297
  cycle, 304
  hydrides, 300
  properties of, *291*, 299, *385*, 393
  uses of, *291*, 299
Carbon dioxide
  manufacture, 302
  molecular shape, 139
  preparation of, *292*, 300, *334*, 344
  properties of, *293*, 301
  test for, *292*, 300
  uses of, 301

Carbon monoxide
  poisonous nature, 305
  preparation of, *295*, 304
  properties of, *296*, 305
  uses of, 305
Carbon tetrachloride, *see*
  Tetrachloromethane
Carbonates
  preparation of, 302
  properties of, *65*, 74, *294*, 303
  test for, 303
Carbonic acid, 301
Carboxylic acids, *421*, 440
Cast iron, 405
Catalyst, *243*, 249
  in reversible reactions, 266
Cathode, 205
Cathodic protection, 216
Cations, 205
Cell
  Daniell, *199*, 214
  simple, *198*, 213
Change
  chemical, *32*, 35
  physical, *32*, 35
Charles' law, 178
Chemical change, *32*, 35
Chlorides
  preparation of, *366*, 372
  properties of, *112*, 122, *365*, 373
  tests for, *365*, 373
Chlorine
  isotopes of, 105
  manufacture, 376
  preparation of, 374
  properties of, *366*, 374
  tests for, 374
  uses of, 376
  water, *368*, 375
Chromatography, 24
  separation of ink by, *17*
  separation of sugars by, *419*, 437
Clark's process, 400
Cleavage of crystals, *87*, 94
Coal, 435
Coke, 299
Combustion, *37*, 43
Compounds, 34
Condensation polymerisation, *424*, 447
Conservation of energy, law of, 276
Conservation of mass, law of, *38*, 44,
  91, 98
Constant composition, law of, *92*, 99

Contact process, 360
Coordinate bonding, 136
Co-polymer, 447
Copper, 409
  effect of heating in air, *30*, 33, *37*, 43
  purification by electrolysis, 209
  uses of metal and compounds, 410
Copper(I) compounds, *390*, 410
Copper(II) compounds, 410
  conversion to copper(I) compounds, *390*, 410
  sulphate, effect of heat on, *49*, 56
Corrosion of metals, *40*, 44, 215
Covalent bonding, *125*, 132
Covalent compounds, differences from ionic, *126*, 137
Cracking, *417*, 435
Crude oil, *417*, 434
Crystal
  cleavage of, *87*, 94
  growth, *16*, 24
  structure, *127*, 139, 144
Crystallisation, *15*, 23
  fractional, *221*, 226
  water of, *49*, 56, *222*, 228
Cycle
  carbon, 304
  nitrogen, 329
Cyclohexane, 433

Dalton's atomic theory, 99
Dative bonding, 136
Decrepitate, 319
Dehydrating agent, *342*, 359
Deliquescence, *222*, 228
ΔH convention, 277
Depolymerisation, *424*, 447
Destructive distillation, *290*, 298
Detergents, *423*, 443
Diamond, *290*, 297, 299
Diatomic molecules, 181
Diffusion, *89*, 96
Dinitrogen oxide, *311*, 321
Dinitrogen tetraoxide, *309*, 320
Direct combination, *32*, 79, *337*, 350
Displacement
  of halogens, *190*, 202
  of hydrogen, *189*, 202
  of metals, *189*, 202
Dissociation, thermal, *255*, 261
Distillate, 24
Distillation, *16*, 24
  destructive, *290*, 298

  fractional, *19*, 25
Double decomposition, *68*, 78
Downs cell, 395
Drying agent, 359
Dynamic equilibrium, *255*, 262

Efflorescence, *222*, 228
Electrical energy
  and thermal energy, *275*, 286
  from chemical changes, *198*, 213
Electricity
  conduction by electrolytes, *125*, 130, *190*, 203
  conduction by metals, *30*, 33, 143, 210
  from chemical reactions in voltaic cells, *198*, 213
Electrochemical series, *187*, 200
  and chemical activity, 201
  and discharge of ions, 202
  and displacement from solution, *189*, 202
Electrode, 187
Electrolysis, *190*, 203
  of aqueous solutions, *193*, 206, 208
  of copper(II) sulphate solution, *194*, 206
  of dilute sulphuric acid, *193*, 206
  of lead(II) bromide, *192*, 205
  of sodium chloride solution, *195*, 207
Electrolyte, *191*, 203
Electronegative elements, 201, 229
Electronic structure, 103
  and bonding, 127
  and periodic table, *110*, 117
Electrons, 103
  arrangement in atoms, 103
Electroplating, 209
Electropositive elements, 201
Electrovalent bonding, *125*, 127
Elements, 33
  classification into metals and non-metals, *30*, 33
  differences between elements, compounds and mixtures, *31*, 34
Empirical formulae, *153*, 161
  from oxidation numbers, 163
  of hydrated salts, *154*, 166
Endothermic reactions, 35, *270*, 277
Energy, *270*, 276
  in photosynthesis, 45
Enthalpy, 277
  and electrical energy, *275*, 286

Enthalpy (cont.)
of a displacement reaction, *274*, 286
of combustion, *270*, 280
of fusion, 279
of hydration, 283
of neutralisation, *272*, 282
of precipitation, *274*, 285
of solution, *273*, 283
of vaporisation, 280
Enzymes, *419*, 437
Epsom salt, 398
Equations, *155*, 167, *176*, 183
calculating reacting masses from,
169
ionic, 168
Equilibrium
dynamic, *255*, 262
effect of catalyst on, 266
effect of concentration change on,
257, 264
effect of light on, *259*
effect of pressure change on, *260*,
265
effect of temperature change on,
258, 265
in the Contact process, 267
in the Haber process, 266
Le Chatelier's principle, *257*, 264,
266
Esterification, 442
Esters, *422*, 442
Ethanal, 439
Ethane, *415*, 427
Ethane-1,2-diol, *416*, 431, 448
Ethanoic acid, *421*, 441
Ethanol, *420*, 437
fractional distillation of aqueous
solution, *19*, 25, 440
Ethene, *416*, 429
molecular shape, 139
Ethyl acetate, *see* Ethyl ethanoate
Ethyl ethanoate, *422*, 442
Ethylene, *see* Ethene
Ethylene glycol, *see* Ethane-1,2-diol
Ethyne, *416*, 432
Evaporation, *14*, 22
Exothermic reactions, 35, *270*, 277

Faraday
constant, 211
laws of electrolysis, *196*, 210
Fehling's solution, *418*, 436
Fermentation, *420*, 437

Fertilisers, 329, 330, 394
Field-ion microscope, 95
Filtrate, 24
Filtration, *15*, 24
Fluorine, *see* Halogens
Formulae
empirical, *153*, 161
from oxidation numbers, 163
graphic, 425
molecular, 165
determination of, *176*, 181
Fountain experiment, *312*, 322, *364*,
372
Fractional crystallisation, *221*, 226
Fractional distillation, 25
of crude oil, *417*, 434
of ethanol and water, *19*, 25, 440
of liquid air, 345
Fractionating column, 25
Frasch process, 351
Freezing point, *21*
effect of impurity on, *21*, 27
Fructose, *418*, 436
Fuels, 282, 356, 434, 439

Galvanic couples, *199*, 214
Galvanising, 215
Gases
diffusion of, *89*, 96
ideal gas equation, 178
laws, *173*, 177
motion of molecules in, *89*, 96
noble, 46, 127
Gay–Lussac's law, *173*, 179
Giant molecules, *see* Macromolecules
Giant structures, *127*, 139, 141
Glass, 307
Glucose, *418*, *420*, 435
Graphic formulae, 425
Graphite, *290*, 297, 299
Group, 114, 117
trends down, 123

ΔH convention, 277
Haber process, 324
Half-life, 107
Halides, *370*, 378
Halogens
preparation of, 376
properties of, *369*, 376
reactivity trend in, 378
Hardness of water, *387*, 398

Hardness of water (cont.)
  removal of hardness, *387*, 399
Heat changes, *270*, 276
Homologous series, *414*, 426
Hydrated salts, *222*, 228
Hydration energy, 284
Hydrides, 119, 394
Hydrocarbons, *368*, 375, 427
  in crude oil, 434
Hydrochloric acid, *365*, 372
  uses of, 372
Hydrogen
  bonds, 229
  displacement from acids, *64*, 73, *189*, 202
  from electrolysis of water, *50*, 57
  manufacture of, 60
  preparation of, *52*, 58
  properties of, *50*, 58, *367*, 375
  test for, *50*, 58
  uses of, 59
Hydrogen chloride
  manufacture of, 372
  preparation of, *364*, 371
  properties of, *364*, 372
  tests for, *364*, 371
  uses of, 372
Hydrogen halides, *370*, 378
Hydrogen peroxide, *333*, *335*, 347
Hydrogen sulphide
  and air pollution, 356
  preparation of, *337*, 351
  properties of, *338*, 352
  tests for, *337*, 352
Hydrogencarbonates
  distinction from carbonates, *295*, 304
  preparation of, 303
  properties of, *295*, 303
  tests for, 303
Hydrogensulphates, *69*, 80, 361
Hydrogensulphides, 353
Hydrogensulphites, 356
Hydrolysis, 83
  of esters, *422*, 443
  of salts, *72*, 83
  of starch and sugars, *419*, 436
Hydrophilic, 443
Hydrophobic, 443
Hydroxide ion, bonding in, 136
Hydroxides, *334*, 346
Hydroxyl group, bonding in, 135
Hygroscopic substances, *222*, 229

Ice, 230
Ideal gas equation, 178
Immiscible liquids, *19*, 25
Impurities
  effect on boiling point of liquids, *20*, 26
  effect on freezing point of liquids, *21*, 27
  effect on melting point of solids, *21*, 27
  removal of, *15*, 23, *221*, 226
Indicators, *62*, 72
Inhibitor, 251
Ink, separation of dyes in, *17*, 24
Insoluble base method, *69*, 78
Intermolecular forces, 140, *222*, 229
Iodine, *see* Halogens
Iodine clock reaction, *241*
Ion, 128
  evidence for, *125*, 130
  exchange, 400
Ionic bonding, *125*, 127
Ionic compounds, differences from covalent, *126*, 137
Ionic equations, 168
Ionic radii, 377, 392
Iron, *382*, *384*, 404
  compounds, *388*, 404
  conversion to steel, 406
  interconversions of iron(II) and iron(III) compounds, *388*, 405
  manufacture, 405
  protection of, 215
  rusting of, *40*, 44
  uses of, 405
Iron(II) sulphide, *337*, 351
Iron(III) chloride, hydrolysis of, *365*, 373
  preparation of, *366*, 374
Isomerism, 426
Isotopes, 104
  radioactive, 107

Joule (unit of energy), 278

Kinetic theory, *89*, 96
Kipp's apparatus, 351

LD process, 407
Lampblack, 299
Lattices, giant
  atomic, 141
  ionic, 139

Lattices, giant (cont.)
  metallic, 141
  molecular, 140
Law of conservation of energy, 276
Law of conservation of mass, *38*, 44,
  91, 98
Law of constant composition, *92*, 99
Law of multiple proportions, *93*, 99
Le Chatelier's principle, *257*, 264
  and the Contact process, 267
  and the Haber process, 266
Lead, 407
  compounds, *389*, 408
  uses of metal and compounds, 409
Light, effect on substitution reactions
  in organic chemistry, *415*, 428
Lime water, *42*, 300
Limekiln, 267, 302
Liquefaction of air, 345
Liquids
  determination of boiling point of, *20*,
    26
  motion of molecules in, *90*, 96
  separation if immiscible, *19*, 25
  separation if miscible, *19*, 25
Lithium, *386*, 392
  compounds, 394
  uses of, 395
Litmus, *63*, 72

Macromolecules, 144, *423*, 445, 446
Magnesium, *382*, *384*, 396
  compounds, *111*, 119, 397
  uses of metal and compounds, 398
Maltose, *418*, 436
Margarine, 431
Mass number, 103
Melting point, effect of impurity on,
  21, 27
  of a pure solid, *21*
Mercury cathode cell, 395
Metals, 382
  differences from non-metals, *30*, 33,
    410
  displacement from solution, *189*, 202
  reaction with acids, *64*, 73, 75, 393
  reaction with air, *382*, 391, 393
  reaction with water, *383*, *386*, 391
  reactivity series, *382*, 391
Methane, 427
  molecular shape, 138
Methylated spirit, 439
Miscible liquids, *19*, 25

Mixtures, 34
  separation of, *14*, 22
Molar enthalpy of fusion, 279
Molar enthalpy of vaporisation, 280
Molar mass, 159
Molar volume, *152*, 161, *174*, 182
Mole, 159, 166
Molecular formulae, 165
  determination of, *176*, 181
Molecular lattice, 140
Molecular mass, relative, 106, 164,
  175
Molecular shapes, 138
Molecule, 101
  bonding in, *125*, 132
Monoclinic sulphur, *336*, 349
Monomer, 446
Motion of particles, *89*, 96
Multiple proportions, law of, *93*, 99

Neutral oxides, 345
Neutralisation, 68, 77, 82
Neutron, 103
Nitrates
  preparation of, 328
  properties of, *317*, 328
  tests for, *317*, 328
Nitric acid
  manufacture of, 327
  preparation of, 326
  properties of, *315*, *317*, 326
  uses of, 327
Nitrogen, 318
  cycle, 329
  manufacture, 319
  preparation of, *309*, 318
  properties of, *309*, 318
  uses of, 319
Nitrogen dioxide, *309*, 319
Nitrogen oxide, *310*, 320
Noble gases, 46, 127
Non-metals, differences from metals,
  30, 33, 410
Nucleus, 102
Nylon, *424*, 447

Open Hearth process, 406
Organic,
  acids *421*, 440
  chemistry, 414, 425
  compounds, action of heat on, 414
    structure of, 138, 139, 425
Ostwald process, 327

Oxidation
  in electrolysis, 205
  in terms of electron transfer, 145
  in terms of oxidation number, 149
  in terms of oxygen and hydrogen,
    53, 59
  number, 129, 147, 163
Oxides, *111*, 122, 393
  types of, *333*, 344
Oxidising agents, 145
  tests for, 454
Oxonium ion, 81
Oxygen, 343
  in dissolved air, *48*, 55
  manufacture, 345
  preparation of, *40*, *333*, 343
  properties of, *333*, 343
  test for, *333*, 343
  uses of, 345

pH, *63*, 73
PVC, 447
Particles
  motion of, *89*, 96
  packing in solids, 94, 97, 139
  size of, *88*, 95
Peptide linkage, 445
Percentage composition of
  compounds, 167
Period, 114
  trends across, *111*, 118
Periodic table, *110*, 113, 464–5
  and chemical properties, *111*, 119
  and physical properties, *111*, 118
Permanent hardness of water, *387*,
  400
Peroxides, 345
Perspex, *424*, 447
Petroleum, *417*, 434
Phosphate fertilisers, 330
Phosphorus, *38*, 330, *334*, 344, *367*,
  375
  compounds, *111*, 122
Photosynthesis, *42*, 44
Physical change, *32*, 35
Plastic, 448
Plastic sulphur, *336*, 350
Polyester, 448
Polymerisation, *423*, 446
  addition, *423*, 446
  condensation, *424*, 447
Polymers, 446
Polystyrene, *423*, 447

Polythene, 446
Potassium, *382*, 392
  compounds of, 394
  uses of, 394
Proteins, *423*, 445
Proton, 102
Purification of compounds, *15*, 23,
  221, 226
Purity, tests for, *20*, 26

Qualitative analysis, 451
Quantitative analysis, 454

Radicals, 162
  bonding in, 135
Radioactivity, 106
Rate of reaction, 246
  effect of catalyst on, *243*, 249
  effect of concentration change on.
    240, 248
  effect of light on, *246*, 251
  effect of particle size on, *239*. 248
  effect of pressure change on. 249
  effect of temperature change on.
    242, 249
  iodine clock reaction, *241*
Reacting volumes, *173*, 179
Reactivity series, *382*, 391
Reactivity trends, in metals. *383*. *386*.
  392
    in non-metals, *370*, 378
    in the periodic table, *111*. 119. 123
Redox reactions, 146
Reducing agents, 146
  tests for, 454
Reduction
  in electrolysis, 205
  in terms of electron transfer. 146
  in terms of oxidation number. 149
  in terms of oxygen and hydrogen.
    53, 59
Relative atomic mass, 105. *152*. 158.
  466
Relative molecular mass, 106, 164, *175*
Residue, 24
Respiration, *42*, 44
Reversible reactions, *254*, 261
  effect of catalyst on, 266
  effect of concentration change on,
    257, 264
  effect of light on, *259*
  effect of pressure change on, *260*,
    265

Reversible reactions (cont.)
  effect of temperature change on, *258*, 265
Rhombic sulphur, *336*, 348
Ring compounds, 433
Rock salt, purification of, *17*, 24
Rusting
  conditions needed for, *40*, 44
  prevention of, 215

Saliva, *419*, 437
Salts, 75
  acid, *69*, 79
  hydrated, *222*, 228
  hydrolysis of, *72*, 83
  names of, 77
  normal, 79
  preparation of, *67*, 77
  solubility in water, 77
Saponification, *422*, 445
Saturated compounds, 427
Saturated solution, *219*, 225, *256*, 263
Sea water
  dissolved solids in, 55
  salt from, *14*
Seeding, *219*, 225
Separating funnel, *19*, 25
Separation of mixtures, *14*, 22
Shapes of molecules, 138
Siemens–Martin Open Hearth process, 406
Silicon, 307
  compounds, *111*, 119, 307
  uses of silicon(IV) oxide, 307
Soap, *422*, 443
  reaction with hard water, *387*, 398
Soapless detergents, *423*, 443
Soda lime, 427
Sodium, *382*, 392
  carbonate, manufacture, 396
  compounds, *111*, 119, 394
  hydroxide, manufacture, 395
  manufacture, 395
  uses of metal and compounds, 394
Solids
  motion of particles in, *90*, 97
  packing of particles in, 94, 97, 140
  structure of, *127*, 139, 144
Solubility
  curves, *219*, 226
  effect of temperature change on, *14*, 22, *219*, 225
  of air in water, *48*, 55

of common bases and salts in water, 77
  of gases, *224*, 233
  of liquids, *223*, 232
Solution, *14*, 22, *219*, 225
  saturated, *219*, 225, *256*, 263
  standard, 454
  supersaturated, *219*, 225
Solvent, *14*, 22, 225
  other than water, *223*, 232
  water, *222*, 231
Spectator ion, 169
Stalactites, 398
Stalagmites, 398
Standard solution, 454
Standard temperature and pressure, 179
Starch, *418*, 435
States of matter, *90*, 97
Steel, 405
  manufacture of, 406
Strong acids, *71*, 82
Structure
  atomic, 102
  crystal, *127*, 139, 144
  determination, 144
  diamond, 298
  electronic, 103
  giant atomic, 141
  giant ionic, 139
  giant metallic, 141
  graphite, 298
  molecular, 140
  sulphur, *336*, 349
Sublimation, *18*, 25
Substitution reaction, 429
Sugars, *418*, 435
Sulphates
  preparation of, 361
  properties of, 361
  test for, *339*, 361
Sulphides
  preparation of, *337*, *338*, 350, 353
  properties of, *337*, 351
  test for, 354
Sulphites, *339*, 356
Sulphur
  allotropes, *336*, 348
  compounds, 120, *337*, 350
  extraction, 351
  properties of, *31*, *336*, 348
  uses of, 350

Sulphur dioxide
  and air pollution, 356
  preparation of, *334*, *339*, 344, 354
  properties of, *339*, 354
  tests for, 354
  uses of, 356
Sulphur trioxide
  preparation of, *341*, 357
  properties of, 121, *341*, 358
  uses of, 358
Sulphuric acid
  manufacture, 360
  preparation of, *341*, 358
  properties of, *341*, 358
  uses of, 359
Sulphurous acid, *339*, 355
Supersaturated solution, *219*, 225
Suspension, 227
Symbols, 101, 466
Synthesis, *31*, 34, *50*, 57

Temporary hardness of water, *387*, 400
Terylene, 432, 448
Tetrachloromethane, 306
Thermal decomposition, 262
Thermal dissociation, *255*, 261
Thermoplastic, *423*, 448
Thermosetting plastic, 448
Tin plating, 215
Titration, *68*, 78, 454
Transition metals, the special properties of, 404
Transition temperature, 349

Universal indicator, *63*, 73
Unsaturated compounds, 429
  tests for, 432

Valency, 129, 133
Van der Waals' forces, 140
Vinegar, 442
Voltaic cells, *198*, 213

Voltameter, *193*
Volume
  atomic, 118
  molar, *152*, 161, *174*, 182
Volume strength of hydrogen peroxide solution, 348
Volumetric analysis, 454

Washing soda, *222*, 228, 400
Water
  air dissolved in, *48*, 55
  as a solvent, *222*, 231
  as an electrolyte, 204, 208
  composition of, *50*, 57
  distilled, 24
  domestic supplies, 56
  electrolysis of, *50*, 57
  hardness of, *387*, 398
  hydrogen bonding in, 229
  molecular shape, 139
  of crystallisation, *49*, 56, *222*, 228
  polar nature of, *222*, *223*, *224*, 229, 232, 234
  pollution, 55
  properties of, 56, *333*, 344, *369*, 378, *383*, *386*, 391, 392
  removal of hardness from, *387*, 399
  sources of, 49, 54
  synthesis of, *50*, 57
  tests for, 56
Weak acids, *71*, 82
Wood charcoal, *290*, *291*, 298
Wrought iron, 405

X-ray diffraction, 144

Yeast, *420*, 437

Zinc, *382*, *384*, 403
  compounds, 403
  uses of metal and compounds, 403
Zincates, *334*, 345, 347, 403